TEXTILE DYEING

TEXTILE DYEING

Dr. N. N. Mahapatra

B.Sc (Hons), B.Sc (Tech) (Bom)M.Sc (Chem), Ph.D (Chem),
M.B.A (IMM, Cal) C.Col FSDC (UK),
CText FTI (Manchester), FRSC (UK) Int Trg (Australia), Sen Mem,
AATCC (USA), FAIC (USA) FIC, FTA, FICS, FIE, FIIChE, MISTE (INDIA)

WOODHEAD PUBLISHING INDIA PVT LTD

New Delhi

Published by Woodhead Publishing India Pvt. Ltd.
Woodhead Publishing India Pvt. Ltd.,
303, Vardaan House, 7/28, Ansari Road,
Daryaganj, New Delhi - 110002, India
www.woodheadpublishingindia.com

First published 2018, Woodhead Publishing India Pvt. Ltd.
© Woodhead Publishing India Pvt. Ltd., 2018

Woodhead Publishing India Pvt. Ltd. ISBN: 978-93-85059-26-1
Woodhead Publishing India Pvt. Ltd. e-ISBN: 978-93-85059-91-9

Typeset by Allen Smalley, Chennai

Contents

Preface

The book "Textile Dyeing" is my 7th book . During my student days (1981–84) in UDCT (now known as ICT), Matunga, Mumbai, I was an admirer and follower of my guide Late Prof V. A. Shenai. During that time he had already written more than ten books. In his books he mentioned his practical knowledge in working as a technical consultant. So that was the reason his books were familiar in the Academic circle both in India and abroad.

Similarly I have also shared my 25 years of experience working in the shop floor production areas working in big corporate houses in India and abroad. My aim was to connect the Industry with Academics.

Textile Dyeing can be done in many ways where the substrate can be Fibre, Tow, Yarn, Fabric, Garments, Denims, etc. Accordingly we have different types of dyeing in India. I have tried my best to include all types of process used in the Industry Dye houses and Process Houses.

I am thankful to Mr. Subhash Bhargava, FSDC (UK), M.D. of Colorant Ltd. for giving sufficient scope, cooperation and help in writing this book.

I am also thankful to my ex colleague, Mr. Chetan Mulani (a leading Dyeing expert) for giving me few practical inputs.

I am also thankful to ATE, Mumbai, India, the leading Textile Dyeing manufacturer an supplier for arranging the machine photographs and machine parameters etc.

I am also thankful to Mr. Punit Makharia, M.D., and Mr. Gautam Makharia, Jt. M.D., of Shree Pushkar Chemicals & Fertilisers Ltd., Mumbai for allowing me to give the final touches to the book.

I hope this book is unique in nature and it will be very helpful to the textile students, supervisors, CEO, etc of the Textile Dyeing Industry.

Lastly, I want to thank my wife Seemani and both my daughters and son in law (Sanchitta, Nittisha and Nishant) for providing me timely support.

I want to dedicate this book to my Father in Law, Mr. Srichandra Parija and mother in law Mrs. Anjali Parija who inspired me to complete this project.

Dr. N. N. Mahapatra
C.Col FSDC (UK), CText FTI (Manchester), FRSC (UK), FAIC (USA)
Business Head (Dyes)
SHREE PUSHKAR CHEMICALS & FERTILISERS LTD.
301-302, 3rd Floor, Atlanta Centre, Near Udyog Bhawan,
Sonawala Road, Goregaon (E), Mumbai - 400 063, India

1
Introduction

Early evidence of dyeing comes from India where a piece of cotton dyed with a vegetable dye has been recovered from the archaeological site at Mohenjo-daro (third millennium BC). The dye used in this case was madder, which, along with other dyes such as Indigo, was introduced to other regions through trade.

Contact with Alexander the Great, who had successfully used dyeing for military camouflage, may have further helped aid the spread of dyeing from India.

Dyeing was a hand process since ages. Dyers used large containers or vats for dyeing relatively short lengths of fabric or small quantities of fibre and yarn. Until the development of synthetic dyes, all dyes were natural. Fastness varied widely among the natural dyes. The processes required to achieve certain colours were long and involved and carefully guarded by dyers. In the latter half of the 19th century, research into a better understanding of the chemistry of natural dyes led to the development of synthetic dyes. By approximately 1900, synthetic dyes had replaced natural dyes in almost all applications.

Developments during the industrial revolution increased the amount of fibre, yarn or fabric dyed in a vat. Dye boxes, vats and jigs were the primary pieces of machinery used to dye industrial quantities of fabric in batch process. However, in the 20th century, researchers and dyeing technologists developed continuous methods so that undyed fabric would enter and dyed fabric would exit the machine. Other 20th-century developments which took place were increase production rates, improve dye quality and lower costs.

Dyeing is the process of imparting colours to a textile material in loose fibre, yarn, cloth or garment form by treatment with a dye.

1.1   Dye types

For most of the thousands of years in which dyeing has been used by humans to decorate clothing, or fabrics for other uses, the primary source of dye has been nature, with the dyes being extracted from animals or plants. In the last 150 years, man has produced artificial dyes to achieve a broader range of colours, and to render the dyes more stable to resist washing and general use. Reactive dyes combine directly with the fibre, resulting in excellent colourfastness. The first ranges of reactive dyes for cellulose fibres were introduced in the mid-1950. Today, a wide variety is available.

Different classes of dye are used for different types of fibre and at different stages of the textile manufacturing process. Many different dyeing processes exist.

1. In batch processes, a predetermined quantity of fibre or length of yarn or fabric is dyed.

2. In continuous methods, very long lengths of yarn or fabric are dyed.

3. In skein dyeing, yarns are wrapped around poles and vat dyed.

4. In package dyeing, yarn is wound on to perforated tubes through which dye is pumped.

5. In beam dyeing, yarn or fabric is wrapped around a large perforated beam through which dye is pumped.

6. In beck and jet dyeing, fabric is loosely twisted into a long loop and circulated around reels and guide rollers into and out of the dyebath.

7. Jig dyeing is a batch process in which the fabric is under tension during dyeing as it passes through the dyebath and winds on to one roller, reverses direction and passes back through the dyebath and on to the other roller.

8. In continuous dyeing, the dye is padded on to the fabric, excess dye is removed by rollers and steam heat sets the dye.

1.2   Dyeing process

Dyeing is a method which imparts beauty to the textile by applying various colours and their shades on to a fabric. Dyeing can be done at any stage of the manufacturing of textile, such as fibre, yarn, fabric or a finished textile product including garments and apparels. The property of colour fastness depends upon two factors: selection of proper dye according to the textile material to be dyed and selection of the method for dyeing the fibre, yarn or fabric.

Following are the dyeing methods adopted in reactive dyes:

1. *Isothermal dyeing methods.* In this method, first add salt at 40°C and then raise temperature to 60°C. Then add dye solution and continue dyeing at that temperature. Then add alkali and run for 45 min and then check shade.

2. *Exhaust dyeing methods.* In this method, first add pre-dissolved dye at 40°C and run for 10 min. Raise temperature to 60°C. Then add salt in three instalments and continue dyeing, followed by adding soda ash slowly at 60°C and then check sample. Results are at par with both the methods. Soda ash alone is always the preferred alkali but to reduce the amount of handling needed, a mixture of soda ash and

caustic soda or a formulated liquid alkali may be used for fixation of the dyes.

3. *Grafting dyeing method.* While dyeing turquoise shade/royal blue shade, glauber salt must be used as an electrolyte (salt) for the dyeing process. Dyes like turquoise blue H2GP must be dyed by this method. In this method, after alkali addition, the dyeing continued for 60 min at 60°C. Raise the temperature to 80°C and continue for 40 min give good results.

1.3   Dyes

Dyes are used for colouring the fabrics. Dyes are molecules which absorb and reflect light at specific wavelengths to give human eyes the sense of colour. There are two major types of dyes: natural and synthetic dyes. The natural dyes are complex mixtures of components derived extracted from natural substances such as plants, animals or minerals. Synthetic dyes are made in a laboratory. Chemicals are synthesized for making synthetic dyes. Some of the synthetic dyes contain metals too.

All the dyes are not fixed to the fibre during the process of dyeing. Table 1.1 gives the percentage of unfixed dyes for different textiles. The reactive dye that is used for cotton shows the poorest rate of fixation. As half of the textile-fibre market is cotton, the problems of coloured effluents stems from dyeing cotton with reactive dyes.

Table 1.1. Percentage of unfixed dye for various dye types and applications

Fibre	Dye type	Unfixed dye (%)
Wool and nylon	Acid dyes/reactive dyes for wool	7–20
	Pre-metallized dyes	2–7
Cotton and viscose	Azoic dyes	5–10
	Reactive dyes	20–50
	Direct dyes	5–20
	Pigment	1
	Vat dyes	5–20
	Sulphur dyes	30–40
Polyester	Disperse	8–20
Acrylic	Modified basic	2–3

1.3.1   Synthetic dyes

Synthetic dyes are classified based upon their chemical composition and the method of their application in the dyeing process.

1.3.2   Basic/cationic dyes

Basic dyes are water soluble and are mainly used to dye acrylic fibres. They are mostly used with a mordant. A mordant is a chemical agent which is used to set dyes on fabrics by forming an insoluble compound with the dye. With mordant, basic dyes are used for cotton, linen, acetate, nylon, polyesters, acrylics and modacrylics. Other than acrylic, basic dyes are not very suitable for any other fibre as they are not fast to light, washing or perspiration. Thus, they are generally used for giving an after treatment to the fabrics that have already been dyed with acid dyes.

1.3.3   Direct dyes

Direct dyes colour cellulose fibres directly without the use of mordants. They are used for dyeing wool, silk, nylon, cotton, viscose, etc. These dyes are not very bright and have poor fastness to washing although they are fairly fast to light.

1.3.4   Chrome dyes

Chrome dyes are acidic in character. Sodium or potassium bichromate is used with them in the dyebath or after the process of dyeing is completed. This is done for getting the binding action of the chrome. They are mostly used for wool which gets a good colour fastness after treatment with chrome dyes. They are also used for cotton, linen, silk, viscose and nylon but are less effective for them.

1.3.5   Vat dyes

Vat dyes are insoluble in water and cannot dye fibres directly. However, they can be made soluble by reduction in alkaline solution which allows them to affix to the textile fibres. Subsequent oxidation or exposure to air restore the dye to its insoluble form. Indigo is the original vat dye. These dyes are the fastest dyes for cotton, linen and viscose. They are used with mordants to dye other fabrics such as wool, nylon, polyesters and acrylics.

1.3.6   Reactive dyes

Reactive dyes react with fibre molecules to form a chemical compound. These dyes, they are either applied from alkaline solution or from neutral solutions which are then alkalized in a separate process. Sometimes heat treatment is also used for developing different shades. After dyeing, the fabric is washed well with soap so as to remove any unfixed dye. Reactive dyes were originally used for only cellulose fibres but now their various types are used for wool, silk, nylon, acrylics and their blends as well.

1.3.7   Disperse dyes

Disperse dyes are water insoluble. These dyes are finely ground and are available as a paste or a powder that gets dispersed in water. These particles dissolve in the fibres and impart colour to them. These dyes were originally developed for the dyeing of cellulose acetate but now they are used to dye nylon, acetate and acrylic fibres too.

1.3.8   Sulphur dyes

Sulphur dyes are insoluble and made soluble by the help of caustic soda and sodium sulphide. Dyeing is done at high temperature with large quantities of salt so that the colour penetrates into the fibre. After dyeing, the fabric is oxidized for getting desired shades by exposure to air or by using chemicals. Excess dyes and chemicals are removed by thorough washing. These dyes are fast to light, washing and perspiration and are mostly used for cotton and linen.

1.3.9   Pigment dyes

Although pigments are not dyes in a true sense, they are extensively used for colouring fabrics like cotton, wool and other man-made fibres due to their excellent light fastness. They do not have any affinity to the fibres and are affixed to the fabric with the help of resins. After dyeing, the fabrics are subjected to high temperatures.

1.4   Dyeing methods

Colour is applied to fabric by different methods and at different stages of the textile manufacturing process.

1.4.1   Loose stock dyeing

In loose stock dyeing method, the fibre is dyed even before it is spun. This method is used to dye fibres. In this process, the staple fibres are packed into a vessel and then dye liquid is forced through them. Although the dye solution is pumped in large quantities, the dye may not penetrate completely into the fibres and some areas may be left without dyeing. However, the following blending and spinning processes mix up the fibres in such a way that it results in an overall even colour. Polyester, acrylic, cotton and viscose are dyed in fibre form.

1.4.2   Top dyeing

Top is the combed wool and gilled polyester tops. In this method, the fibre is dyed in the stage just before the appearance of finished yarn. Polyester and wool tops are loaded into a vessel and then dye liquid is forced through them.

1.4.3   Piece dyeing

In this method, small batches of constructed natural-coloured fabric are dyed according to the demands for a given colour. The constructed fabrics are piece dyed for the flexibility they provide. The textile manufacturer can dye the whole fabric in batches according to the fashion demands of the time thus avoiding wastage and resultantly loss. There are several methods prevalent or piece dyeing.

1.4.4   Beck dyeing

It is used for dyeing long yards of fabric. The fabric is passed in rope form through the dyebath. This rope of the fabric moves over a rail on to a reel which immerses it into the dye and then draws the fabric up and forward and brings it to the front of the machine. This process is repeated many times until the desired colour intensity is obtained.

1.4.5   Jigger dyeing

It is similar to the process of beck dyeing with a slight variation. The fabric in jigger dyeing is held on rollers at full width rather than in rope form as it is passed through the dyebath.

1.4.6   Pad dyeing

Padding is also done while holding the fabric at full width. The fabric is passed through a trough having dye in it. Then it is passed between two heavy rollers which force the dye into the cloth and squeeze out the excess dye. Then it is passed through a heat chamber for letting the dye to set. After that it is passed through washer, rinse and dryer for completing the process.

1.4.7   Jet dyeing

Fabric is placed in a heated tube where jets of dye solution are forced through it at high pressures. The fabric too moves along the tube. The solution moves faster than the cloth while colouring it thoroughly.

1.4.8   Solution pigmenting or dope dyeing

This is a method applied for dyeing the synthetic fibres. Dye is added to the solution before it is extruded through the spinnerets for making synthetic filaments. This gives a colourfast fibre as the pigments are used which are the fastest known colours.

1.4.9   Garment dyeing

Garment dyeing is applied to finished products such as apparels and garments. When the finished textile products such as hosiery or sweaters are dyed, it is called garment dyeing. A number of garments are packed loosely in nylon net and put into a dyestuff filled vessel with a motor-driven paddle. The dye is thrown upon the garments by the moving paddles' effect.

1.4.10   Yarn dyeing

When dyeing is done after the fibre has been spun into yarn, it is called yarn dyeing. In this method, the dyestuff penetrates the fibres to the core of the yarn. There are many forms of yarn dyeing: skein (hank) dyeing, package dyeing, warp-beam dyeing and space dyeing.

1.4.10.1   Skein (hank) dyeing
The yarns are loosely arranged in skeins or coils. These are then hung over a rung and immersed in a dyebath in a large container. In this method, the colour penetration is the best and the yarns retain a softer, loftier feel. It is mostly used for bulky acrylic and wool yarns.

1.4.10.2   Package dyeing
The yarns are wound on spools, cones or similar units and these packages of yarn are stacked on perforated rods in a rack and then immersed in a tank. In the tank, the dye is forced outward from the rods under pressure through the spools and then back to the packages towards the centre to penetrate the entire yarn as thoroughly as possible. Mostly, the carded and combed cotton which are used for knitted outerwear is dyed through this method.

1.4.10.3   Warp-beam dyeing
It is similar to package dyeing but more economical. Here, the yarn is wound on to a perforated warp beam and then immersed in a tank for dyeing it applying pressure.

1.4.10.4   Space dyeing

In this method, the yarn is dyed at intervals along its length. For these, two procedures such as knit–deknit method and OPI space–dye applicator are adopted. In the first method, the yarn is knitted on either a circular or a flat-bed knitting machine and the knitted cloth is then dyed and subsequently it is deknitted. Since the dye does not readily penetrate the areas of the yarn where it crosses itself, alternated dyed and undyed spaces appear. The OPI space–dye applicator technique produces multicoloured space-dyed yarns. The yarns are dyed intermittently as they run at very high speeds through spaced dyebaths. They are continuously subjected to shock waves produced by compressed air having supersonic velocities.

1.4.11   Bale dyeing

This is a low cost method to dye cotton cloth. The material is sent without scouring or singeing, through a cold water bath where the sized warp has affinity for the dye. Imitation chambray and comparable fabrics are often dyed this way.

1.4.12   Batik dyeing

This is one of the oldest forms known to man. It originated in Java. Portions of the fabric are coated with wax so that only un-waxed areas will take on the dye matter. The operation may be repeated several times and several colours may used for the bizarre effects. Motifs show a mélange, mottled or streaked effect, imitated in machine printing.

1.4.13   Beam dyeing

In this method the warp is dyed prior to weaving. It is wound on to a perforated beam and the dye is forced through the perforations thereby saturating the yarn with colour.

1.4.14   Burl or speck dyeing

This is done mostly on woollens or worsteds, coloured specks and blemishes are covered by the use of special coloured links which come in many colours and shades. It is a hand operation.

1.4.15   Chain dyeing

This is used when yarns and cloth are low in tensile strength. Several cuts or pieces of cloth are tacked end-to-end and run through in a continuous chain in the dye colour. This method affords high production.

1.4.16   Cross dyeing

This is a very popular method in which varied colour effects are obtained in the one dyebath for a cloth which contains fibres with varying affinities for the dye used. For example, a blue dyestuff might give nylon 6 a dark blue shade, nylon 6, 6 a light blue shade, and have no affinity for polyester area unscathed or white.

1.4.17   Jig dyeing

This is done in a jig, kier, vat, beck or vessel in an open formation of the goods. The fabric goes from one roller to another through a deep dyebath until the desired shade is achieved.

1.4.18   Piece dyeing

The dyeing of fabrics in the cut, bolt or piece form is called piece dyeing. It follows the weaving of the goods and provides a single colour for the material, such as blue serge, a green organdie.

1.4.19   Random dyeing

Colouring only certain designated portions of the yarn. There are three ways of doing this type of colouring.

Skeins may be tightly dyed in two or more places and dyed at one side of the dye with one colour and at the other side with another one. Colour may be printed on to the skeins which are spread out on the blanket fabric of the printing machine.

Cones or packages of yarn on hollow spindles may be arranged to form channels through which the yarn, by means of air-operated punch, and the dyestuff are drawn through these holes by suction. The yarn in the immediate area of the punch absorbs the dye and the random effects are thereby attained.

2.1   Loose stock dyeing

In Asian countries, it is called fibre dyeing, and in Europe and the United States, it is called stock dyeing. Both are the same. Initially grey cotton fibre was spun into grey yarn and then woven into grey fabric and it was dyed in jigger, winch, etc., under atmospheric pressure and at boiling temperature using direct dyes, vat dyes, sulphur dyes, naphthols and reactives. Fibre dyeing was given prominence when polyester was used. In 1969, the first polyester fibre dyeing was done at Kiran Spinning Mills.

Stock dyeing or yarn dyeing is another less expensive process used in textile mills to colour fabrics. Due to the nature of this material, it is unable to outperform solution-dyed materials. In the stock-dyeing process, the colour is applied after the basic white fibres have been spun and tightly woven together. Once the material has been fabricated, the colouring is applied via force to penetrate the porous fibres as deeply as possible. This method, regardless of how much dye is used or the amount of pressure applied, is not able to achieve consistent colouring throughout the fibres. Thus, the colouring tends to reflect that of a radish (Fig. 2.1).

Materials that do not have thorough colouring are, unfortunately, more susceptible to fading and tearing, as the fabric weakens more rapidly when exposed to intense sunlight. In current market, it has been observed that most polyester materials are stock dyed because the process to make it is less costly than the solution-dying process. Textiles may be dyed at any stage of their development from fibre into fabric or certain garments. Stock dyeing is done in the fibre stage, and top dyeing in the combed wool sliver stage.

2.1.1   Definition of stock dyeing

Stock dyeing refers to dyeing a staple fibre before it is spun. There are two methods. The older and widely practiced procedure is that of removing the packed fibre from the bales and then packing the stock in large vats and circulating dye liquor through the mass of fibre at elevated temperatures.

Figure 2.1. Staple fibre packed into a career and placed into a vat for stock dyeing

The newer method, bale dyeing, which is applicable to wool and all types of man-made fibres, is that of splitting the bale covering on all six sides, placing the entire bale in a specially designed machine (covering and straps need not be removed), and then forcing the dye liquor through the bale of fibre. This latter method obviously saves time and labour costs.

Raw stock dyeing is about taking a bale of fibre in natural fibre form with specific deniers and cut lengths to the dyeing process. The fibre is then broken up by hand or by auto loader into a dyeing machine. After the raw-stock-dyeing process is completed, it is removed from the dyeing vessel by an internal carrier holding the fibre and then delivered on to a mobile skid carrier or into a holding truck with wheels. These carriers are then taken to the drying line areas in which the next process occurs.

The fibre is now dyed, soaking wet and needs to be extracted and re-opened. Water is extracted from the fibre and is constantly being re-opened in this process. Once this process is complete, the dried/opened fibre goes to re-baling to send to the next phase of textile processing.

Although the dye liquor is pumped through the fibre in large quantities, there may be areas where the dye does not penetrate completely. However, in subsequent blending and spinning operations, these areas are so mixed with the thoroughly dyed fibre that an overall even colour is obtained. In stock dyeing, which is the most effective and expensive method of dyeing, the colour is well penetrated into the fibres and does not crock readily. Stock-dyed fibre does not spin as readily as undyed fibre because it loses some of its flexibility, but lubricants add in the final rinsing overcome most of this difficulty.

2.1.2   Dyeing of loose acrylic fibres

It is available in various denier like 1.5×44, 2.0×64 and 1.2×44 mm. It all comes in non-shrinkable form. It comes in bales of 200–250 kg and in both indacryl and supacryl forms. Following are the steps involved in the dyeing process of acrylic loose fibre.

1. *Lot packing.* After opening the bales, loose fibre is filled in the carrier. It should be filled properly with manual loading or stamping machine using water continuously. It is very crucial. If loading of fibre is not uniform, then there will be air gaps in between the fibre package. Thus dyeing channel will occur, which result in the uneven dyeing of the fibre.

2. *Dye selection.* Normally in industries, cationic dyes and basic dyes are used for acrylic fibre dyeing. It is not manufactured in India. It was previously imported from European countries. But nowadays it has been imported from China. The main distributers in India are the Jalan Brothers from Mumbai, New Delhi and Ludhiana. Big companies like Vardhaman directly import acrylic dyes.

There are two factors involved in the process of selecting dyes for acrylic dyeing.

1. *F* value: dyestuff saturation factor
2. *K* value: combination factor

Such values are mentioned in the shade card. The *F* value is a dye constant which is used to calculate the saturation concentration in percentage of a given dye or combination of dyes obtainable on a given acrylic fibre. But the *K* value indicates the compatability of cationic dyes. It is expressed as a figure between 1 and 5. It is used to select suitable dye combinations, particularly for fibre dyeing. Dyes with a low *K* value tend to exhaust more rapidly in combination than those with a high *K* value. The *K* values of dyes to be used in combination should not differ by more than 0.5 to 1. This will avoid unevenness in acrylic fibre dyeing.

2.1.3   Dyeing procedure

The glass transition temperature of acrylic fibre is between 65°C and 85°C depending on the fibre type. The dyeing rate should be very slow after this temperature (1°C per minute). Cooling is also very crucial. It should be cooled at 0.5°C till 60–70°C, and then it should be drained.

There are also two other factors which should be taken into consideration while dyeing acrylic fibres. This information should be provided by our acrylic fibre manufacturers. They are as follows:

1. *Fibre saturation value (S value).* It indicates the maximum amount of dye which an acrylic fibre can take up. It is very helpful in deciding the recipe and holding time for dark and heavy dark shades.

2. *Rate of exhaustion (V value)*. It indicates which fibre has got faster rate of exhaustion or slower rate of exhaustion. Then accordingly the dyeing process parameter/cycle can be decided by the dyer.

In acrylic hank dyeing, common salt/Glauber's salt is used for retarding and migration. But in fibre dyeing, it should not be used because it has to be added in large quantities into the dyeing machine. Main problem is after a long period of time, the impurities/dust in common salt get deposited in the centrifugal pump/impeller, and it should be avoided.

Instead of salt, retardar should be used like Retargal ALS (Archroma, India) and Rematard RDB (Archroma, India). Both are good levelling and retarding agents. These retarders slow down the rate of dye uptake of acrylic dyes in the critical temperature range and promote level dyeings. The retarder should be cationic in nature. It is required to be added in light/medium shades in more quantity but in dark and heavy dark shades lesser quantity.

2.1.4   Cotton fibre dyeing

Nowadays cotton mélange and polyester/cotton-fibre-dyed yarns are in demand. Cotton fibre used for dyeing in high temperature–high pressure (HTHP) Dalal, India and Fongs, China dyeing machines is available in carded or combed form. The qualities preferred for cotton fibre dyeing in India is J-34 (Punjab, Haryana), Sankar 6 (Gujarat) and H-4 (Madhya Pradesh). It is classified based on area. Mélange has got good demand in India and abroad. Cotton fibre is dyed and mixed with different percentages of grey (or white) fibre to make mélange yarn. The shade development and shade approval are critical. Specialized persons are involved in this job. They have to meet the required standards, as the clients are very fussy about the shades and quality of mélange yarn. Cotton fibre is dyed and the dyed yarn is made into one lot of 5–10 tons. As the supplier wants 5 or 10 tons in one lot, he will prefer fibre dyed rather than yarn dyed because he will get 5-ton dyed yarn lots having lot-to-lot variation. But care has to be taken for fibre contamination. Planning should be proper and adequate precautions are required. Then the other type of yarn is made like blending dyed cotton fibre with dyed polyester fibre and then dyed polyester/cotton yarn is made. The advantage for the user or buyer is that he can get 5–10 tons in one shade without any shade difference if the polyester/cotton-dyed yarn is made through fibre spinning route instead of yarn-dyeing route.

Stock dyeing refers to the dyeing of the fibres, or stock, before it is spun into yarn. It is done by putting loose, unspun fibres into large vats containing the dye bath, which is then heated to the appropriate temperature required for the dye application and dyeing process. Stock dyeing is usually suitable for woollen materials when heather-like colour effects are desired. Wool fibre dyed black, for example, might be blended and spun with undyed (white) wool fibre to produce soft heather-like shade of grey yarn.

Tweed fabrics with heather-like colour effects such as Harris Tweed are examples of stock-dyed material. Other examples include heather-like colours in covert and woollen cheviot. Pill polyester fibre is more similar to wool in some of its technological properties. The conditions of pre-treatment and particularly the dyeing operation must be suitable for the relatively susceptible wool fibre in the blending process.

2.1.5   Dyeing of fire retardant textile in loose staple fibre dyeing

Normally 1.7 d (denier) × 44 mm Trevirs CS fibres are available for dyeing. The CS fibre is loaded in the carrier. Then the loaded carrier goes into the HTHP dyeing machines. It is dyed at 120°C at a pH of 4–5 with a levelling agent like Lyogen DFT (of Clariant) is added. Dark shades are reduction cleared with caustic soda and sodium hydrosulphite (1 and 2 gpl, respectively) and undergo hot wash at 85–90°C. The fibres are acid neutralized by acetic acid and antistatic agent Sapcotex F (Henkel)–0.4% (owg). Then the carrier is offloaded. The dyed fibres go for hydroextraction and dried in radio frequency (RF) dryer/steam dryer. Total dyeing time is about 4–4.5 h. With a 40°C wash, there are no discernible differences between normal polyester and Trevira CS fibres. At 60°C with the same dyeing similar values are achieved on flame retardants as on non-modified types.

2.1.6   Exhaust dyeing of viscose loose fibre

Nowdays the dyestuff technology has advanced and viscose fibre is being dyed using vat dyes, reactive dyes and sulphur dyes. The types of dyes are selected as per the end-use. For example, 100% dyed viscose yarn is being used in carpets in Belgium, Middle East, Turkey, Egypt and some European countries. The shades made out of dope-dyed viscose get rejected due to less brightness/lustre. Hence some units like RSWM, Gulabpura (Raj), Reliance Chemotex, Udaipur (Raj), RTM and Bhawanimandi (Raj) have switched over to own dyed viscose. The 100% own dyed viscose is very difficult to spin into yarn. But nowadays the spinning technicians have overcome this problem. By the support of fibre dyehouse like better washing, better antistatic finish and uniform drying (use of RF dryer) for better lustre, reactive highly exhaust (HE) dyes are recommended. Viscose fibre is dyed in HTHP dyeing machines. The material/liquor ratio in the dyeing machine should be maintained between 1/7 and 1/10. If water is hard, it is better to add EDTA . For example, in a 500-kg-capacity dyeing machine, the loading will be 550 kg (maximum).

Viscose is semi-synthetic fibre. It is a purified form of cellulose and hence unlike cotton scouring is not required directly start dyeing. Before dyeing, water should be checked. The soft water should have a total dissolved solid between 100 and 200 ppm. Water hardness <50 ppm. pH should be neutral (7). To avoid uneven dyeing of viscose fibre, it should be opened properly with hand and manually loaded (no stamping machine should be used). Loading density should be uniform to avoid channelling and uneven dyeing.

Following are the steps for dyeing viscose loose fibre:

1. Setting of dyebath: use sequestering agent/dyebath conditioner and Tata salt and run for 15 min.
2. Dissolve HE dyes outside and filter and add half quantity through addition tank 3. Run for 10 min at 50°C.
3. Then add half quantity of dyes at 60°C and run for 5 min.
4. At 60°C add Tata salt (as per depth of shade) in three to four instalments through stock tank.
5. Raise the temperature to 85°C at 1°C/min and run for 30 min.
6. Then cool to 70°C and add soda ash (as per depth of shade) in two instalments and run for 30 min at 85°C.
7. Cool to 60°C and drain.
8. Washing with plain water and drain.
9. Hot wash at 95°C and drain.
10. Then acid wash and drain.
11. Soaping at 95°C and drain.
12. Hotwash at 95°C and drain.
13. Cold wash with plain water.
14. Dye fixing at 40°C with acetic acid and drain.
15. Washing with plain water.
16. Finish with cationic softener/antistatic finish at 50°C and drain.

Total time required for the process varies from 6 to 7 h per batch. The dyed fibre goes to hydroextraction. Wet fibre opener is avoided as chance of fibre damage is there. Then it goes to steam dryer/RF dryer. The fibre is dried and goes to spinning with 12–13% moisture. In shades like royal blue, turquose blue and any bright shades Glauber's salt should be used. A suitable levelling agent would ensure greater uniformity. In many dyehouses, bifunctional reactive dyes are used to dye viscose fibre at 60°C and better levelling. The average dyeing cost is Rs 11/kg for light shade, Rs 15/kg for medium shade, Rs 20/kg for dark shade and Rs 30/kg for extra dark shade.

2.1.7   Polyester loose fibre dyeing

Polyester staple fibres are available in various deniers like 1.5, 2.0, 3.0 and in various staple lengths like 38, 44 and 51 mm. Polyester hollow fibres are of 6, 8 and 10 denier. No scouring is required before starting of polyester fibre dyeing. Following are the steps for polyester fibre dyeing.

1. Fibre loading
2. Fibre dyeing
3. Reduction cleaning
4. Soaping
5. Finishing
6. Hydroextraction
7. Drying

The dyeing machines are HTHP type (such as Dalal, Fongs, Thies, etc.). It is of vertical type. It is made of SS316. The capacities of the machine used in production are from 50, 100, 500, 1000 kg, etc. Each machine has got SS perforated carrier. If the carrier capacity is 100 kg, it can contain 120 kg fibre for dyeing. Sometimes after the first loading, the fibre carrier goes for a hot wash, so the loaded fibre shrinks and goes down and again about 10% more fibre can be reloaded into the carrier. The loading of the fibre is done by two methods.

a. Manual loading
b. Stamping machine

In india, manual loading is very common. Two workers get into the fibre carrier and two workers from outside put the staple fibres into the carrier. Then it is wet using a water pipe. Those two persons inside the fibre carrier go on pressing the grey fibres by their legs only. The similar process is done by two mechanical stamping legs which works mechanically and a water sprinkler is attached. The fibre carrier is kept on a rotating station which keeps on rotating. Only two persons have to put the fibre inside the carrier time to time.

After the polyester fibre is loaded, it goes into the HTHP dyeing machine with the help of a crane. The lid of the machine is manually closed by the dyeing machine operator. For polyester fibre dyeing, only S type of disperse dyes are used (high sublimation fastness dyes). Processes involved in the polyester fibre dyeing are as follows:

1. Dissolve the colour in bucket or side colour tank with stirring using slight warm water and little acetic acid.
2. Water is filled inside the machine via tank or directly.

3. The fibre carrier is put inside the machine.
4. Colour is added from the side tank along with levelling agent.
5. Material/liquor ratio is between 1/7 and 1/10 depending on the type of dyeing machine used.
6. Temperature is raised rapidly (2°C/min) to 100°C and hold for 10 min.
7. Again temperature is raised (1.5°C) to 130°C. Hold accordingly to the depth of the shade.
8. High temperature discharge is done (to avoid any oligomers deposition).
9. Reduction clearing with caustic/hydro or reduction clearing agents at 80°C for 20 min.
10. Hot wash at 80°C for 10 min.
11. Soaping at 80°C for 20 min.
12. Hot wash.
13. Cold wash.
14. Finishing treatment. Antistatic finish is applied (2%) on weight of the fibre and run for 20 min.
15. Unloading. Then the dyed cake from the carrier goes to the hydroextractor which removes the maximum water through centrifugal action. Lastly it goes to steam dryer or RF dryer where the fibre is dried properly.

Polyester fibre is dyed at 130°C and high temperature drain is done, to avoid oligomers. Oligomers are cyclic trimers present in polyester fibre. During the dyeing of heavy dark shades at 130°C, these oligomers come out from the polyester fibre and get deposited on the dyed fibre, the machine carrier and the inside of the vessel. It is present in the form of white powders and it gives problem in polyester/viscose spinning as a sticky mass on the card wire, blowroom lattice, drawframe, etc. The dyehouse has to take care in controlling these oligomers. Regular machine cleaning is a must. Dyed polyester fibre is blended with dope-dyed viscose from Grasim, Nagda (Madhya Pradesh) and made polyester/viscose yarn in synthetic spinning system.

2.2   Top dyeing

One step nearer to the finished yarn than stock dyeing is what is called top dyeing in the worsted industry. Top is wool that has been combed to take out the short fibres, then delivered from the combs in a rope like form about 1.25″. The top is wound on perforated spools and the dye liquor is circulated through it. Very even dyeing is possible with this method.

Top dyeing is also the dyeing of the fibre before it is spun into yarn and serves the same purpose as stock dyeing – that is, to produce soft, heather-like colour effects. The term top refers to the fibres of wool from which the short fibres have been removed. Top is thus selecting long fibres that are used to spin worsted yarn. The top in the form of sliver is dyed and then blended with other colours of dyed top to produce desired heather shades. Dyeing of different types of tops is as follows.

1. *Polyester tops*. These tops are available from fibre manufacturers such as Reliance, IOC and Indorama. The denier used for polyester top is 2.5 and low pill. If not tops, they supply polyester tow. In some units, they have tow to top convertor. According to their requirement, the mills make their own tops. TBL tops are also available. It is also called trilobal polyester tops. Reliance can supply grey polyester tops or dope-dyed black polyester tops. The weight of polyester tops is between 8 and 8.5 kg each. It is available in 1.5, 2 and 2.5 denier.

2. *Wool tops*. These tops are supplied directly to the spinning mills. But in some big units, they purchase greasy raw wool and have a scouring range where they do the scouring and then make wool tops. It is a tedious job. Wool tops are normally imported from Australia and New Zealand. But Indian wool is also available. It should be noted that Indian wool, as is well known, is suitable for coarse counts and is slightly harsher. Imported wool, particularly Merino wool, is fine and softer. Wool tops are available from 11.6 to 27 μ.

3. *Viscose tops*. Similarly Grasim, Nagda supplies grey viscose tops and dope-dyed viscose tops in various shades (more than 1000 shades). Viscose grey tops can be dyed in a top dyeing machine. The machine is HTHP. Only the carrier is different. For staple fibre, it is called fibre carrier. For yarn, it is called yarn carrier, and for tops, it is called top carrier.

4. *Silk tops*. Normally muga silk tops are available in brownish colour which need not be dyed. For dyeing, we have mulberry silk tops and tassar silk tops.

5. *Blended tops*. It may be polyester/silk or polyester/ramie or polyester/flax tops.

6. *Cashmere/angora/camel hair tops*. It is a very delicate and costly blend. All are animal fibres.

2.2.1   Top dyeing process

Figure 2.2. Top dyeing machine

There are HTHP dyeing machines of both horizontal and vertical types (Fig. 2.2). Fong's, Thies, Then, Dalal and Hisaka are the machine manufacturers. Normally vertical dyeing machines are used for top dyeing. There are top carriers. According to the capacity of the dyeing machines (from 10 to 1000 kg dyeing machines), each machine has different size of top carrier. There are perforated spindles. Inside the perforated spindles, the polyester tops or wool tops or any tops are loaded manually. The loading of polyester or wool tops is done in a similar way we load the yarn package into the yarn carrier. Then the loaded carrier goes inside the HTHP dyeing machines. According to the dyeing cycle, the grey tops are dyed in various shades. After dyeing, the dyed tops are removed from the top carrier. The dyed tops are fed manually into the hydroextractors where maximum water is removed from the dyed tops. The hydroextractor runs under the centrifugal principle. Then the moisture tops are dried inside the RF dryer or it passes through the back washer. For dark and heavy dark shades, the back washer is used. It consists of three to four troughs, and the sliver is removed and passes through the different troughs consisting of soap, washing, antistatic finish and passes through the steam dryer. The dried and finished sliver collected in can are fed into the gill box, where again tops are formed called gilled tops. This is a long process.

The various counts used are 2/80 and 2/90 Nm for Terewool yarns. Blends of polyester fibres and wool have established themselves particularly in the field of ladies' and men's outerwear because of their good shape retention, easy-care properties and suitability for a wide variety of fashion goods. The low pill polyester fibre is more similar to wool in some of its technological properties.

The conditions of pre-treatment and particularly the dyeing operation must be suitable for the relatively susceptible wool fibre in the blend (e.g., as compared with the polyester or cellulosic fibre). For this, the normal high-temperature dyeing method (120–135°C) cannot be used. Also the methods of after washing or intermediate washing must be adapted to suit the wool. The difficulties caused by the wool fibre in the blend can be avoided by dyeing the two types of fibres of the blend separately in the form of loose stock or tops, and then blending the dyed fibres. Another possibility would be that of pre-dyeing the polyester fibre alone, to blend the undyed wool and to dye this blend yarn or blend fabric in the same shade or in a different shade called as Terewool package dyeing or Terewool piece dyeing.

2.2.2   Worsted spinning

As compared to European countries and North American counties, worsted fabric processing is very less in India. Similarly worsted spinning is also of very less volume in India as compared to synthetic and cotton spinning. Most of the textile units are composite units. They have both worsted spinning and worsted fabric processing such as Raymond, Digjam, Dinesh Woolen Mills, OCM, Reliance, Reid and Taylor. The leading worsted yarn sellers are Jayashree Textiles (Aditya Birla group) and Uniworth, Raipur.

The worsted culture is entirely different from synthetic and cotton culture. Following are the fibres and blends used in worsted processing: (a) polyester tops, (b) wool tops, (c) viscose tops, (d) acrylic tops, (e) silk tops (tussar, muga and mulberry), (f) flax, ramie, pineapple fibres. The popular blends used are poly/wool 65/35, 55/45, 70/30, 40/60, poly/wool/viscose 65/25/10, poly/viscose 65/35, 55/45, wool/viscose and all wool varieties. The preferred blend ratio is 55/45 polyester/wool. This blend yarn is also being used increasingly for knit goods. The yarns often differ markedly from each other in construction and blend ratio. For instance, staple fibre yarns with a blend ratio between 65/35 and 80/20 are used for knit goods. Yarns with a polyester continuous filament core which is covered with wool are also in use. Furthermore, there are blend yarns which contain basic dyeable polyester fibres that have reduced tendency to pilling. Among the fancy items, acrylic, silk, flax, ramie and pineapple are used in 10–15% along with poly/wool blends. The dyed polyester and wool tops go to the worsted spinning department where it is spun into blended yarn of different counts.

2.2.3   Synthetic spinning

We can also make polyester/wool or viscose/wool in synthetic spinning. The wool tops are dyed in HTHP dyeing machine. But here instead of top carrier, the wool tops are loaded in a loose fibre perforated carrier. After the wool tops are dyed, it goes for hydroextraction and drying. Later the dried wool tops go to the cutter where the dyed wool sliver are cut into various staple lengths like 38–51 mm. Then the dyed wool fibres are blended with dyed polyester staple fibre and spun into blended yarn in synthetic spinning system. The khaki shade which is used in the military is made in this way. It is a wool/viscose 55/45, 2/30 or 2/40 count. Here the viscose staple fibre used is khaki dope dyed (N-1700) shade manufactured by Grasim Industries, Nagda (Madhya Pradesh). Only the wool tops are dyed in khaki shade.

2.3   Gel dyeing

There are three types of online colouration such as tow dyeing, dope dyeing and gel dyeing. Gel dyeing, sometimes also called mass dyeing, is a continuous tow dyeing method where soluble dyes are applied to wet-spun synthetic fibres such as acrylic or modacrylic fibres in the gel state. The gel state means that the material is not yet at full crystallinity or orientation. The gel dyeing method is applied after extrusion and coagulation, but it happens before drawing and drying. In other words, passing a wet-spun fibre that is in the gel state (not yet at full crystallinity or orientation) through a dye bath containing dye with affinity for the fibre. This process provides good accessibility of the dye sites. In the wet spinning of acrylic fibre, dope is prepared first for extrusion into a solvent solution, then the drawing, rinsing, oiling and drying stages follow for eventual densification. The process is completed with subsequent crimping, crimp setting and cutting. Before being dried for densification, acrylic fibre remains gel-swollen or in other words, spongy, a state which allows the absorption of cationic dye liquor. In short, gel dyeing is the inclusion of the dyeing and fixing processes after the rinsing stage of spinning.

In the gel dyeing, the wet-spun filaments like acrylic and modacrylic can be dyed while they are still in the coagulating bath. The fibres are still soft at this stage and the fibre is highly absorbent due to the high porosity so the penetration of dye is high. Since the material is coloured in the chemical bath after extrusion through the spinneret, the gel dyeing can be applied only for fibres manufactured by the wet spinning process. In gel dyeing, acrylic fibres are coloured with cationic dyes. Before being dried for densification, fibre remains spongy, a state that allows the absorption of cationic dye liquor. The initial set-up is much simpler than that of solution dyeing, as the fibre is extruded and then coloured. Gel dyeing is different from dope dye which is also known as solution dye or spun dye or mass dye. Dope dye is a dyeing method used for synthetic fibres/manufactured fibres such as acrylic, nylon/polyamide and polyester. To produce this kind of synthetic fibres, the liquid chemical solution (polymer solution) that makes up the fibre is passed through tiny holes in a spinneret like water being passed through shower head. Traditional fibre-dyeing methods require two process steps. In the first step, fibre is produced as uncoloured/undyed, and in the second step, the colour is applied to the surface of the fibre with wet-dyed process. In dope-dyeing technology, fibres are coloured with pigment or insoluble dyes. During this process, the dye (a masterbatch colourant) is added to the liquid fibre solution (polymer melt) before the fibres are extruded through spinnerets for making synthetic filaments. This way, the colour is mixed into the solution thoroughly. Fibres and filaments are already coloured coming out of the spinnerets. After extrusion, the colour pigments become a part of the fibre, they are bonded in the fibre. So the fibres have the colour all the way through the the cross section of the monofilament with excellent colour fastness. In the end, non-soluble colours are locked in the yarn. It is also different from tow dyeing where batches of high-bulk acrylic yarns of up to 15,000 lb have been successfully produced by the tow-dyeing route. The method of dyeing is called the 'atoz' process. It include loader, hot water system, press, dyeing machine with closed basket, centrifuge, automatic unloading system and robot for loading tow in hot air dryer belt (included padder system to add antistatic chemical before dryer), tow unloading robot that takes the tow from dryer belt either in storage trolley for conditioning or directly to a bale press. Advantages of the method include: colouration is economical; the yarn is free from flattening, tangling and other disturbances; and fancy effects, such as marls, can be produced easily. Limitations include: it is not economic to produce quantities of less than 1000–1500 lb by the method; extreme care is required to avoid contamination of yarn with fibres of different colours; and considerable forward planning is necessary.

2.3.1   Production of gel-dyed acrylic fibres

Gel dying is a good method for production of coloured fibres. Their major advantage is that they are colour-fast and the end product is more uniformly dyed in textile industries. The colour-fastness against such natural elements as sunlight, washing, etc., is unparalleled among man-made fibres. Depending on the customer's demand, the fibres are produced with a variety of dTex and Deniers.

The acrylic fibres are produced using monomer acrylonitrile as the major monomer (over 85%), methylacrylate or vinyl acetate as the second comonomer (<15%) and a third comonomer for dyeability, by solution polymerization and wet spinning using SNIA technology. In both dull and semi-dull products, titanium dioxide (max. 0.5%) is used. In addition, in the process of production, the fibres are submerged in finish (finish of about 0.5%) prior to packing. Consistent with SNIA technology, the production process for the product starting with the mixing of monomers as feedback input to the reactors for polymer solution in dimethylformamide in the chemical area through to fibre production, drawing, finishing and packaging is continuous and that at each stage, samplings and controls are conducted and finally the product is graded and released in accordance with the test results. Bright, dull and semi-dull products are normal. The product is packaged in bales weighing 350 ± 20 kg and then put in covered polypropylene jute and rod guards. The bales measure $120 \times 110 \times 73$ mm. About 20–22 tons of the product can be transported in a 40-foot container.

2.3.2   Uses

These fibres are used in a spinning and yarn production process for use in textile, blanket and carpet industries.

2.3.3   Advantages of gel dyeing

- Dyeing economy: lower energy, water and dye requirements and material handling
- Better fastness levels: light, rubbing and hot processing
- Availability of shrinkable dyed acrylic in box-cut permits various innovative blends on the cotton-spinning system
- Best for uniform dyeing and big lots (lot size >150 million tonnes [MT] even is possible)
- Tow breaking and spinning performance of gel-dyed tow and fibre are the best, and waste generation is lower compared to acrylic dyed by any other method
- Most environment friendly process among all acrylic dyeing methods
- Water requirement is five times less compared to conventional tow dyeing
- No extra energy requirement and negligible dyestuff losses
- Spinners do not have to follow strict pollution norms
- Consistency in shade reproduction
- The gel dyeing has the lowest cost of colouration. It is economical. It requires dyes of high quality
- It is highly flexible since the small batches can be dyed economically and to change the dyeing shade is quite easy. The switching of colour is quick. From pale to deep, any hue is possible with cationic dyestuff
- The gel dyeing spends little water, thus it is very environment friendly

Gel dyeing of acrylic fibres, also known as 'coloured fibres', is done in the gel state of the dyes. The corresponding raw materials, the fibre quickly colouring dye. Gel dyeing has many technical advantages: the traditional chemical fibre dyeing time, up to more than 100 min, while the new technology of production requires only a transient gel stained fibres; than the traditional dyeing process of saving pre-treatment, dyeing, drying and other multi-channel processes, and thus lower energy costs. As dyes on fast, high colour fastness, colourful, capability sun, sweat soaked, washed, etc., so has been welcomed by customers. Recently, Daqing Petrochemical Company plant newly developed acrylic 'gel dyeing of acrylic fibres' smooth in the 10,000-ton industrial production line to enlarge the production, this is the first time, the plant received two textile enterprises in Jiangsu and Hebei, the order, quickly made a response. According to market user information, the new development and production of acrylic fibre gel dyed into the six different shades like light yellow, light beige, light grey, camel, reddish-brown and dark brown. Produced samples with user supplied not only consistent with standard sample, but also some of the new products, more than 28% shrinkage rate, the product colour in full compliance with user-supplied fibre colour and Poor's criteria, and all products were less than an aberration, the product all the targets to achieve superior quality Acura requirements

The gel dyeing of acrylic fibre development and production of true representation of the acrylic fibre plant in Daqing Petrochemical Company is 'to help others help themselves' business philosophy. Daqing Petrochemical Company of acrylic works makes use of its technical superiority and advanced waste water treatment conditions, a positive for the downstream the sake of the textile users, not only to help the textile printing and dyeing enterprises to reduce costs, and more importantly, for users to save a coloured textile waste water treatment process links, eliminating the need for small-scale textile enterprises downstream dyeing the formation of environmental pollution problems, reflects the Petrochemical Company's 'mutual benefits and common development' values.

The gel-dyed acrylic fibres are produced in a variety of companies including Thai Acrylic, Pasupati, AKSA, Zayandeh Rood Fiber Company and Polyacryl Iran Corporation. Pasupati is the only producer of superior quality gel-dyed fibres in India with a capacity of 12,000 MT per annum. Gel dyeing is the most advanced technology where fibre is dyed online. Pasupati's gel-dyed acrylon fibre is far more superior than others in terms of its vibrant colours, brightness, excellent shade uniformity, soft and silky feel, excellent sunlight fastness, washing fastness, rubbing fastness and perspiration fastness properties. Pasuapti's gel-dyeing process is backed by strong quality control process which is supported by most modern robotized dye kitchen of CIR Italy, and computerized colour-matching system (CCM) of NOSELAB. Pasupati supplies all type of gel-dyed fibre, tow and tops, suitable for processing on all types of cotton and worsted spinning system to produce wide range of bulked and non-bulked dyed yarns for hosiery and other textile applications.

2.3.4   Gel dyeing

Passing a wet-spun fibre that is in the gel state (not yet at full crystallinity or orientation) through a dye bath containing dye with affinity for the fibre. This process provides good accessibility of the dye sites (Fig. 2.3).

We can express gel dyeing as follows:

- In the wet spinning of acrylic fibre, dope is prepared first for extrusion into a solvent solution, then the drawing, rinsing, oiling and drying stages follow for eventual densification. The process is completed with subsequent crimping, crimp-setting and cutting.
- Before being dried for densification, acrylic fibre remains gel-swollen or in other words, spongy, a state which allows the absorption of cationic dye liquor. In short, gel dyeing is the inclusion of the dyeing and fixing processes after the rinsing stage of spinning.

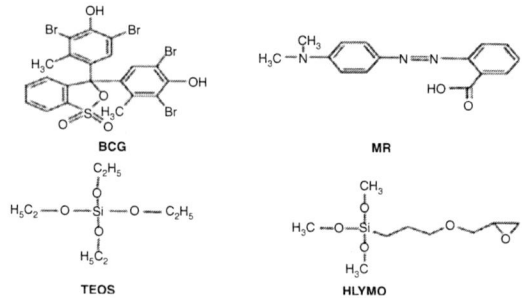

Figure 2.3. Structure on gel dyes

2.3.5   Application of gel dyeing in fibre

This dye is used for hydrophobic fibres such as:

- Regenerated fibres (e.g., cellulose acetate)
- Synthetic fibres (e.g., polyamides, polyester, polyacrylonitrile)

2.3.6   Properities of gel dyeing

Sodium dodecyl sulphate polyacrylamide gel electrophoresis (SDS-PAGE) describes a technique widely used in biochemistry, forensics, genetics and molecular biology to separate proteins according to their electrophoretic mobility (a function of the length of a polypeptide chain and its charge) and no other physical feature. SDS is an anionic detergent applied to protein sample to linearize proteins and to impart a negative charge to linearized proteins. In most proteins, the binding of SDS to the polypeptide chain imparts an even distribution of charge per unit mass, thereby resulting in a fractionation by approximate size during electrophoresis.

2.3.7 Preparing acrylamide gels

The gels typically consist of acrylamide, N,N'-methylenebisacrylamide (bisacrylamide), SDS and a buffer with an adjusted pH. The solution may be degassed under a vacuum to prevent the formation of air bubbles during polymerization. Alternatively, butanol may be added to the resolving gel after it is poured, as butanol removes bubbles and makes the surface smooth. A source of free radicals and a stabilizer such as ammonium per sulphate and N,N,N',N'-tetramethylethylenediamine (TEMED) are added to initiate polymerization. The polymerization reaction results in a gel because of the added bisacrylamide, generally about 1 part in 35 relative to acrylamide, which can form cross links between two polyacrylamide molecules. The ratio of acrylamide to bisacrylamide can be varied for special purposes. The acrylamide concentration of the gel can also be varied, generally in the range from 5 to 25%. Lower percentage gels are better for resolving very high-molecular weight proteins, while much higher percentages are needed to resolve smaller proteins. Determining how much of the various solutions to mix together to make gels of particular acrylamide concentration is possible (Fig. 2.4).

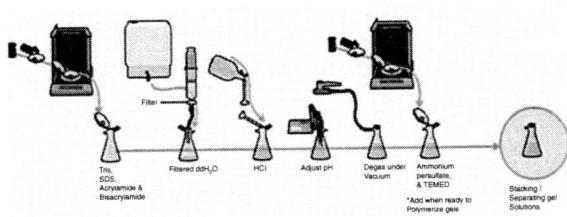

Figure 2.4. Preparing gels

Gels are usually polymerized between two glass plates in a gel caster, with a comb inserted at the top to create the sample wells. After the gel is polymerized, the comb can be removed and the gel is ready for electrophoresis.

2.3.8 Composition

Chemical buffer stabilizes the pH value to the desired value within the gel itself and in the electrophoresis buffer. The choice of buffer also affects the electrophoretic mobility of the buffer counterions and thereby the resolution of the gel. The buffer should also be unreactive and not modify or react with most proteins. Different buffers may be used as cathode and anode buffers, respectively, depending on the application. Multiple pH values may be used within a single gel, for example, in DISC electrophoresis. Common buffers in SDS-PAGE include Tris, Bis-Tris or imidazole.

Counterion balances the intrinsic charge of the buffer ion and also affects the electric field strength during electrophoresis. Highly charged and mobile ions are often avoided in SDS-PAGE cathode buffers, but may be included in the gel itself, where it migrates ahead of the protein. In applications such as DISC SDS-PAGE, the pH values within the gel may vary to change the average charge of the counterions during the run to improve resolution. Popular counterions are glycine and tricine. Glycine has been used as the source of trailing ion or slow ion because its pKa is 9.69 and mobility of glycinate is such that the effective mobility can be set at a value below that of the slowest known proteins of net negative charge in the pH range. The minimum pH of this range is approximately 8.0.

When acrylamide (C_3H_5NO; molecular weight: 71.08) dissolved in water, slow, spontaneous autopolymerization of acrylamide takes place, joining molecules together by head on tail fashion to form long single-chain polymers. The presence of a free-radical-generating system greatly accelerates polymerization. This kind of reaction is known as vinyl addition polymerization. A solution of these polymer chains becomes viscous but does not form a gel, because the chains simply slide over one another. Gel formation requires linking various chains together. Acrylamide is a neurotoxin. It is also essential to store acrylamide in a cool dark and dry place to reduce autopolymerization and hydrolysis.

The SDS ($C_{12}H_{25}NaO_4S$; molecular weight: 288.38) is a strong detergent agent used to denature native proteins to unfolded, individual polypeptides. When a protein mixture is heated to 100°C in the presence of SDS, the dee tergent wraps around the polypeptide backbone. It binds to polypeptides in a constant weight ratio of 1.4 g SDS/gram of polypeptide. In this process, the intrinsic charges of polypeptides become negligible when compared to the negative charges contributed by SDS. Thus polypeptides after treatment become rod-like structures possessing a uniform charge density, that is, same net negative charge per unit length. The electrophoretic mobilities of these proteins will be a linear function of the logarithms of their molecular weights. Without SDS, different proteins with similar molecular weights would migrate differently due to differences in mass/charge ratio, as each protein has an isoelectric point and molecular weight particular to its primary structure. This is known as native PAGE. Adding SDS solves this problem, as it binds to and unfolds the protein, giving a near uniform negative charge along the length of the polypeptide.

Ammonium persulphate (APS; $N_2H_8S_2O_8$; molecular weight: 228.2) is a source of free radicals and is often used as an initiator for gel formation. An alternative source of free radicals is riboflavin, which generated free radicals in a photochemical reaction.

The TEMED ($C_6H_{16}N_2$; molecular weight: 116.21) stabilizes free radicals and improves polymerization. The rate of polymerization and the properties of the resulting gel depend on the concentrations of free radicals. Increasing the amount of free radicals results in a decrease in the average polymer chain length, an increase in gel turbidity and a decrease in gel elasticity. Decreasing the amount shows the reverse effect. The lowest catalytic concentrations that will allow polymerization in a reasonable period of time should be used. APS and TEMED are typically used at approximately equimolar concentrations in the range of 1–10 mM.

Bisacrylamide ($C_7H_{10}N_2O_2$; molecular weight: 154.17) is the most frequently used cross-linking agent for polyacrylamide gels. Chemically it can be thought of as two acrylamide molecules coupled head to head at their non-reactive ends. Bisacrylamide can cross link two polyacrylamide chains to one another, thereby resulting in a gel.

2.3.9   Mechanism of gel dyeing

The first step after performing denaturing SDS-PAGE is to disassemble the gel cassette and place the thin (1 mm thick) polyacrylamide gel in a tray filled with water or buffer. The electrophoresed proteins exist as concentrated 'bands' embedded within each lane of the porous polyacrylamide gel matrix. Typically, the proteins are still bound to anionic SDS detergent, and the entire gel matrix is saturated in a particular buffer.

To make the proteins visible, a protein-specific, dye-binding or colour-producing chemical reaction must be performed on the proteins within the gel. Depending on the particular chemistry of the stain, various steps are necessary to hold the proteins in the matrix and to facilitate the necessary chemical reaction. All steps are done in solution, that is, with the gel suspended in a tray filled with one liquid reagent or another.

Given the common constraints of this format, most staining methods involve some version of the same general incubation steps:

1. A water wash to remove electrophoresis buffers from the gel matrix.
2. An acid or alcohol wash to condition or fix the gel to limit diffusion of protein bands from the matrix.
3. Treatment with the stain reagent to allow the dye or chemical to diffuse into the gel and bind (or react with) the proteins.
4. Destaining to remove excess dye from the background gel matrix

Depending on the particular staining method, two or more of these functions can be accomplished with one step. For example, a dye reagent that is formulated in an acidic buffer can effectively fix and stain in one step. Conversely, certain functions require several steps. For example, silver staining requires both a stain-reagent step and a developer step to produce the coloured reaction product.

2.4   Tow dyeing

In some of the industries, there are three types of dyeing followed such as tow dyeing, dope dyeing and gel dyeing. Gel dyeing, sometimes also called mass dyeing, is a continuous tow-dyeing method where soluble dyes are applied to wet-spun synthetic fibres such as acrylic or modacrylic fibres in the gel state. The gel state means that the material is not yet at full crystallinity or orientation. The gel-dyeing method is applied after extrusion and coagulation, but it happens before drawing and drying.

The polyester tow and acrylic tow are formed after the filaments come out of the spinnerets. Then the tow goes for cutting to the cutter at various staple length. Then the staple fibre goes for dyeing which we call fibre dyeing. But polyester tow dyeing is not popular in the industries, where the grey polyester tow is dyed by continuous method in a separate dyeing machine. May be the coloured pigments are added at the polyester polymerization stage where they get different type of dope-dyed polyester tow which is cut into fibres and goes to ring spinning or the tow goes to another machine called tow to top converter where dyed polyester tops are formed and it goes to worsted spinning. But we do not get dope-dyed acrylic fibre. We get dope-dyed viscose fibre. In case of acrylic, we get gel-dyed acrylic and tow-dyed acrylic.

2.4.1   Dyes used in acrylic tow dyeing

There are two types of cationic dyes used for acrylic dyeing.

1. *MAXILON dyes*. It is suitable for CD-polyester (CD-PES), modacrylic (MOD) and meta-aramid fibres. MAXILON® dyes have a good exhaustion rate and high build-up and offer bright colours with good light fastness and wash fastness. Their main application is in exhaust dyeing of yarn and piece material as well as printing.
2. *MAXILON® LIQ*. It is particularly suitable for acrylic (PAN) wet tow dyeing (gel dyeing) and acrylic printing.

Acrylic tow dyeing can be carried out in two ways:

1. *Exhaust tow dyeing*. The acrylic tow comes in the bale form to the dye house. This type of process is carried in fibre dyeing houses

where there are 200–1000 kg vertical HTHP dyeing machines made by Dalal, Fongs, Thies, etc. All these machines have perforated carrier. By a suitable loading or packing arrangement, the acrylic tow from the bale is packed into the steel perforated carrier with alternate stamping and water sprinkling arrangements. Then the tow packed carrier goes inside the HTHP dyeing machine. The acrylic tow dyeing is carried out using basic dyes or cationic dyes similar to the exhaust dyeing process of acrylic fibre and yarn-dyeing route. After the acrylic tow is dyed, the tow goes to hydroextractor where maximum water is removed from the tow. Then the tow goes for drying in RF dryer or steam conveyer dryer. The dyed acrylic tow goes to tow to top converter, and dyed acrylic tops are made and go for worsted spinning process to make dyed acrylic yarn.

2. *Continuous dyeing of acrylic tow (pad-steam method)*. This process is very popular in industries. The tow dyeing machine is manufactured by Serracant, Spain and Fleissner, Germany. The procedures involved are as follows:

   a. Padding of the tow with dye liquor
   b. Fixation of dye by steaming
   c. Washing off and softening using backwasher
   d. Drying

The recipe for padding liquor is

   a. X g/L acrylic liquid dye
   b. 2–10 g/L non-ionic, acid-resistant thickener (locust bean gum)
   c. 6–12 g/L non-ionic padding auxiliary
   d. Adjust pH 4–4.5 with acetic or tartaric acid
   e. Padding temperature 30–50°C

Pick up percentage varies from 50 to 80% machine to machine, steaming temperature varies 100–115°C and steaming time 30–50 min.

The dyed tow has to be further processed satisfactorily to remove all residues of chemicals and thickeners used in dyeing. The tow is therefore treated on suitable back washers (2–4 baths) with the overflow at a temperature of 50–70°C. In the washing tanks, a good detergent and a dispersing agent are required.

After washing, the tow is softened by a suitable cationic softener in the last tank. The drying of the tow is carried out in a perforated dryer with overfeed at 110–130°C. Finally conditioning should be not <24 h. Then the dyed acrylic tow goes to tow to top converter machine where dyed acrylic tops are made. The tops are reading for worsted spinning.

2.4.2   Advantages of dyed acrylic tow

1. It is best suited for excellent running on stretch breaking tow to top convertor machines.
2. Minimum waste generation.
3. Highest productivity in grey as well as dyed form due to the perfect combination of its tenacity and elongation parameters.
4. It is well-accepted worldwide as an ideal raw-material for producing ready to use dyed on cone high bulk/non-bulk acrylic yarns for hosiery and textile applications.
5. Pasupati Acrylon, India, supplies acrylic tow in a broad range of k-tex ranging from 110–130 k-tex, making it versatile for processing on all types of stretch breaking tow to top convertor machines with best efficiency.

The characteristics and properties of acrylic tow are summarized in Tables 2.1 and 2.2.

Table 2.1. Characteristics of acrylic tow

Denier	0.9–15
Weight per meter (k-tex)	110–130
Lustre	Bright/semi-dull
Characteristics	Regular, shrinkable
Packing	HDPE bales of 400 ± 10 kg

Note. HDPE, high-density polyethylene.

Table 2.2. Properties of acrylic tow

Properties	Unit	Type						
Fineness	Denier	0.9	1.2	1.5	2	3	5	7
Tenacity	g/denier	>3.5	>3.2	>3.0	>3.0	>2.8	>2.6	>2.5
Weight per meter	k-tex (NKT)	108 ± 3	109 ± 3	110 ± 3	122 ± 4	122 ± 4	122 ± 4	122 ± 4
	k-tex (OKT)	–	–	–	–	132 ± 5	132 ± 5	–
Elongation	(%)	32 ± 5	33 ± 5	38 ± 7	40 ± 7	40 ± 7	40 ± 7	40 ± 7
Shrinkage	(%)	<2.5	<2.5	<2.5	<3.0	<3.0	<3.5	<3.5

2.4.3   Conclusion

Acrylic tow dyeing by continuous method is not popular in India. Very few units are using this technology. In India, acrylic hank dyeing and fibre dyeing are very much in demand. Acrylic package dyeing is also not much in demand in India. Polyester tow dyeing also should be started in the same manner.

2.5   Dope dyeing

The process of introducing dye (or pigment) into the molten plastic or solution from which yarns are produced to manufacture coloured yarns rather than the usual white ones is called dope dyeing. As no extra dyeing process is required, it is the least impact dyeing process. Dye is added to the solution before it is extruded through the spinnerets for making synthetic filaments. A process called solution pigmenting or dope dyeing has been used for man-made fabrics ranging from rayon through saran and glass fibres. In dope dyeing, dye is added to the spinning solution before it is extruded through the spinnerets into filaments. This method also gives a greater degree of colourfastness. Effective results have been obtained by this method.

This is the method of dyeing in which the dye or pigment is added to the spinning solution before the extraction of filaments/fibres through the spinnerets. The addition of dye or pigment to the spinning solution before it is forced through the spinnerets.

- Solution dyeing also called mass pigmentation and dope dyeing.
- This method of dyeing is generally used in the production of man-made fibres so as to save a great deal of money and time.
- Most olefin fibres are solution dyed.

2.5.1   Process

The procedure is used for other man-made fibres that are difficult to dye and have special requirements for colour fastness.

- Solution dyeing is the method of colouring that is actually part of the manufacturing process of man-made fibres. In this method, the colouring agent is added to the liquid spinning solution of man-made fibre before it is extruded from a spinneret.
- The liquid spinning solution is sometimes called as fibre dope. Hence the term 'dope dyeing'. The colour becomes part of the fibre itself and is thus permanent.

2.5.2   Merits and demerits

2.5.2.1   Merits

- Solution dyed colours are practically fade proof under all common conditions of use. Their fastness to light is outstanding.
- Fabrics made from solution-dyed yarns are thus well suited from draperies, automotive fabrics and other applications where long sun light exposure is anticipated.
- Almost perfect colour reproduction consistency from dye lot to dye lot can be achieved by solution dyeing.

2.5.2.2   Demerits

- Solution-dyed materials are available in only a limited range of colours.
- Economically unfeasible.
- Solution-dyed yarns are not widely used in fashion apparel fabrics because the range of available colours is too limited to satisfy the great variety of colours needed for the fashion market.

Solution dyeing, also known as dope or spun dyeing, is the process of adding pigments or insoluble dyes to the spinning solution before the solution is extruded through the spinneret. Only manufactured fibres can be solution dyed. It is used for difficult-to-dye fibres, such as olefin fibres, and for dyeing fibres for end-uses that require excellent colourfastness properties. As the colour pigments become a part of the fibre, solution-dyed materials have excellent colourfastness to light, washing, crocking (rubbing), perspiration and bleach. Dyeing at the solution stage is more expensive, as the equipment has to be cleaned thoroughly each time a different colour is produced. Thus, the variety of colours and shades produced is limited. In addition, it is difficult to stock the inventory for each colour. Decisions regarding colour have to be made very early in the manufacturing process. Thus, this stage of dyeing is usually not used for apparel fabrics.

2.5.3   Package-dyed yarns

Package dyeing is a mature technology that is one of the most commonly used methods of yarn dyeing. This eight-step process involves dyeing natural yarns that have been wound on perforated spools that are dyed in a pressurized tank.

2.5.3.1   Package dyed advantages

- Small lot sizes are common and possible using package dyeing.
- Custom colour and shade matching are possible due to the smaller lot sizes than dope dyeing.

- These factors make package dyeing good for flexibility and colour development.

2.5.3.2   Disadvantages

- Colour uniformity is difficult to achieve from lot to lot, creating waste or product variation.
- Colourfastness is not as good as solution dyeing, leading to bleeding or transfer of dye onto other materials during subsequent operations as well as shade changes during downstream processing.
- Fading and changes to colour more likely from package-dyed yarns.
- Eight-step process has a longer lead time than solution-dyed yarn.
- Energy and water usage costs are very high for package-dyed yarn.

While package dyeing is adequate for most materials, thick, high twist yarns and threads may not allow for good dye penetration, and materials exposed to outdoor environments are likely to experience fading and shade changes.

2.5.4   Dope-dyed yarns

Doped-dyed yarns are created by adding a masterbatch colourant to the polymer melt in spinning or extrusion. This results in fibres and filaments that are fully impregnated with pigment coming out of the spinnerets in a one-step process.

2.5.4.1   Advantages

- Dope-dyed yarn is highly resistant to ultraviolet (UV) fade and shade changes.
- Dope-dyed yarns are fully uniform in colour and typically do not vary from lot to lot.
- Dope-dyed yarns are colourfast, resistant to multiple washings and mild bleach solutions.
- Energy costs for dope-dyed yarns are the same as natural yarns.
- More environmental friendly since no water is used in the dope-dyeing process.

2.5.4.2   Disadvantages

- Large lot sizes needed for dope-dyed yarns, usually 4500 kg minimum.
- Fewer colours available, custom colours impossible without very large lot quantities.

- Depending on the fibre being produced, potential for slightly lower tenacity for dope-dyed yarns.

For primary colours such as black, yellow, green, red and blue where precise colour matching is not required, dope-dyed yarns are recommended. Dope-dyed yarns are also ideal for use in producing bonded polyester and nylon sewing threads for both indoor and outdoor applications, since there is little possibility of dye bleeding and UV fading. The environmental impact of package dyeing with high water usage and energy costs is also leading many manufacturers to change to dope-dyed yarns to increase their sustainability scores.

They hence have good fastness to light, washing, hot pressing, dry cleaning, bleeding, perspiration, crocking and also sea water. In case on man-made fibres it is possible to mix pigments thoroughly with the fibre solution so that the fibre is coloured as it is formed and the pigment forms a part of the fibre. The colours thus produced are relatively permanent. In addition to being insoluble in water, pigments have no affinity for fibres. Technically pigments are not dyes, because they are insoluble in water or in the solvents typically used in dyeing.

2.5.5   Viscose dope dyeing

It means mass colouration pigments are added during viscose manufacturing stage. Total production of Grasim industries, Nagda (Madhya Pradesh) viscose staple fibre is 400 tons/day. Of this, 200 tons/day is dyed and 200 tons/day is grey fibre. Biggest consumer of dope-dyed viscose fibre in India is Sangam Spinners, RTM, CTM, BTM and RSWM. Dope-dyed viscose is used by dyed polyester/viscose spinning units. It is manufactured from wood pulp, which is imported in fine quality. The wood pulp is chemically treated and made into viscous solution where colour pigments are added and dyed viscose tow is made. Then it goes to cutter where it is cut into different staple length. Clariant offers a broad range of colourants for viscose dope dyeing.

Following are the pigments used in dope dyeing:
1. Vernafil violet O
2. Vernafil yellow A2GN
3. Vernafil red AGN
4. Vernafil blue ARN
5. Vernafil yellow AGX
6. Vernafil blue FRVD
7. Vernafil rubine BFVD
8. For black shade, they are using carbon black pigment

These pigments are manufactured by Colourchem, Mumbai or imported by Clariant, Switzerland. For a very long time dope-dyed viscose was used by each and every spinning mill in India making polyester/viscose, cotton/ viscose, etc., yarn. The dyeing charges of dope-dyed viscose were as follows: pastel shades – Rs 8/kg, light shades – Rs 12/kg, medium shade – Rs 18/kg, dark shade – Rs 22/kg, extra dark – Rs 28.5/kg, black shade – Rs 22.35/kg. Rate of grey viscose is Rs 85/kg.

2.5.5.1   *Advantage of dope-dyed viscose*

1. Washing fastness is good, rating 4–5.
2. Perspiration fastness is good, rating 4–5.
3. Light fastness is good, rating above 5.
4. Fastness to organic solvent, rating 4–5.
5. Sublimation fastness, rating 3–4, 4, 4–5 (depending on shade) at 150°C, 180°C, 210°C, respectively.
6. Runnability in spinning department is very good.
7. About 5000 shades range.
8. Over 120 shades manufactured every month.

2.5.5.2   *Disadvantage of dope-dyed viscose*

1. Brightness/lustre is less.
2. Prone to thermomigration.
3. Monopoly in standard shades minimum quantity to be ordered 3 tons and non-standard shades minimum quantity to be ordered is 5 tons.
4. During summer, the working in spinning mills is disturbed.
5. Difficult to manufacture in very light shades – less than 0.1% pigment depth.

Viscose is based on natural cellulose and is used for different textile and non-woven applications. It is pressed through spinnerets and is coagulated in a bath of sulphuric acid. Dope dyeing means that the colourants are added to the viscose before the spinning process takes place. With Clariant's pigment preparations, an excellent product quality could be obtained by economic processing in terms of longer filter life cycles and improved colour yields. The properties of pigments used are as follows:

- Flowable paste
- Easy incorporation and dispersion in water-based systems
- Maximum particle size <1 μm (ultrafine dispersed)
- High tinting strength and high transparency
- Especially low sedimentation tendency
- Ultra-fine dispersed pigments with excellent sedimentation behaviour and transparency

2.5.6   Polyester dope dyeing

Dope-dyed polyester, on one hand, is available only in a limited number of colours and shades. The required shade has often to be developed through R&D work or fibre has to be often dyed. With dyed fibre fastness to rubbing and washing will not be as good as dope-dyed fibre. Carbon black is used 1.5–2.5% depth to produce dope-dyed black. It is cheaper than own dyed black polyester. It can be available in fibre and filament form. The overall fastness is good as compared to own dyed black polyester. It is manufactured by Reliance, IOC, Indorama and imported from China. Dope-dyed polyester is available only in a limited number of colours and shades.

Apart from Black, Reliance has developed a wide range of dope-dyed polyester fibres in 1.4, 2.0, 6.0 and 15 denier in regular and special cut length. They are available in special quality pigments and dyes for a wide range of applications in apparel and automotives. The range includes olive green, khaki green, navy blue, sky blue, dark grey and chocolate brown, red and parrot green. All the Recron dope-dyed polyester fibres have a very high shade consistency over long yardage, as well as unbeatable washing, light, heat and sublimation fastness.

The IOC is also having a complete range of dope-dyed polyester fibre. It can directly go for spinning of yarn or blended with Grasim Nagda dope-dyed viscose and spun into dyed polyester/viscose yarn. Dope-dyed fibres have been produced starting from PET bottle wastes. Dope dyeing has been carried out both by the incorporation of pigments during drying before melt extrusion and by masterbatch addition. Fibres are produced using different pigments.

2.5.7   Polypropylene dope dyeing

Density of polypropylene (0.91 g/cc) is much lower than that of polyester (1.38 g/cc). Diameter of polypropylene fibre is therefore proportionately higher than polyester fibre of the same denier. As a result, thicker, bulkier yarns and loftier fabrics and more comfortable carpets are made with the former for a given count of yarn and area density of fabric. As polypropylene molecular chains have no polar functional groups (active sites for chemical bonds or dye sites) and relatively high crystallinity (50–65%), dye molecules cannot be chemically attracted to the fibres. Adsorbed dye molecules which interact with the fibre surface by weak van der Waals force are easily washed away because of polypropylene's hydrophobic properties. However, polypropylene fibre is dope dyed (spun dyed) by the manufacturer in virtually unlimited colour choices. Polypropylene is dope dyed and is available in an extensive range of colours and shades. It is therefore much easier to achieve colour and shade matching by mixing a minimum number of shades of fibres.

2.5.8   Nylon dope dyeing

It is made before the extrusion process, the solution of polymer is mixed with pigment, and extruded yarns are coloured and with excellent colour fastness. Nylon dope-dyed yarn is considered as an eco-friendly yarn. No dyeing process is needed after knitting/weaving process. This helps to save water and energy consumption, avoid from dyeing pollution and reduce CO_2 emission.

2.5.9   Acrylic dope dyeing

Daqing Petrochemical has successfully tested its dope-dyed acrylic fibre production line. The new technique replaces the traditional acrylic dyeing process of chemical dyes with a new process that gives higher finished colour fastness and more protection from sunrays. In the new production technique, the entire process is carried out without any waste water discharge. Dope-dyed acrylic fibre is widely used in the making of umbrellas, tents, beach umbrellas and other outdoor leisure items.

2.6   Flock dyeing

Historians claim that flocking can be traced back to circa 1000 BC, when the Chinese used resin glue to bond natural fibres to fabrics. Fibre dust was strewn onto adhesive coated surfaces to produce flocked wall coverings in Germany during the middle ages. In France, flocked wall coverings became popular during the reign of Louis XIV of France.

Look around you. The home you live in. The paper you write on. The car you drive. From the carpet you stand on, to the chair you sit in, flock has become an integral part of everyone's daily life. Flocking fascinates because a textile, velvety or brush-like surface may be applied to almost any material. Flocked products are everywhere – flock is found on T-shirts, in packaging for perfumes, car glove boxes, car head liners, floor coverings, model railway landscapes, eyeliner brushes, scrubbing pads, etc.

Consumers are always looking for something different and unusual. Suppliers seek the same thing – a special item or product that will increase their market share or generate new business. Developing something different is always a top priority and is the driving force behind the recent resurgence of printer interest in learning about flocking.

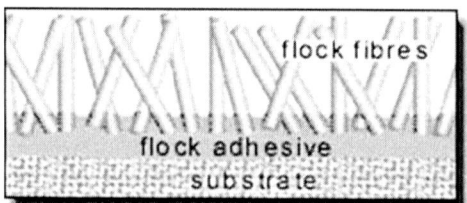

Figure 2.5. Flocking

Flocking for decoration is not new. Similar methods were used in the middle ages to attach fibre dust to sticky surfaces. It was in the 1970s, with the advent of improved technologies and adhesives, that flocking became a popular decoration method. In the 1980s and early 1990s, flocking popularity faded away and few printers used the process. While flocking is not the most widely used decorating process, the average person is aware of its velvet or suede feel (Fig. 2.5).

Over the last several years, however, inquiries about the process have begun to increase, and flocking is once again in demand as a decorating method. Even though flocking may not be most decorators' first choice process at present, it is used widely in many industrial applications. Flocked surfaces reduce water condensation, act as good thermal insulators and have been used in the automotive industry for years for such items as glove compartment boxes, door mouldings and window trim.

2.6.1   What is flocking?

The flocking process involves applying short monofilament fibres, usually nylon, rayon or polyester, directly onto a substrate that has been previously coated with an adhesive. Flocking is defined as the application of fine particles to adhesive coated surfaces. Nowadays, this is usually done by the application of a high-voltage electric field. In a flocking machine, the 'flock' is given a negative charge, while the substrate is earthed. Flock material flies vertically onto the substrate attaching to previously applied glue. A number of different substrates can be flocked including textiles, fabric, woven fabric, paper, PVC, sponge, toys and automotive plastic.

The majority of flocking done worldwide uses finely cut natural or synthetic fibres. A flocked finish imparts a decorative and/or functional characteristic to the surface. A variety of materials that are applied to numerous surfaces through different flocking methods create a wide range of end products. The flocking process is used on items ranging from retail consumer goods to products with high technology military applications.

The diameter of the individual flock strand is only a few thousand of a centimetre and ranges in length from 0.25 to 5 mm. Adhesives that capture the fibres must have the same flexibility and resistance to wear as the substrate. The process uses special equipment that electrically charges the flock particles causing them to stand-up. The fibres are then propelled and anchored into the adhesive at right angles to the substrate. The application is both durable and permanent. Flock can be applied to glass, metal, plastic, paper or textiles. Flock design applications are also found on many items such as garments, greeting cards, trophies, promotional items, toys and book covers.

2.6.2   Application methods

Decorative flocking is accomplished by using one of four application methods:

- Electrostatic,
- beater bar/gravity,
- spraying and
- transfers.

The electrostatic method is perhaps the most viable flocking method, especially for the printer doing more than an occasional flocking job.

Flocking material can also be sprayed using an air compressor, reservoir and spray gun similar to spraying paint. It is primarily used when large areas require flocking. It is an untidy process, because some of the flocking fibres become airborne. Flocking is also applied by printing an adhesive onto a substrate, and then rapidly vibrating the substrate mechanically, while the flock fibres are dispensed over the surface (Fig. 2.6).

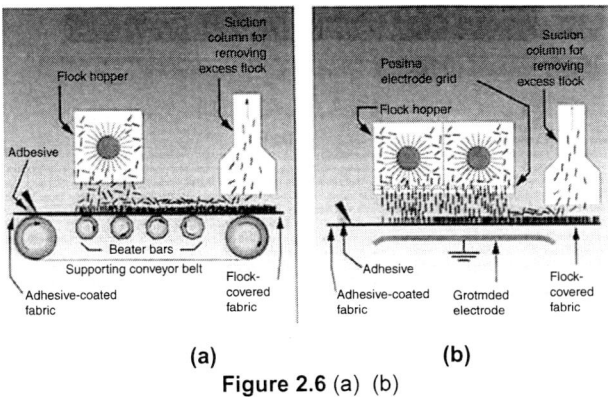

Figure 2.6 (a) (b)

The vibration promotes the density of fibres, which is critical to good fibre coating, and causes the flocking fibres to adhere to the adhesive and pack into a layer. This process is called a beater bar or gravity flocking system and is basically a mechanical process. With this process, the flocking fibres are randomly adhered to the surface of the substrate, and each fibre adheres to the adhesive at a different depth, creating an irregular flocked surface. Since the fibres adhere to the surface of the adhesive, rather than penetrate or imbed in it, some fibre shedding occurs. Loose flocking fibres generated during production also have a tendency to migrate, so many of these systems are installed in a separate area to prevent fibre contamination of the factory.

A much easier way to add flocking to your operation is to purchase and apply standard flock transfers. Another choice is a patented process, where the transfers are multi-coloured and very detailed, and are produced without ink.

Flocking is a value-added alternative decorating method for achieving that unusual look. It only costs slightly more that producing a standard screen print, and in conjunction with textiles, it is certainly less expensive than embroidery. This process uses dyed fibres of nylon that are charged with high voltage and driven into the design. This process eliminates fibre shedding, which is common with most other flocked images. Standard flock transfers can also be made in-house.

Basically the pre-press is virtually the same as it is for screen printing with only a few differences. The screens are made from coarser mesh counts and the line thickness of some of the artwork may need to be increased. The adhesive is printed, and a crystalline powder is applied to the transfer and then is sent through the dryer at a low temperature. This removes moisture from the adhesive, but does not melt the crystalline powder. Excess powder is removed and the transfer is sent through the dryer a second time to chemically bind the powder to the adhesive.

2.6.3   The flock

Flock can be made from natural or synthetic materials such as cotton, rayon, nylon and polyester. There are two types of flock – milled and cut. Milled flock is produced from cotton or synthetic textile waste material. Because of the manufacturing process, milled flock is not uniform in length.

Cut flock is produced only from first quality filament synthetic materials. The cutting process produces a very uniform length of flock. Lengths can be obtained from 0.3 to 5.0 mm and 1.7 to 22 dtex in diameter.

Advantages of flocked surfaces are:

- pleasant appearance
- fibred grip
- persistence against bounces and scratches
- good sliding effect on even surfaces and many other specialized properties

The fineness of the flock, length of fibres and adhesive coating density determine the softness of the flocking. It should be noted, however, that fine or short flock is difficult to work with, since it has a tendency to ball-up during processing.

Milled cotton flock has the advantage of being the lowest in cost and the softest, but has the least abrasion and wear resistance. Rayon is a little bit better on wear resistance and nylon is the best. For cut flock, rayon is the least expensive with the least wear resistance. Cut nylon is the best grade of flock and produces a good feel, but is also the most expensive. Cut polyester is basically used for industrial applications such as automobile window seals, glove compartments and roofing. Besides cutting or milling, flock manufacturing includes several other steps.

After cutting, the flock is cleaned of oils that accumulated during processing. It is vat dyed to any number of colours, and then chemically treated to enable the fibres to accept an electrical charge. Since the fibres are all dielectric, a certain amount of conductivity must be present for electrostatic flocking process to occur. When the process is complete, the fibres are spin dried and then oven dried to a specific moisture content. Note that flocking fibres are never totally dried, since moisture content adds to their conductivity. Finally, the flock is packaged in moisture proof bags that maintain proper humidity.

2.6.4   The adhesive

Flock adhesives are available in both a single and a two-part catalyzed system. There are also plastisol and water-based adhesives. Many of the adhesives have the consistency of plastisol ink. Care should be exercised to select a stencil emulsion or film that is compatible with the adhesive to be printed. Proper application of the adhesive is the most important part of the process.

A very heavy deposit of adhesive is required, but at the same time the adhesive should not be 'squeegeed' through the substrate. Care should also be exercised not to apply a too thin coating.

2.6.5   The electrostatic system

Electrostatic flocking equipment is available in three configurations:

- an automatic carousel for multicolour flocking,
- a single station flocking unit that usually attaches to one station of a garment press and
- a portable hand-held unit for lower volumes.

The cost of the equipment varies from hundreds or a few thousand dollars for hand-held units to tens or hundreds of thousands of dollars for automatic multicolour systems. All of the equipment operates using the same basic procedure. In flocking, the electrical charge is generated by the use of two electrodes:

- a high voltage, direct current grid connected to a power generator and
- a grounded substrate

An electrostatic charge is generated that propels the fibres at a high velocity onto the adhesive coated substrate. This causes the flocking fibres to penetrate and imbed in the adhesive at right angles to the substrate. This forms a high density uniform flock coating. Controlling the electrical field by increasing or decreasing either the applied voltage or the distance between the electrodes and the substrate controls the speed and thickness of the flocking.

2.6.6   Hand held units

The hand held units are comprised of a metal plate, a generator and a flocking head. The metal plate must be grounded, and it can be placed where convenient. It is the equivalent of the plate on a textile press. The generator creates the electrostatic charge, and is wired to a canister that contains the loose fibres. A metal screen is mounted halfway inside the canister opening. The open end of the canister is then passed over the adhesive coated substrate, drawing flock fibres from the canister through the screen. The electrostatic charge propels the fibres towards the grounded metal plate. The adhesive coated substrate intercepts the fibres and flocking occurs. The substrate is then cured in a conventional textile dryer, and the loose fibres are removed by shaking, vacuuming or by using compressed air.

2.6.7   The environment

Having a controlled atmosphere for flocking operations is another essential ingredient for success. Ideally, the flocking area should have a relative humidity of 60% and a temperature of 20°C (68°F). A small variation in temperature or a change in the percentage of relative humidity can result in a three- to four-factor change in the conductivity or electrical sensitivity of the flock and the substrate. Flocking fibres are very sensitive to humidity and temperature conditions.

When a new batch of flock fibres is opened, the fibres will give off or receive moisture based on the surrounding environment. Less than 30% relative humidity in the production area will lead to fibres that would not accept a charge. Relative humidity in excess of 65% causes the flock to stick together and flow poorly through the metal screen or plate. For best results, the flocking operation should be located in an atmospheric controlled room.

2.6.8   Uses of flock fibres

Flocking is used in many ways. One example is in model building, where a grassy texture may be applied to a surface to make it look more realistic. Similarly, it is used by model car builders to get a scale carpet effect. Another use is on a Christmas tree, which may be flocked with a fluffy white spray to simulate snow. Other things may be flocked to give them a velour texture such as t-shirts, wallpaper, gift/jewellery boxes or upholstery.

Besides the application of velvety coatings to surfaces and objects there exist various flocking techniques as a means of colour and product design. They range from screen printing to modern digital printing to refine for instance fabric, clothes or books by multicolour patterns. Presently, the exploration of the flock phenomenon can be seen in the fine arts.

Flocking in the automotive industry is used for decorative purposes and may be applied to a number of different materials. Many rally cars also have a flocked dashboard to cut down on the sun reflecting through the windscreen. A view on the present state-of-the-art of flocking can be found in the first international exhibition 'Flockage: The Flock Phenomenon' in the Russell-Cotes Art Gallery & Museum in Bournemouth.

In the photographic industry, flocking is one method used to reduce the reflectivity of surfaces, including the insides of some bellows and lens hoods. It is also used to produce light-tight passages for film such as in 135 film cartridges.

Flock consists of synthetic fibres that look like tiny hairs. Flock print feels somewhat velvet and a bit elevated. The length of the fibres can vary in thickness which co-determines the appearance of the flocked product. Thin fibres produce a soft velvety surface, thicker fibres a more bristle-like surface.

The flocking process is used on items ranging from retail consumer goods to products with high technology military applications. Today we see flock in every day applications from fine upholstery to durable carpeting; from fashion fabrics to children's shoes; from Christmas ribbon to engineered automotive parts. The applications are practically endless.

3.1   Package dyeing

It is a yarn colouration process. Yarns wound onto perforated poly propylene tubes or stainless steel dye spring fit over perforated pipes in package dyeing machines called as high temperature and high pressure (HTHP) which are available in vertical or horizontal type of dyeing machines manufactured by Fong's, Thies, Then, Cubotex and Dalal.

The term package dyeing usually denotes for dyeing of yarn that has been wound on perforated cores. The dye flows through to the yarn package with the help of the deliberate perforations in the tube package. Once full exhaustion is brought about, the carrier of coloured yarn is consequently removed from the vessel.

The package which is wound may be of polyester, cotton, viscose or acrylic. The weight of the package varies from 500 g to 2 kg each. The other names for package dyeing is cop, cheese and cone dyeing. Following are the steps followed in package dyeing.

6. Soft package winding. In package dyeing even the most expensive dyeing equipment will not produce perfect results without properly wound dye packages. In fact, with the right kind of package your job of yarn dyeing is already half done. In other words, each and every yarn dye package has to conform to a prescribed density, and this density has to be uniform from inside to outside and from tip to toe of the package. Similarly, packages of identical density and uniformity have to be produced on any spindle of a winding machine at any given time, meaning total reproducibility. The density of spun yarn dye packages recommended by leading dyeing vessel manufacturers are:

    a. For cotton: 420 g/L
    b. For cotton/polyester: 460 g/L

Since such packages, as compared with those intended for use in warping creels or knitting creels, feel spongier, they are universally referred to as 'soft packages'.

7. Package dyeing machine. Package dyeing machines are the most widely used now a days for dyeing of almost all types of yarns, due to economical, automatic and accurate dyeing results. The term package dyeing usually denotes for dyeing of any type yarn wound on the compressible dye springs/perforated solid dyeing tubes or cones. Yarn dyeing in package form is done at high temperature and under high pressure, with the packages mounted on hollow spindles. These spindles are fixed on the dyeing carriers, which is inserted into the dyeing vessel after closing the lid of the machine, the dyeing liquor is forced through the packages in two-way pattern (inside to out and outside to in) and goes on circulating throughout the vessel and yarn. Heat is applied to the dye liquor to achieve the dyeing temperature, time–temperature and flow reversal are controlled through a programmer.

A series of technical developments in the recent years has resulted into package dyeing being developed into a highly sophisticated as well as an economic process. Latest design package dyeing machines are amenable to accurate control and automation. These features would likely to lead to increases in the application of package dyeing.

The term package dyeing usually denotes for dyeing of yarn that has been wound on perforated cores. This helps in forcing the dye liquor through the package. With the start of dyeing cycle, the dye liquor goes on circulating throughout the vessel and tank. This happens till all the dye is used up or fully exhausted. The dye flows through to the yarn package with the help of the deliberate perforations in the tube package. Once full exhaustion is brought about, the carrier of coloured yarn is consequently removed from the vessel. A large centrifuge removes excess water from the packages. Finally the yarn is dried using an infra red drying oven. The image shows the process working of a package dyeing machine.

3.1.1   Working process of package dyeing machine

The material to be dyed is wound on the dye springs, perforated plastic cheeses or steel cones and loaded in the carrier spindles, which are compressed and bolted at the top to make a uniform and homogeneous dyeing column. The liquor containing dye chemicals and auxilliaries is forced through with the help of pump, and circulated through the material from inside–out and is reversed periodically so that each and every part of the material get the same and uniform treatment. The dyeing cycle is controlled through a micro computer and different chemicals may be added through the injector pump or colour kitchen at any stage of dyeing.

In case of fully flooded machines, the liquor expands with the rise in temperature (approximately 5% volume increases from 30°C to 130°C temperature) is taken back in the expansion tank through a back cooler. This extra water is then again injected to the dyeing vessel through an injector pump. Expanded volume of the dye liquor is thus remains in continuous circulation in the system. Any type of addition can be done to the machine through the injector pump. The quantity and time of injection can be controlled through the programmer.

In case of air pad machines, the air above the liquor acts as a cushion, which is compressed with the increase in liquor volume, the pressure is controlled by pre-set pressure control valve. In air pad, machines have an advantage, that entire dye liquor participate in dyeing and dye exhaustion is perfect. In case some addition has to be done in air pad machines, if the machine temperature is <80°C, the liquor is taken back by back transfer valve to addition tank and injected back to machine vessel. If the machine temperature is >80°C then cooling has to be done to bring down the machine temperature.

Air pad technology is possible in all types of machines such as vertical kier, horizontal kier and tubular dyeing machines. The material after dyeing is washed and finished properly in the same machine and taken out hydro extracted or pressure extracted in the same machine and dried subsequently.

3.1.2   Advantages of package dyeing machine

Package dyeing methodologies have been subjected to intensive research and development. As a result package dyeing machine has evolved into a very sophisticated apparatus. It offers a number of advantages (Fig. 3.1).

Figure 3.1. Package dyeing machine

- Considerable reduction in yarn handling
- Compatible to automatic control, in the process leading to reproducible dyeings
- Open to large batches
- High temperature dyeing a possibility
- Low liquor ratios, giving savings in water, effluent and energy
- Uniform and high rates of liquor circulation that leads to level application of dyes. Machinery totally enclosed resulting in good working conditions at the dye house.

3.1.3   Types of package dyeing machines

Different types of package dyeing machines are

1. Vertical kier dyeing machines
2. Horizontal kier dyeing machines
3. Tubular dyeing machines

3.1.3.1   *Vertical kier dyeing machines*

These machines have a vertical cylindrical dyeing kier, in which material loaded into carriers with vertical perforated spindles, is dyed. The machine could be fully flooded or air-pad type. These are high pressure machines and suitable up to 135°C temperature dyeing.

3.1.3.2   *Horizontal HTHP dyeing machines*

These machines are similar to vertical type machines in which the cylindrical dyeing kier is in horizontal position. The dyeing carriers with vertical spindles are used in these machines, which are inserted into the machine via trolleys. These machines are erected at the ground level and hence do not need an overhead hoist as well as platform, thus making the dyehouse design and layout is simple.

3.1.3.3   *Tubular HTHP dyeing machines*

These machines may be of vertical or horizontal type, and have one or many tubes acting as small dyeing vessels, each with a single individual spindle. The spindle is taken out of the tube, loaded and then inserted back into it. These machines can be operated either fully loaded tubes or to partial loads by using dummies. Since all individual tubes in a machine are connected and serviced by a main pump, it is also possible to operate as many tubes as required and disconnecting others.

These machines can be erected at ground level and hence do not need a platform or hoist. These machines are most flexible as for as the capacity variation is concerned, without altering the material to liquor ratio.

In its simplest form a package-dyeing machine is a vessel capable of containing packages of textile material through which heated dye liquor is passed by means of a circulation pump. Later developments accelerated by the need to dye polyester at temperatures above the boil lead to enclosing and strengthening such vessels so that they could operate up to 140°C at pressures around 70 psi (4.95 kg/cm^2).

Accessories were added to allow samples to the extracted without depressurizing the whole system and to inject dyes and chemicals from out with the main circulation circuit. Later still, simple controls of time and temperature were replaced with fully automatic programmes based on sophisticated microprocessors that reduced operator involvement in the dyeing process to a minimum and elevated limits of accuracy sophistication previously unattainable levels.

3.1.4   Liquor flow

We expect a main circulating pump to deliver 30 L of liquor per kilogram of thread at 1.26 kg/cm^2 pressure which usually means the bath is pumped through the thread load up to four to five times per minute. Exceptionally, cheese-bleaching machines need only deliver half of the discharge to be effective. Many of us, when faced with an unlevel cheese dye lot, blame our troubles on poor liquor flow which, because the dyeing process, by necessarily, is unobservable and because there is no instrument to read out the flow and is hard to prove one way or the other. But a small one, however, can interpret the evidence available to him, for example, here are a few tips on how to ascertain whether or not abnormal liquor flow is the source of unlevelness.

1. Check the in–out and out–in pressure gauges and compare the readings with your past experience. Your Dalal and Staffi machines with their modest pumps should register a pressure differential of around 0.5 kg/cm^2. If the differential is significantly lower than this value, liquor may be freewheeling or channelling through a badly seated carrier, a sprung cap or a loosely loaded column of cheeses.
2. Likewise pressure differential higher than 0.5 kg/cm^2 could indicate that something is causing unduly high back pressure, for example, very dense cheeses.
3. Unlevelness on a number of cheeses which represent one spindle or multiples of one spindle might indicate poor sealing of the number(s) of spindles involved.
4. Unlevelness on a number of cheeses that represent one complete layer as horizontal cross section of a carrier load of cheeses may mean that the machine has (levelled out) for sampling or during a

power failure exposing the top most layer of cheeses to oxidation or differential dye uptake.

5. Loss of air-pad pressure in one way low liquor dyeing can cause reduced liquor flow.

3.1.5   Open expansion tank

This tank is sized so that the top row of cheeses is exposed when liquor is levelled bag to the expansion tank from the kier by gravity. The tank feeds the suction side of the secondary pump, which normally discharges into the main pump housing via the non-return valve. The expansion tank is an invaluable aid to level dyeing as it allows controlled additions of chemicals and redip dyes, when pressurized.

Extraction rate from the expansion tank is usually 5–25 L/min with the pump running at a pressure of around 3.6 kg/cm². It is important that the right balance between expanding main kier liquor and expansion tank injection rate is struck otherwise liquor flow may be affected. This balance is obtained by drilling out the orifice plate on the cooled liquor return from the main kier to the expansion tank.

The efficiency of the back cooler or condenser is also important since if the temperature in the expansion tank is allowed to rise about 80–85°C, the adversely secondary pump may cavitate thus affecting the flow characteristics of the dyeing system. If the liquor is over cooled, energy is wasted in reheating it in the main kier and of course cooling water volumes are unnecessarily high.

3.2   Hank dyeing

The yarns are loosely arranged in skeins or hanks (*skein* is the length of a yarn wound in a loose coil). Then they are hung over a bar and immersed in a dyebath. In this method, the colour penetration is the best and the yarns retain a softer feel. The process of dyeing yarn in the form of hanks or skeins is carried out by both hand and machinery. It is also called skein dyeing. Skein dyeing consists of immersing large, loosely wound hanks (skeins) of yarn into dye vats that are especially designed for this purpose. Soft, lofty yarns, such as hand knitted yarns are usually skein dyed. Skein dyeing is the most costly yarn-dye method.

3.2.1   Manual method

In previous years, hanks were dyed manually with two persons holding the stick. Hank dyeing is a simple but time-consuming process. First, the skein of yarn is looped over a hook and washed in water, opening the fibres to receive the dye. It is then dipped into the dye for up to 48 hours, washed and redipped. This procedure is repeated for several times. Once the desired colour is achieved, the yarn is steamed to fix the dye to the fibres. Because it does not use as many chemicals as other forms of dyeing, hank dyeing is less damaging to the material. The final dye colours are also usually richer than those achieved by other dyeing methods.

3.2.2   Why hank dyeing?

Although package dyeing route is much simple and easy to follow for dyeing of different kinds of yarn, the hank route is popular for certain qualities. There are two reasons for this:

1. The quality and nature of the product do not allow to follow package route.
2. The package route is non-practical and non-economical.

3.2.3   Product quality

Certain kinds of yarns which are so delicate that their physical properties such as strength, appearance, texture and construction cannot be retained while following the package route in which yarn is processed under stretched conditions at elevated temperatures. The examples of such products are natural silk, viscose rayon, voluminous yarns, high bulk acrylic, shrink resist treated wool and its blends, cashmere, its blends with wool and nylon, nylon and hand knitting yarns of different fancy constructions.

3.2.4   Economy of process

In case of mercerized cotton yarns, the yarn mercerizing is done in the hank form and material has to be neutralized before dyeing. Therefore, it is not economical to convert these hanks to packages for dyeing.

Process sequence for yarn dyeing in hank form:

1. Singeing (for mercerized cotton yarns)
2. Reeling
3. Mercerization (for mercerized cotton yarns)
4. Dyeing
5. Hydro extraction or squeezing
6. Drying
7. Hank to cone winding

3.2.5   Types of different hank dyeing machines

3.2.5.1   Roller hank dyeing machine

These machines are based on Hussong type of hank dyeing machines in design. These machines are very simple in design and operation. In the basic design of machine, there is rectangular stainless steel (SS) tank with proper heating mechanism (either a heating coil or direct steam injection through a perforated steam pipe) and have a false bottom above the heating coil. The tank is fitted with a temperature indicator and control sensor. There is an array of SS rollers/poles which are fitted on a suitable frame. The rollers can rotate in clockwise or anticlockwise directions with the gears arrangement. The machine is fitted with a stop/start system to control the roller movement.

The hanks loaded on rollers can be lowered in the tank or lifted out by lifting or lowering the frame with the help of hydraulic system. The speed (RPM) of rollers can be varied by changing the gears and rotation cycles are controlled by timers which change the direction of movement of motors. The dyeing vats have the filling and draining facility of dye liquor. The dyes and chemicals are added directly into the dye vat.

Operation of the machine

The hanks to be dyed are loaded onto the roller bars by avoiding overlapping of hanks, in lifted position. The dyebath is kept ready with proper liquor level and mixed homogenously, preferably with an external circulation pump. The hanks are lowered into the dyebath and movement is started, approximately 40–50% part of the hanks remain the liquor and rest is exposed to atmosphere. The material is lifted up for addition of colours, chemicals, salt, alkali, etc., or for raising the temperature of dyebath (in case of direct steam injection). In case of indirect heating, the dyebath temperature may be increased at set rate of heating, but in case of direct heating, the temperature rise is step by step. The sample may be drawn at the completion of dye cycle and checked for shade, without stopping the machine. All operations can be done in the same machine.

Advantages and disadvantages of roller hank dyeing machines

1. Roller dyeing machines are simple in design and operation.
2. Cost of the machine is not very high.
3. Almost all types of dye classes such as direct, reactive, vat, naphthols, pigments and soluble vats can be applied to cotton, mercerized cotton or rayon filaments.
4. All types of yarns can be processed such as cotton, mercerized cotton, viscose filament or even silk can be processed. The machines with simple modifications can be used to produce tie and dye effects in bigger lots particularly in cellulosic substrates.

5. The machine capacity depends upon the number of rollers, which could be from 1 to even 10 rollers, and each roller can handle from 2 to 4 kg of yarn.
6. Only 40–50% of yarn takes part in dyeing and rest is exposed to atmosphere. Therefore hydrosulphite consumption is high in these machines.
7. The machines are open type. Therefore maximum dyeing temperature is 95–98°C.
8. The liquor ratio increases with heating in case of direct injection of steam.
9. Liqour ratio is high, generally 1:15 to 1:20; therefore, chemicals, steam and water consumption are comparatively high.
10. The yarn entanglement takes place during running which reduce the hank to cone winding efficiency.
11. Only limited automation is possible.
12. The dyeing is carried out under atmospheric pressure and there is no liquor pressure. Therefore, the colour penetration in hard twisted material is poor.

3.2.5.2   Cabinet type hank dyeing machines

These machines are designed to have liquor flow parallel to the yarn and they work with low liquor ratios to save substantially in consumption of thermal energy, water and chemicals. The material to be dyed in hanks is loaded on the sticks, generally two sticks are used one at the top and other at the bottom. The bottom stick is used to avoid the yarn entanglement during the liquor flow. The carriers are loaded with the sticks having hanks. The sticks are supported into slots made in the body of the carrier. The loaded carriers are then put into the dyeing machine and door is closed.

In a second variant of the machine, the machine is open at the top and career is loaded from the top, where a rubber gasket is provided around the periphery on which the machine career rest, and vessel is sealed by the weight of the career and material, thus making closed circuit. The liquor circulation and flow remain the same as in ordinary cabinet machine. The dye liquor in these machines moves while the material remains stationary. The flow direction is reversed frequently to ensure level dyeing. A cabinet dyeing machine consists of following main components:

- Main dyeing vessel made of SS
- Carriers for loading of hanks with SS sticks for loading of hanksAn internal axial pump or an external pump for liquor circulationAn indirect heating system (with internal or external heat exchanger)
- An addition tank with injector pump

- Preparation tank
- An internal or external sampling device

The advantages and disadvantages of cabinet type hank dyeing machines are as follows:

1. *Pressure dyeing cabinet for hanks.* The GMW-HDD model is specially designed for dyeing of polyester yarn/polyester cotton at a working temperature of up to 125°C and working pressure of up to 2.5 kg/km².

2. *Low liquor ratio steam and power consumption.*
   a. Automatic air-pad pressurization operator at 0.4 bar static pressure, which is obtained by a compressed air in the area above the liquor and this act as an expansion chamber for the increase of liquor volume during temperature rise or addition of dyes and chemicals and occupies approximately 20–25% of internal volume.
   b. Due to the fitment of steam coil at the bottom of the chamber eliminates the flooded liquor to the top of the chamber as compared with the traditional cabinets. Hence, the only flooded volumes are the hank racks, impellor and centre partition.
   c. The air-pad operation acts as an insulation, hence there is low heat dispersion, thus saves energy and steam.
   d. Total elimination of external piping and heat exchanger by use of centre partition for full width liquor circulation as compared to traditional cabinets.
   e. Liquor ratio varies from as low as 05 to 6:1 for Malai Dori (polyester blend), 9 to 10:1 for mercerized cotton and 12 to 14:1 for acrylics.

3. *Low investment and running costs.* The GMW-HDD requires very low initial investment as compared to HTHP package dyeing which requires high working height ceiling, electric overhead cranes, etc., and does not require hank to cone and then cones to hank binding, thus saving high labour and machinery costs.

4. *Multiple dyeing options.* The GMW-HHD gives an option of dyeing all types of yarns in hank form including wool, HB acrylics, cotton, polyester, polyester blends, Malai Dori, etc.

3.2.5.3   *Spray hank dyeing machine*

The ABEP is a new technical evolution which extends the wide range of products that Loris Bellini manufactures for special applications on delicate yarns on hanks:

- Cashmere, alpaca, mohair, wool and blend yarns
- Natural silk yarns and blends with viscose and wool
- Filament viscose yarns
- Mercerized cotton yarns
- Acrylic knitting yarns, very-high bulk type

ABEP solves the problems relevant to the common arms dyeing machines:

- Hank loading and unloading operations
- Regulation of the flow rate when working at variable load
- Possibility of VAT dyeing

The technical features and advantages are as follows:

- Pressurized operation by air cushion, up to a maximum temperature of 102°C at sea level, which is independent to atmospheric pressure variations. This allows VAT dyeing.
- Extractable arms to make easy the hanks loading/unloading operations.
- Magnetic flow meter, feeding back the inverter, to work at a constant pre-set liquor flow-rate value no matter how many arms are in operation.
- Structure entirely made of high thickness plate in SS.
- Possibility to exclude arms down to 50% of the nominal installed capacity to reduce the total output per batch.
- Electro-galvanic mirror-polishing spray arms with internal liquor distributor engineered to guarantee a constant and even liquor flow rate along the entire length of the arms. Dyeing liquor distributor from circulation pump to sprayer arms designed for a perfectly even flow rate on all sprayer arms independently from their central or lateral position.
- AISI 316 liquor self-lubricating no-maintenance internal arms rotating mechanism, fault free controlled through proximity limit switches by our Leonardo PC software.
- Liquor level control to work at a constant liquor ratio, by setting the liquor volume depending on the number of arms in operation: high level of colour repeatability and reduction of colour corrections.
- Automatic control of liquor level and static pressure by continuous electronic probing sensor.
- Hank rotation system by rotating movement of arms: perfect material conservation for high-speed hank-to-cone winding.
- No drippings on the hanks, due to a special design of internal roof.

3.2.6   Dyeing of cotton in hank form

In the initial stages, cotton yarn was dyed in hank in a wooden vat manually. Then hank dyeing machines were started. The machine capacity varies from 50 to 300 kg. Then cabinet-type hank dyeing machines were started. After the cotton hank is dyed, it goes for hydroextractor, then is dried in chamber dryer/radio frequency (RF) dryer. Then hank to cone is made in a winding machine.

3.2.7   Dyeing of wool in hank form

Hank dyeing is mainly done for carpet wool yarn dyeing. It is dyed in cabinet hank dyeing machine. The biggest centre is Bhadohi, Banaras (Uttar Pradesh), Panipat and Amritsar. For carpet yarn dyeing, you can use acid milling dyes from Clariant. Cabinet hank dyeing machine is originally from Mezzera, Italy. But nowadays Suraj Industries, Ludhiana, is also making cabinet hank dyeing machine of various capacities such as 3, 50, 100, 300, 500 kg, etc. Previously Stafi used to make hank dyeing machine of carrier type. These machines are not HTHP dyeing machine. It is dyed at atmospheric pressure. Wool hanks are dyed at 98–100°C. The weight of wool hanks are ranging from 100 to 200 g. Cabinet dyeing machine looks like a steel almirah. The door is opened. Inside the vessel, there are steel rods where the wool hanks are hanged and all the rods are put inside the vessel. Then dyeing is started and completed as above. After dyeing, the dyed hanks are hydroextracted and sent for drying to RF dryer or chamber dryer or tray dryer.

3.2.8   Dyeing of viscose in hank form

Due to the above problem in package dyeing, many local dyers in Panipat, Bhadohi and Banaras are dyeing viscose in hank form using cabinet hank dyeing machines, which convert hank to cone and sell in the market.

3.2.9   Dyeing of acrylic hanks

Following are the processes adopted in hank dyeing.
  1. *Bulking.* It is required to give the yarn a softer feel and better dye penetration. It is carried out with saturated steam. The bulking machine is like a vacuum steamer. In this machine, the grey acrylic hanks are hanged on rods/sticks and kept on trolleys. These trolleys are kept inside the bulking machine (similar to autoclave) temperature is 100–105°C for 20–30 min with an intermediate vacuum. The

maximum shrinkage (20–25%) takes place in the grey yarn under the influence of steam. Steam penetration should be uniform and drops of condensation must be avoided. Each batch should be bulked in one operation.

2. *Dyeing.* The bulked yarn is packed on the rods and loaded in the cabinet dyeing machine of various capacities such as 50, 100, 200, 500 kg, etc. This cabinet dyeing machine is like a steel almirah having doors. It is dyed at normal pressure and temperature between 100°C and 105°C. It can also be dyed with Stafi hank dyeing machine, where the carrier is taken out from the machine and loaded outside with rod/stick attachments. Then the loaded carrier goes inside the hank dyeing machine with the help of electrical hoist.

3. *Hydroextraction.* The dyed yarn in hank form is put inside the hydroextractor which removes the maximum water. This machine works on the centrifuge principle.

4. *Drying.* The dyed yarn goes to chamber dryer. The hanks can be kept on trays and put inside the machine chamber. The temperature is 100–105°C. By the help of steam, the pipes are heated and the blower/fan circulate the hot air inside. In some places, instead of a chamber dryer, a cemented room is made where steam pipes and fan arrangements are there. The hanks get dried overnight. The latest technology is RF drying of acrylic hanks. RF dryer is supplied by Stalam, Italy, and Mong Astray Field, Pune, but care must be taken while drying acrylic hank in RF dryer because the fibre is very prone to fire.

5. *Checking and packing.* After drying, the hanks are checked by checker in case of any shade variation from hank to hank or within the hank. Then it is packed in bales. In some cases, balling machine is there, where the knitting ball is made from the hank and directly supplied to the market which is sold as hand knitting yarn where used by ladies to make sweaters by hand or machine. In some places, there is hank to cone attachment winding machine, where cone is made and sent to circular knitting machine of making t-shirts/jerseys, etc. During making the grey yarn, some fancy yarn is also made by mixing polypropylene fibre or sparking nylon or wool fibre.

3.2.10   Polyester hank dyeing

In the early 1980s, sewing thread (count 3/15, 2/40, 3/20) was dyed in hank form in Jayashree Textile, Rishra (West Bengal) and RSWM, Bhilwara. About 70–100 g, hank was made and loaded in fibre carrier and dyed in HTHP Dalal or Stafi dyeing machine. At 130°C after dyeing of the hanks, it was hydroextracted and then dried in chamber dryer/tray dryer. It was sold in hank form, due to uneven dyeing in hank form and difficulties in making hank to cone.

3.3   Space dyeing

Space dyeing is a technique used to give yarn a unique, multi-coloured effect. While a typical skein of yarn is the same colour throughout, a skein of space dyed yarn is two or more different colours that typically repeat themselves throughout the length of the yarn. Space dyed yarn is sometimes referred to as dip dyed yarn. The secret to space dyeing yarn involves the use of a special chemical called a mordant. The purpose of a mordant is to help permanently fix the dye to the yarn after the space dyeing process. Since different colours of dye require different types of mordants, this makes it possible to dye the same skein of yarn many different colours.

Space dyed yarn can be dyed in either coordinating or contrasting shades. Space dyeing yarn in coordinating colours, such as various neutral tones or assorted shades of blue, provides a subtle yet sophisticated look. Space dyed yarn in contrasting shades, such as yarn that is purple, red and blue, offers a funkier feel. Spaced dyed yarn is most commonly used for knitting and crocheting. When space dyed yarn is made to make a knitted or crocheted item, the resulting project features uneven horizontal stripes that produce a collage-like effect. However, the size of the stripes in the finished piece depends on what size of yarn is used. As you might expect, thicker yarn produces thicker stripes. There are various methods of space dyeing on different forms of yarn.

3.3.1   Hank or skein printing

The ready to dye yarn in hank or skein form is mounted on a stand or laid on a suitable surface and then different colours are applied either by brush or by spraying at fixed places to achieve the desired multicolour effect. Depending on the type of substrate and the dyestuffs used, the hanks are then processed for suitable dye fixation followed by finishing. Though this method imparts controlled yarn colouration at specified points and desired length, the process is cumbersome and time consuming and is mainly suitable for short length fabric.

3.3.2   Space dyeing in package form

Depending on the machinery available and the effect desired, the yarn printing in package form is done by two methods (Fig. 3.2).

1. *Manual operation.* Colours are injected by hand at different places in the package using a syringe filled with the dye solution. The skill set developed by the technicians enable them to insert the needle in the package horizontally, vertically or at various angles and release the colour while taking the needle out so that the required amount of dye is applied at the specified place in the package. Another variant of this method is dip-stick process. In this case, different places of the yarn package are dipped in different dye solution so that multiple colours get applied at various places and with varying intensity which after unwinding of yarn provides a unique fancy dyed effect.

2. *Machine processing.* The ready for dyeing yarn is pre-treated as per the standard application process for the specified substrate and dyed in a single package form in a machine which can dye up to eight colours. Each colour has a different feed tank as well as injector pump. The colour is injected at a fixed place at high pressure and the excess colour is collected through vacuum. This creates beautiful patterns of dyed patches on package with good sharpness. Subsequently the coloured yarn package is given adequate dye fixation treatment followed by washing off of unfixed superficial dye and finishing with a yarn lubricant/softener.

Figure 3.2. Space dyeing machine

3.3.3   Knit–deknit process

This process involves first knitting the yarn on a circular or flat-bed knitting machine into a tubular fabric followed by printing using engraved rollers and then unravelling the knit to produce a space dyed yarn. As the printed colour does not readily penetrate the areas of the yarn where it crosses itself, alternately dyed and undyed spaces appear. In another method, the pre-dyed knitted material is overprinted with different colours and subsequently deknitted to produce contrast effect of overprinted and base dyed areas.

This imparts a unique 'micro-spaced' design, owing to the limited dye diffusion during the printing stage. The unravelled printed yarn is rewound onto cones and used for subsequent knitting or carpet weaving. Since this process achieves limited dye penetration into the yarn, as the pressure exerted by the printing rollers is insufficient to insure adequate dye diffusion, incomplete and non-uniform coloration may result. Also, there is a characteristic curl to the un-ravelled yarn due to pre-knitting which affects cut pile of carpets.

3.3.4   Continuous or warp yarn printing

In this process, multiple strands of yarn are continuously printed at spaced intervals with different colours. These yarns usually have long spaces of each colour spaced at about 3–7 inches. The yarn is dyed as singles or piled form and the colour is applied either by air jets or by dye troughs. Given below are few methods of continuous yarn printing for achieving space dyeing effect based on the information gleaned from various patented processes.

One of the methods describes yarn dyeing at intervals along its length as it runs at very high speed through spaced dye baths. Another method specifies systems where yarns are taken from wound packages on a creel and colour is applied, either by lick rollers or by a spinning disc applicators. Such warp printing imparts a 'long spacing' design which is generally preferred for man-ufacturing of tufted carpets.

One process indicates passing of yarns between a pair of cylinders around the surface of which are mounted rows of small dye applicator pads. The lower cylinder is dipped into a trough of dye liquor which gets picked up on the surface of the pads and when the opposing pairs are in the raised posi-tion, the yarn passing between them gets printed.

Another method proposes space dyeing by running yarn through the nip of printing rollers which have patterned grooves on the outer periphery. In this apparatus, the printing rollers apply dye to the yarn only when the lands of the two rollers are aligned across the nip. Another process incorporates printing rollers with grooves about their peripheries to provide a gear-like appearance. Thus, when the yarn is deflected into the nip, that portion of the yarn which passes through the nip gets intermittently printed to achieve closely spaced marks or dots and thus further vary the dye pattern applied to the yarn.

The advantage of space dyeing process over conventional solid dyeing is not only in terms of the intended special effect but also in terms of reduced consumption of dyestuff and chemicals, avoidance of printing thickener and considerable reduction in effluent. However, the challenges associated with space dyeing process include process to avoid mixing or overlapping of col-ours along the length of yarn and on the adjacent yarns which otherwise tend to result in visible streaks on the face of the fabric affecting its appearance.

Another issue pertains to tendency of dye migration during the subsequent heat setting operation resulting in blurring of adjacent colour bands affecting its appearance. Further, there is a chance of dye dripping or contact mark off between different yarn layers resulting in unwanted patches. Moreover, the dye overspray from the various colours being applied often mix together in a single collection system, which result in added costs for replacement of dye as well as for waste handling and disposal. This has necessitated further development of an in-line process for controlled, efficient and repeatable space dyeing of yarn.

Space dyeing is a technique used to give yarn a unique, multi-coloured effect. When woven or knitted together for apparel, home furnishing and carpet making, beautiful patterns can emerge depending on the length and variation of each colour block. Adequate use of dyes and auxiliary chemical combined with specialized machines can create finished product as some of the best in the world. Presently the 'space dyeing' segment occupies a miniscule (<1%) share in the textile coloration space. However, considering its potential to offer unique and customized print design effects, it has great potential to grow by many folds in the near future.

Resist dyeing

Resist dyeing has been very widely used in Eurasia and Africa since ancient times. The first discoveries of pieces of linen were from Egypt and date from the fourth century. Cloth, used for mummy wrappings, was coated with wax, scratched with a sharp stylus and dyed with a mixture of blood and ashes. After dyeing the cloth was washed in hot water to remove the wax. In Asia, this technique was practiced in China during the T'ang dynasty (618–907), in India and Japan in the Nara period (645–794). In Africa, it was originally practiced by the Yoruba tribe in Nigeria, Soninke and Wolof in Senegal.

Resist dyeing (resist dyeing) is a term for a number of traditional methods of dyeing textiles with patterns. Methods are used to 'resist' or prevent the dye from reaching all the cloth, thereby creating a pattern and ground. The most common forms use wax, some type of paste made from starch or mud, or a mechanical resist that manipulates the cloth such as tying or stitching. Another form of resist involves using a chemical agent in a specific type of dye that will repel another type of dye printed over the top. The most well-known varieties today include tie dye and batik.

Resist dyeing is a widely used method of applying colours or patterns to fabric. A substance that is impervious to the dye blocks its access to certain areas of the fabric, while other parts are free to take up the dye colour. Tie dyeing involves pinching areas of cloth and tying them tightly with thread before dyeing. Removal of the thread reveals small circular marks in the original fabric colour. Complex patterns can be built up by repeating the process using another dye colour. In applied resist dyeing, the pattern is marked on to the cloth with a substance such as paste or wax. After dyeing and removal of the resist substance, the pattern is revealed in the original fabric colour. This process can be repeated several times.

Cultures around the world have developed an array of resist dye techniques. Dyeing provides rich colours but once the fabric has been coloured in a dark shade, lighter colour patterns will not show up. To allow lighter colours to come through, areas have to be blocked from receiving dye. Any of these techniques of blocking the dye are referred to as a 'resist'. Sometimes these techniques have arisen independently; sometimes the techniques have been passed across cultures through trade and exchange. In many cases, the origins have been lost to time, leaving only rich and remarkable textile traditions. Resist techniques can be seen in the most expensive and treasured textiles, but also in relatively humble objects.

The exhibition is organized by technique to bring together examples from around the globe. The objects are grouped into three main categories of resist methods: mechanical, chemical and ikat. While specific techniques may vary widely, they rely on a few basic principles. The dye can be resisted using mechanical means by tying, stitching or folding. Alternately the resist can be chemical, generally paste or wax. The third category, ikat, refers to textiles in which the resist is applied to the threads before weaving. Ikat is generally a mechanical resist technique, in which the threads are wrapped and bound.

4.1   Mechanical resist

shibori, bandhani, tie-dye

Tie dye is a technique that has become familiar to many Americans because of brightly coloured t-shirts popular in the 1960s and 1970s. The technique of tying off sections of cloth or garments before treating it with dye has been around for centuries. Japan and India are among the many parts of the world with long traditions of tie dye. While most of the examples of mechanical resist techniques in the collection are variations of tying and binding with thread, other methods such as clamping and pole wrapping can also be used. While these techniques have been practiced for centuries and are performed by highly experienced artisans, there is always an element of randomness and chance to the results. The subtle variations in shade and pattern are intrinsic to the beauty of the handmade pieces.

This technique is used extensively in India, where it is known as 'bandhani' from which we get the word 'bandanna' – a silk neck cloth that was originally tie dyed. Various methods are used to mark out a pattern on the fabric before tying. In one of the most traditional methods, now used less frequently, the dampened fabric is placed over a pattern block of raised pins. The cloth is pinched between the thumb and index finger at each point and tied with waxed thread. Another way is to block print the design of dots using a medium that washes out in water, such as soot or red ochre. Sometimes a thin sheet of plastic pierced by holes is placed over the fabric and the fugitive solution spread over it. This leaves a pattern of small dots on the fabric. It is also possible to roughly mark out the pattern and tie by eye. The ties are often not removed before the cloth is sold, to show that it has been hand dyed and not mechanically printed.

A tie-dying method called 'lehariya' is used in India for turban cloths. Fine cloth such as muslin is folded concertina fashion and tied tightly at intervals. It is dipped quickly in dye of a pale colour. Some areas are then unrolled and the process is repeated with progressively darker dyes, to build up a range of colours in stripes.

4.2   Chemical resist

batik, adire eleko, tsutsugaki, modrotlac
From the 1920s and 1930s, adire was a main local craft in cities like of Abeo-kuta and Ibadan; the dye pits of Kano are well known. The old technique of starch resist designs reached its peak in popularity up to the early 1970s, and it is still practiced with a few adaptations. In later years, the starch was applied through lace fabrics and dyed in synthetic dyes. The starch is very carefully applied to the cloth and allowed to dry. When immersed in the dye, the starch resists/repels the penetration of the pigment into the fabric. After the dye has been set and the process is completed, the starch is washed out of the cloth.

The use of paste or wax as a resist has developed in many cultures around the world. In the earliest forms, the patterns were created free hand by drawing the wax or paste onto the fabric. Such techniques can be seen in the finest Indonesian batiks and Japanese *tsutsugaki*. As textile printing developed, resists played a critical role in preventing dark colours from spreading into lighter areas. Several cultures developed techniques of print-ing the resist onto the fabric before dyeing. Achieving several colours on a textile demands repeated application of wax or paste before each submer-sion in the dye. Additional colours are created by over dyeing one colour over another.

This technique is called tsutsugaki in Japan where rice paste is used as the resist, and 'batik' in Java where wax is used. Originally the hot wax was applied with a shaped strip of bamboo, but in the 17th century, the invention of the 'canting' (pronounced janting) – a copper crucible with spouts of dif-ferent sizes – meant that the wax could more easily be applied in continuous lines of varying thicknesses, thus improving the fineness of the patterns that could be attempted.

The earliest batiks were monochrome patterns against an indigo back-ground, but multicoloured ones were produced from the 18th century on-wards using methods learnt from expert Muslim dyers in India. Typical patterns represented ancient symbolic designs in complex, symmetrical, in-tertwining layouts, and reflected the social class of the owner through their level of intricacy. Some of the ceremonial garments produced and decorated in this way are among the most superb examples of textile ornamentation known.

In India, beeswax resist was used for part of the fabric colouring process in the production of chintz. Pouncing was used to transfer the pattern in char-coal onto the cotton cloth; a porous bag of loose charcoal powder called a 'pounce' was dusted over a design pricked out onto paper. Then the hot wax was drawn on with a reed pen, following the charcoal guidelines. The textile workers were largely low-caste Hindi family groups, each family skilled in a separate stage of the complex chintz-making process and working in their own small craft workshops (not their own homes). The fabric moved from family to family for each of the many stages 'appearing, like a snail, to make no progress' until the cloth was complete, as a Dutch agent recorded in the 1680s.

In several cultures around the world, paste resist has been traditionally used in the dyeing of textiles. In Nigeria, adire eleko is a method which uses cassava as the base for the paste.

The name, Adire eleko, refers to a cloth produced by Yoruba textile artists using a resist-dye technique. The dye that is most commonly used is indigo – a deep-blue vat dye derived from the indigo plant (*Indigofera*). Designs are produced by painting or stencilling a starch made from cassava flour on the surface of the cloth before dying. The starch is made of the flour mixed with water that is boiled and then strained; a small piece of copper sulphate is added while the mixture is boiling to make the solution last longer.

Ikat, jaspe, adras, kasuri

While it is not known when and where the resist technique first developed, Asia has several cultural regions with a particularly strong ikat tradition. Maritime Southeast Asia, India and Central Asia are all potential candidates for the origin of the technique, but it may also have evolved independently in several locations. Ikat may have spread at an early age through many parts of the Austronesian-speaking world, as technical similarities exist between Indonesian ikat production and that of Madagascar, which was settled by maritime Southeast Asians early in the first millennium CE. As Malagasy weavers also use a distinctly Austronesian version of a horizontal back-strap loom, the technique may have arrived simultaneously with the spread of loom technology. Ikat patterning is probably represented in garments shown in the Ajanta cave paintings of India (from the fifth to the seventh century). A fragment patterned in the technique, kept for centuries in the Horyuji temple at Nara but now in the Tokyo National Museum, was apparently brought there from China during the Tang period (618–907 CE), but was probably produced in Central Asia. Cotton textiles with relatively simple warp ikat stripes were made in Yemen by the eighth or ninth century and were traded to Egypt, where they have survived.

Ikat is a resist dye technique used to pattern textiles. The more common methods of resist dyeing involve covering parts of a fabric to shield the reserved areas from penetration of the dye, as in tie dyeing, where threads are wound around the fabric, or in batik, where wax is applied to the surface of the cloth. The term ikat by contrast is used for a process where prior to weaving, warp (lengthwise yarn) or weft (crosswise thread) or sometimes both are tied off with fibre knots that resist absorbing colour and are then dyed. To facilitate the pattern tying, the threads are set up on a frame. They are then grouped into bunches of several threads to be tied at once; this results in the creation of knot units from which the overall pattern is built up. Resist ties are removed or new ones added for each colour; their combinations create the design. After dyeing is completed, all resists are opened, and the patterned yarns are woven.

The word ikat comes from the Malay-Indonesian word for "tie"; it was introduced into European sources of textile technology and history in the early 20th century when Dutch scholars began paying attention to the rich textile traditions of the Netherlands Indies, the present-day Indonesia. Depending on whether the tied fibres are applied to the warp or weft, the technique is identified as either warp ikat or weft ikat. A third variety, double ikat, combines both warp- and weft-tied resist. For the pattern to be visible, the resist-dyed thread system has to be the prominent one, so for warp ikat, the weave has to be warp faced, and weft ikat needs a weft-faced structure, which means that either warp or weft is predominantly visible. Plain weave is especially suitable for showing the ikat's design, but for weft ikat, a twill weave may also be used. Double ikat, where the design is built up from both systems, should ideally by woven in a balanced weave, with warp and weft equally visible. All textile fibres may be used for ikat, although silk and cotton are the most common ones. For the seminal study of ikat as a resist dye technique, its history and geographic distribution, see Alfred Bühler (1972).

The word ikat derives from the Indonesian verb *menigikat*, which means 'to bind, tie or wind around'. Clearly the word first applied to Indonesian textiles, but has come to be the general term used to describe any textile made with this technique. The method involves wrapping yarn with a resist before dyeing. When such yarn is woven, the resulting textile will be patterned. The elaborate Central Asian and Indonesian examples required repeated binding and dyeing to achieve the variety of colours and intricacy of design. The patterns created in ikats will have a characteristic raggedness around the edges. The patterns can be created on the warp or the weft, or both. When both the warp and the weft are patterned, the resulting textile is a double ikat.

'Kasuri', which is also known by the Indonesian term 'ikat', takes a different approach and requires extreme accuracy. It is the unwoven warp or weft yarns that are tied and dyed so when the cloth is woven the pattern emerges from the pre-dyed threads. In India, highly valued double ikats called 'patola', in which both warp and weft are dyed, are woven in silk.

5

Fabric dyeing

Warp beam dyeing

Beam dyeing is the much larger version of package dyeing. An entire warp beam is wound on to a perforated cylinder, which is then placed in the beam dyeing machine, where the flow of the dye bath alternate as in the package dyeing. Beam dyeing is more economical than skein or package dyeing, but it is only used in the manufacture of woven fabrics where an entire warp beam is dyed. Knitted fabrics, which are mostly produced from the cones of the yarn, are not adaptable to beam dyeing.

Piece dyeing

The dyeing of cloth after it is being woven or knitted is known as piece dyeing. It is the most common method of dyeing used. The various methods used for this type of dyeing include jet dyeing, jig dyeing, pad dyeing and beam dyeing. Dyeing is done in a closed, tube-like system in which the fabric passes through a fast moving stream of pressurized dye liquor. It is primarily used for fabrics prone to felting (Fig. 5.1).

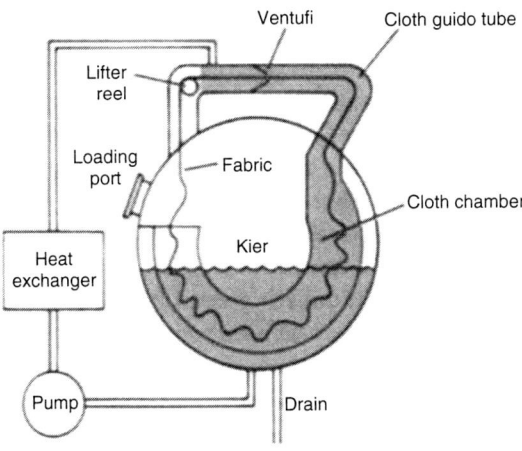

Figure 5.1. Piece jet dyeing

5.1   Jigger dyeing

Dyes use for dyeing fabric in Jig or Jigger dyeing machine which is suitable for woven fabric than knitted fabric because jigger exert considerable length-wise tension on the fabric. This is also particularly suitable for cellulosic fibres because the natural dyes generally do not exhaust well and the jig works with an exceedingly low liquor ratio. It is one of the oldest dyeing machines for dyeing fabric. There are mainly two types of jigger dyeing machine for dyeing woven fabrics with dyes. One is open jigger dyeing machine and other is closed jigger dyeing machine. Normally after the process of dyeing machine, dyed fabric is sent to stenter machine.

The types of jigger dyeing machine are

1. Textile jigger machine for woven fabrics
2. Hydraulic jigger machine for dyed fabric
3. Semi-automatic dyeing jigger for dyeing fabric
4. Automatic jigger textile machine for woven fabrics
5. Semi and jumbo jigger machines show how to dye fabric
6. Manual jigger for natural dyes

The parts of jigger machine are specified as follows: batch roll, hinged cover, fabric unloaded, dye liquor container, guide roller, steam heating coil, main roller, fabric winding, adjustable expander bar, fabric expander bar, dye vat and drain (Fig. 5.2).

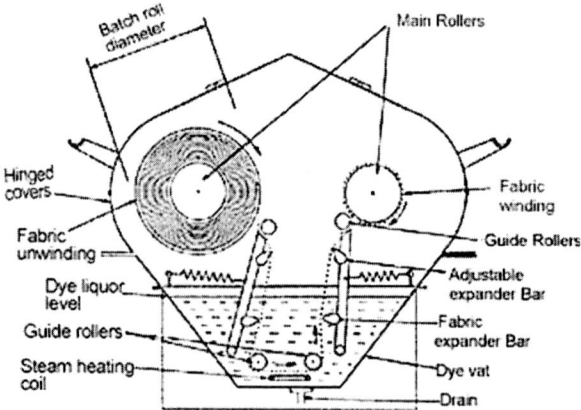

Figure 5.2. Parts of jigger dyeing machine

5.1.1   Advantages of jigger dyeing machine

1. The cloth is dyed in open width form in textile industry
2. The material to liquor ratio is 1:3, which is used for dyed fabric in textile industry

3. Liquor ratio 1:4, which saves huge amount of chemical cost and steam cost
4. Less chemical lose for woven fabric in textile industry
5. No need training to operate how to dye fabric

5.1.2   Disadvantages of jigger dyeing machine

It exerts lot of tension in the warp direction and because of this normally knitted fabrics, woolen, silk, etc., cannot be dyed in jigger dyeing machine because it is only for woven fabric.

5.1.3   Objective of jigger dyeing machine

Dyeing (exhaust) is a chemical process by which textile substrates are immersed in liquor containing dyestuffs and chemicals by which the dyestuffs are transferred to textile substrates in a proper manner. For every shade and batch, quality control (QC) goes shade check, wash fastness, rubbing fastness, water fastness and light fastness, tear test, tensile test, pick-up test, shade listing and tailing.

A dye-trough design ensures minimum possible liquor ratio.

Fabric rope monitors control system for precise calculation and control of the fabric speed and tension.

Uniform dyeing conditions in the dye trough are independent of the batch size.

Water metre is used to control rinsing.

Dosing of dyes and chemicals is dependent on the fabric length passing through the dye bath (Table 5.1).

The key accessories used are wooden plate, sewing machine, A-frame/batcher, eye-protecting glass, hand gloves, rotation station, bucket mixer and filter cloth.

5.1.4   Dyeing process

The dyeing processes involved in jigger dyeing machine are as follows (Fig. 5.3, Table 5.2):

- Fabric inlet: After mercerizing batcher is feed into the inlet of Jigger machine, where the fabric is passed through dye bath.
- Dye + chemical padding: Both chemical and colour is prepared in different tank according to required
- Washing: Washing is done by using detergent and wetting agent
- Striping: Stripping is done by using hydrose (4 g/L) and caustic (15 g/L)

- Mercerizing: Chemical strength is 30 Baume
- De-finishing: De-finishing is done by using soda (8 g/L) and caustic (10 g/L)
- Bleaching
- The fabric is directly pass through dyebath and then batching on another roller, and again this process is start in opposite direction. This system is continued for three to four times. If shade is match, then fabric will be outlet. If not match, dyeing will be continuing.
- Fabric outlet: If shade is ok, then fabric will be outlet, and it becomes ready for finishing.

Table 5.1. Standard machine set-up

Machine set-up parameters	Actual parameter range	Machine set value
Temperature	58–60°C initially 88–90°C later	60°C, 90°C
Chemical level	500–520 L	520 L
Fabric position	Face	Face
Speed	40–80 m/min	40–80 m/min

- pH 8
- M:L = 1:10
- Time: 4 h
- Fabric length: 500–1000 yards
- Route temperature: 57°C

Figure 5.3. Jigger dyeing machine

Table 5.2. Chemicals used in jigger dyeing

Chemicals/materials used	Used in average
Wetting agent	0.25–0.5 g/L
Detergent	0.5–1 g/L
Dye stuff (reactive, sulphur, etc.)	Differs depend on the shade of fabric
Caustic	1.323 g/L
Soda ash	5 g/L
Reducing agent	9–10%
Oxidizing agent	2 g/L
Acid	2 g/L
Compressed air	As required
Salt	20 g/L
Steam, water	As required

Control panel of jigger dyeing machine: From control panel, all types of instruction can be provided to the machinery. This machine has a full manual control panel (Fig. 5.4).

Figure 5.4. Control panel of jigger dyeing machine

The operator controls every process by control switch of jigger dyeing machine.

5.1.5   Cotton jigger dyeing

Processes involved in cotton jigger dyeing are as follows:
1. Half colour – 1 end (chal) at room temperature
2. Add rest half colour – 1 end at room temperature
3. Add half salt – 1 end at room temperature
4. Add rest half salt – 1 end at room temperature
5. Raise temperature up to 60–65°C – 4 ends
6. Add half soda – 1 end at 60°C
7. Add rest half soda – 1 end at 60°C

8. More 2 ends at 60°C
9. Then after treatment process,

Liquor ratio is 1:3.
If Fabric is 250 kg, then liquor will be 750 L (include of wet fabric).
As per thumb rule, if weight of fabric is 250 kg, it contains 250 L water. So extra 500 L to be added. So

$$250 \text{ L} + 500 \text{ L} = 750 \text{ L liquor}$$

5.2   Winch dyeing (beck dyeing)

The winch or beck dyeing machine is the oldest form of piece dyeing machine. The construction is comparatively simple and therefore economical to purchase and operate. It is suitable for practical all types of fabric, especially light weights, which can normally withstand creasing when in rope form as woollen and silk fabric, loosely woven cotton and synthetic fabrics, circular and warp knitted fabrics. This dyeing machine is used for fabrics in rope forms with stationary liquor and moving material. A winch dyeing machine consists of a vat (vessel) that has a curved back. Over the top of vat, extending its length is a horizontal elliptical winch rotated generally by an individual electric motor. In this machine, operations like scouring, bleaching, dyeing and washing can be easily carried out. Winch can be classified as follows:

1. Conventional winch dyeing machine
2. High temperature winch
3. High temperature winch with circulating liquor

Passage of fabric in winch dyeing machine: In loading a winch for dyeing operation, each length of fabric is run over the guide roll and the winch, so that the fabric falls in the vat at the rear end of the machine; the fabric is allowed to run over the rotating winch and get piled at the bottom of the machine; as the bottom is sloped towards the front of the machine, the piled fabric slides along it, acquiring a wavy shape and moves slowly towards the front end of the machine. When the free end of the fabric reaches the front end, it is picked up, passed between the pegs and stitched to the other free end of the fabric. Thus, the entire length of the fabric is now endless loop with much of it in a folded form at the bottom of the fabric winch. The winch is the prime mover of the fabric, but for the greater part of the dyeing period the fabric lies wholly immersed in the dye liquor (Fig. 5.5).

5.2.1   Features and parameters

1. The machine operates at a maximum temperature of 95–98°C
2. The liquor ratio is generally quite high (1:20–1:40)

3. This is a dyeing machine for fabrics in rope form with stationary liquor and moving material.
4. In winch machines, a number (1–40) of endless ropes or loops of fabrics of equal length (about 50–100 m) are loaded with much of their length immersed in folded form inside the dyebath.
5. As for all forms of rope dyeing, the fabric must be fairly resistant to length ways creasing.
6. A perforated separating compartment, positioned at a distance of 15–30 cm from its vertical side, creates an inter space for heating and for adding reagents.
7. Heating can be supplied by means of direct or indirect stem heating.
8. The rope passes from the dyebath over two elevated reels. The first roller is free running (jockey or fly roller) and the second is winch reel.
9. The winch reel not only controls the rate of movement of the fabric rope, but also the configuration of the rope in the dyebath.
10. The winch reel does not grip the fabric positively, but by the weight of the wet fabric and the friction between the reel and fabric.
11. Now-a-days stainless reels with corrugated and broken surface for increase frictional forces are used.
12. The maximum motion speed of the fabric must be approximately 40 m/min.
13. The winch dyeing method is suitable for all fabrics, expects those which tend to originate permanent creases or which could easily distort under the winch stretching action.

A dyeing machine consisting essentially of a dye vessel fitted with a driven winch (usually above the liquor level) which rotates and draws a length of fabric, normally joined end to end, through the liquor.

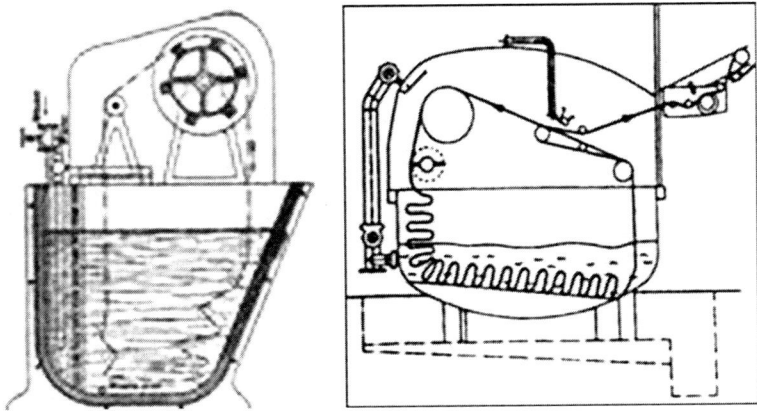

Figure 5.5. Winch dyeing **(beck dyeing)** machine

Winch dyeing machines are a low cost design that is simple to operate and maintain, yet versatile in application proving invaluable for preparation, washing or after treatments as well as the dyeing stage itself. In all winch dyeing machines, a series of fabric ropes of equal length are immersed in the dyebath but part of each rope is taken over two reels or the winch itself. The rope of fabric is circulated through the dyebath being hauled up and over the winch throughout the course of the dyeing operation. Dyestuff and auxiliaries may be dosed manually or automatically in accordance with the recipe method.

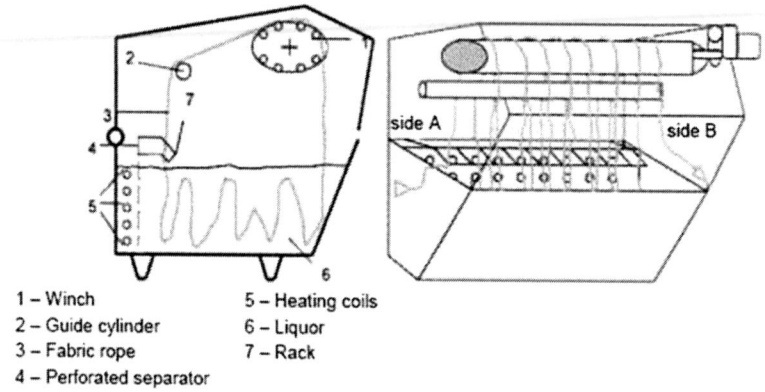

1 – Winch	5 – Heating coils
2 – Guide cylinder	6 – Liquor
3 – Fabric rope	7 – Rack
4 – Perforated separator	

Figure 5.6. Parts of a winch dyeing machine

5.2.2   Description and dyeing method on winch dyeing machine

The basic principle of all winch dyeing machines is to have a number of loops or ropes of the fabric in the dyebath, these ropes are of equal length, which are mostly immersed in the liquor in the bath. The upper part of each rope runs over two reels which are mounted over dyebath. At the front of the machine, above the top of the dye liquor, is a smaller reel, which is called jockey or fly roller (Fig. 5.6).

The fly roller remains free wheeling along with fabric rope. At the back of winch tank is the winch wheel, which pulls the fabric rope from the dyebath over the jockey reel for dropping in the dyebath for immersion. From the dropped location, the fabric rope travels back to be lifted and fed to winch wheel. The dyeing process on winch dyeing machines is based on higher M:L as compared with other dyeing machines. The process is conducted with very little tension. The total dyeing time is lengthier as compared to other machines.

5.2.3   Advantages of winch dyeing machine

1. Construction and operation of winch are very simple.
2. The winch dyeing machines are suitable for types of wet processing operations from desizing to softening.
3. The winch dyeing machine is suitable for practically all types of fabrics, which can withstand creasing in rope form processing.
4. The tension exerted on winch is less than jigger dyeing machine, the material thus dyed is with fuller hand.
5. The appearance of the dyed goods is clean and smooth on winch dyeing machines.

5.2.4   Limitations of winch dyeing machine

1. Batch dyeing operations needs trimming, sewing, opening out the rope, loading and unloading for individual lots separately.
2. Since several lengths of fabric are run over the winch reel into the liquor and sewn end to end. Continuous length processing is not possible in a single batch.
3. Fabric is processed in rope form which may lead to crease marks, particularly in heavy, woven, thin and light synthetics.
4. Most of the machine work under atmospheric conditions.

5.3   Jet dyeing

It was found that in using winch machines, there were some inherent problems. So the jet dyeing machines when they came up in the 1970s were specifically designed to overcome those shortcomings.

In the jet dyeing machine, the reel is completely eliminated. A closed tubular system exists where the fabric is placed. For transporting the fabric through the tube, a jet of dye liquor is supplied through a venturi. The jet creates turbulence. This helps in dye penetration along with preventing the fabric from touching the walls of the tube. As the fabric is often exposed to comparatively higher concentrations of liquor within the transport tube, so little dyebath is needed in the bottom of the vessel. This is just enough for the smooth movement from rear to front. Aqueous jet dyeing machines generally employs a driven winch reel along with a jet nozzle.

Figure 5.7 explains the functioning of a jet dyeing machine:

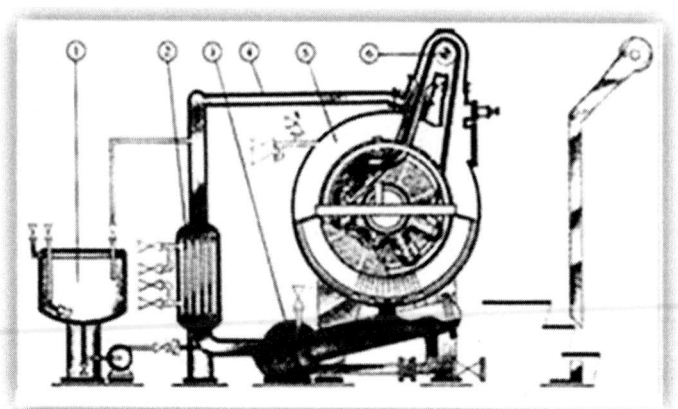

Figure 5.7. Jet dyeing machine

5.3.1   Advantages of jet dyeing machine

The jet dyeing machine offers the following striking advantages that make them suitable for fabrics like polyesters.

1. Dyeing time is short compared to beam dyeing
2. Material to liquor ratio is 1:5 or 1:6
3. Production is high compared to beam dyeing machine
4. Low consumption of water
5. Short dyeing time
6. Can be easily operated at high temperatures and pressure
7. Comparatively low liquor ratios, typically ranges between 1:4 and 1:20
8. Fabrics are handled carefully and gently

5.3.2   Disadvantages jet dyeing machine

- Cloth is dyed in rope form
- Risk of entanglement
- Chance for crease formation

5.3.3   Features of jet dyeing machine

- Capacity: 200–250 kg (single tube)
- Liquor ratio: 1:1 (wet fabrics)

- Dye: 30–450 g/m² fabrics (polyester, polyester blends, woven and knitted fabrics)
- High temperature: Up to 140°C
- Fabric speed: 300 m/min

In this development, there is no fabric drive reel to move the fabric. The fabric movement by only force of water. It is economical, because of low liquor ratio. It is users friendly because comparison with long tube dyeing machine, to control the fabric movement four valves required. In jet dyeing machines and fabric dyeing machine, there is only one valve. Absent of reel, reduce connecting electric power, maintenance of two mechanical seal and breakdown time, if jet pressure and reel speed not synchronized. Nearly 4000 without fabric drive reel machines are saving energy and reducing pollution in 20 countries of this planet.

5.3.4   Types of jet dyeing machine

In deciding the type of dyeing machine, the following features are generally taken into consideration for differentiating. Shape of the area where the fabric is stored, i.e., long shaped machine or J-box compact machine. Type of the nozzle along with its specific positioning, i.e., above or below the bath level. Depending more or less in these criteria for differentiation following types of jet machines can be said to be as developments of the conventional jet dyeing machine.

1. Overflow dyeing machine
2. Soft-flow dyeing machine
3. Airflow dyeing machine

5.3.5   Overflow jet dyeing machine

Overflow jet dyeing machine is used for pre-treatment and dyeing of rope fabrics, with both liquor and materials moving; the architecture and the design of the system and the liquor ratios are similar to the jet machine ones. The main difference is the fabric transport system, driven partly by a motorized reel, and partly by the sequential flow of the liquor. The jet system nozzle, based on a venturi tube, is replaced by a vessel containing the liquor; the liquor enters the straight pipe section and then flows through the transport channel, together with the fabric rope (Fig. 5.8).

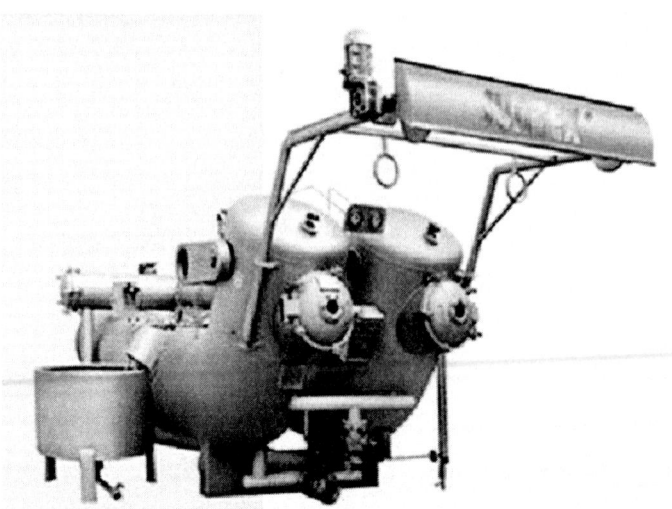

Figure 5.8. Overflow jet dyeing machine

These machines are designed for the use in delicate knitted and woven fabrics that are made up of natural as well as synthetic fibres. They are also extensively used in the production of carpets. The main difference between jet and overflow jet machines is that in jet machines the fabric gets transported by a bath that flows at high speed through the nozzle, while in overflow dyeing machine, it is the gravitational force of the liquor overflow that is responsible for fabric transportation (Fig. 5.9).

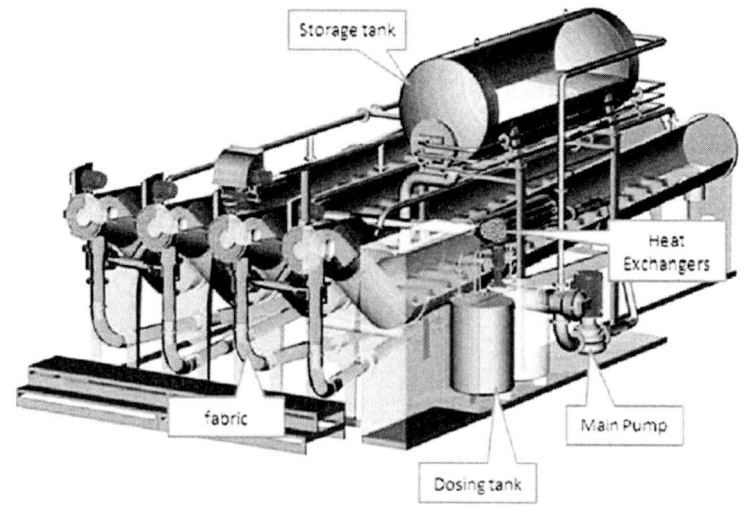

Figure 5.9. Parts of overflow jet dyeing machines

5.3.5.1   Working principle of jet overflow dyeing machine

A typical overflow jet dyeing machine works like this. A winch that is not motor driven usually is located in the top side of the machine where the fabric is hanged. A longer length of textile is made to hang from the exit side of the winch as compared to the inlet side. By applying the force of gravitation, the longer length of textile is pulled downward more strongly than the shorter one. Consequently the fabric is soaked in the bath without any sort of tension. Fig. 5.10 well illustrates the working process and the technical parameters are listed in Table 5.3.

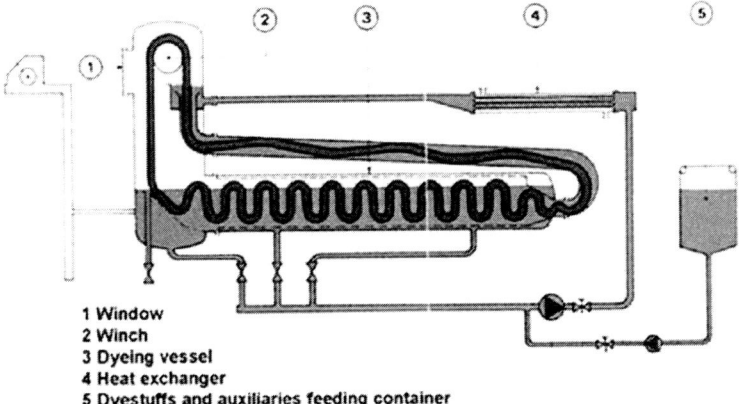

1 Window
2 Winch
3 Dyeing vessel
4 Heat exchanger
5 Dyestuffs and auxiliaries feeding container

Figure 5.10. Schematic diagram of overflow jet dyeing machines

Table 5.3. Technical parameters of overflow jet dyeing machine

Items	SME-150B
Form of the machine	Horizontal tubular jet overflow single pipe
Liquor ratio	1:5 to 1:8
Max fabric capacity (kg)	100–150
Max working temperature	140
Max working pressure (MPa)	0.40
Fabric speed (m/min)	60–300
Heating exchange area ()	5
Main pump power (kW)	14.7

5.3.5.2   Advantages of overflow jet dyeing machine

- No evaporation losses: As the dyeing vessel is closed, there is no evaporation losses stemming from the dyebath. Further, depending on the situation, the temperature may be raised to more than 100°C.

- No build-up of steam condensate in the dyebath: The latest technology implies that the dyebath gets heated by a heat transducer which is steam driven. This technology apart from being very efficient ensures that there is no build-up of steam condensate in the dyebath.
- Low liquor ratios: Dyeing is conducted at relatively low liquor ratios, e.g., 10:1 and may be lesser resulting in substantial savings in water and energy.
- Excellent dye liquor contact: Excellent dye liquor contact with the fabric rope results in better and more improved level dyeings.
- Computer control: The machines are operated by computer and hence, operator error is eliminated.

Ropes of fabric are transported around a dyeing machine by the combination of winch reels and high pressure jets of dye liquor and dyeing is carried out in warm, highly alkaline dyebaths from very high concentrations of salt. The salt is used to attract to dyes to the fibre before they become permanently fixed by the addition of alkali. Around 10–30% of the dye, depending on the specific dye, remains attracted to the fibre but unfixed (on the fibre but not reactively bonded to the fibre) at the end of dyeing and this has to be fully removed so that the fabrics pass colour fastness tests. Unfortunately, the presence of salt makes the removal of unfixed dyed difficult because, in the presence of salt, the dyes prefer to sit on the fibre rather than in the dyebath. The salt has to be diluted with large quantities of water to remove the unfixed dye. Most effluent treatment plants are not designed to remove salt from effluent so large amounts of salt is discharged into water courses which can disrupt the natural balance if there is insufficient dilution. Jet dyeing can be applied to both woven and knitted fabrics and the bleaching, dyeing and wash-off are normally carried out in the same machine using a single, sequential combined process.

5.3.6   Airflow dyeing machine

This is another development of the very popular jet dyeing machines. The main difference between the air flow machine and jet dyeing machine is that the airflow machine utilizes an air jet instead of the water jet for keeping the fabric in circulation. Typically the fabric is allowed to pass into the storage area that has a very small amount of free liquor. This results in a reduction in consumption of water, energy and chemicals.

Figure 5.11 shows how in an airflow machine the bath level is always under the level of the processed textile. Here the fabric does not remain in touch with the liquor (the bath used is below the basket that holds the fabric in circulation). This invariably means that the bath conditions can be altered without having any impact on the process phase of the substrate.

Figure 5.11. Airflow machine

5.3.6.1   Advantages of airflow machine

- Completely separated circuit for liquor circulation without getting in touch with the textile
- Bath less dyeing operation
- Rinsing process offers all the added benefits of continuous processing as it is no longer a batch operation
- Extremely low liquor ratio
- Virtually non-stop process
- Comparatively lesser energy requirement due to faster heating/cooling and optimum heat recovery from the hot exhausted dye liquors
- Reduction in consumption of the chemicals (e.g., salt) dosage of which is based on the amount of dyebath
- Lesser water consumption savings up to 50% from the conventional jet dyeing machines
- Sensitivity towards ecology
- Economical operation
- More safety while dyeing

5.3.6.2   Unique water saving capacity of air flow machine

Water remains a perpetual challenge for the world as the most precious resource and textile dyeing processes are notorious for consuming galons of water. The latest technology of airflow machine surely takes care of such problem.

Airflow dyeing machine can operate at a liquor ratio that is even below 1:5 while a conventional hydraulic dyeing system generally operates with a liquor ratio of about 1:10. It is worthwhile to know that exchanging the liquor ratio of 1:10 from a single 300 kg dye lot to ratio of 1:5, can result in water savings corresponding to an average monthly water consumption of one person in a big country like Germany.

This is the most modern machine used for the dyeing of polyester using disperse dyes. In this machine, the cloth is dyed in rope form which is the main disadvantage of the machine.

In this machine, the dye tank contains disperse dye, dispersing agent, levelling agent and acetic acid. The solution is filled up in the dye tank and it reaches the heat exchanger where the solution will be heated which then passed on to the centrifugal pump and then to the filter chamber.

The solution will be filtered and reaches the tubular chamber. Here the material to be dyed will be loaded and the winch is rotated, so that the material is also rotated. Again the dye liquor reaches the heat exchanger and the operation is repeated for 20–30 min at 135°C. Then the dyebath is cooled down, after the material is taken out.

Metering wheel is also fixed on winch by external electronic unit. Its purpose is to record the speed of the fabric. The thermometer, pressure gauge is also fixed in the side of the machine to note the temperature and pressure under working. A simple device is also fixed to note the shade under working.

5.4   Soft flow dyeing

In the soft flow dyeing machine, water is used for keeping the fabric in circulation. The conception difference of this equipment from a conventional jets that operate with a hydraulic system is that the fabric rope is kept circulating during the whole processing cycle (right from loading to unloading). There is no stopping of liquor or fabric circulation for usual drain and fill steps. The principle working behind the technique is very unique. There is a system for fresh water to enter the vessel via a heat exchanger to a special interchange zone. At the same time, the contaminated liquor is allowed channel out through a drain without any sort of contact with the fabric or for that matter the new bath in the machine.

5.4.1   Key features of soft flow dyeing machine

1. Significant savings in processing time.
2. Savings in water that is around 50%.
3. Excellent separation of different streams results in optimum heat recovery and a distinct possibility of further use or a dedicated treatment.

5.4.2   Principle of soft flow dyeing machine

Textile material can be dyed using batch, continuous or semi-continuous process. Batch processes are the most common method used to dye textile materials. There are three general types of batch dyeing machines:

1. In which fabric is circulated
2. In which dyebath is circulated
3. In which both bath and material is circulated

Jet dyeing machine is the best example of a machine that circulated both the fabric and the dyebath. Jet dyeing is used for knitted fabrics. For Terry towels, soft flow dyeing is use. In jet dyeing machine, the fabric is transported by a high speed jet of dye liquid.

As seen in Fig. 5.12, this pressure is created by venturi. A powerful pump circulates the dyed bath through a heat exchanger and the cloth chamber. Cloth guide tube helps in circulation of fabric.

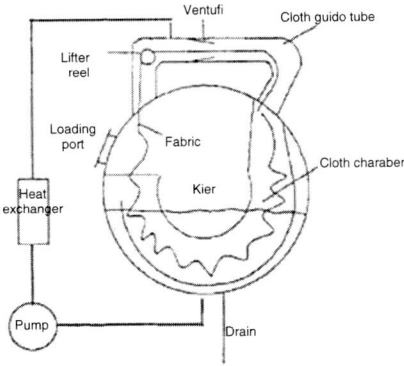

Figure 5.12. Soft flow dyeing machine

5.4.3   Types of soft flow dyeing machine

A few of the commercially popular brands along with their particular technical specifications are discussed here. The categories are not exhaustive as such.

5.4.3.1   *Multi-nozzle soft flow dyeing machine*

The technical features are as follows:

1. Very low liquor ratio: around 1:1 (wet fabric)
2. Can reach high temperature up to 140°C
3. Easily dye 30–450 g/m² of fabrics (woven and knitted fabrics)
4. Number of very soft-flow nozzles
5. No pilling effect
6. Wide capacity

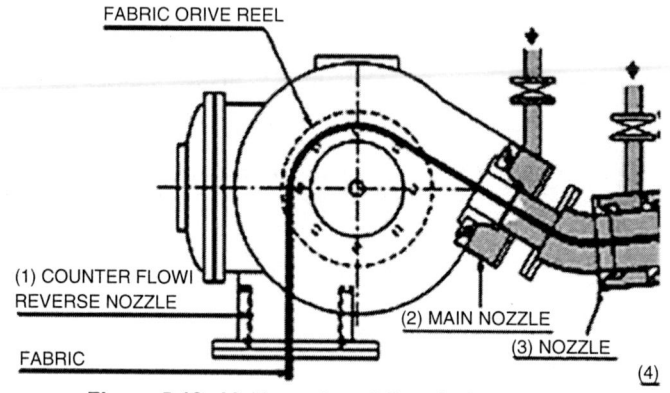

Figure 5.13. Multi-nozzle soft flow dyeing machine

5.4.3.2   *High temperature high pressure soft flow dyeing machine*

The technical features are as follows:

1. Compact body made of stainless steel
2. High efficiency heat exchanger for quick heating/cooling
3. Compact body made of stainless steel
4. Heating rate: around 4°C/min up to 900°C, around 3°C/min up to 135°C, at steam pressure of 6 Bar
5. Cooling rate: around 4°C/min at water pressure of 4 Bar and 15°C
6. Maximum working temp is 135°C
7. Maximum working pressure of 3.2 Bar
8. Control manual as well as automatic
9. Heavy duty stainless steel pump

5.4.4   Soft flow dyeing machine

1. The vigorous agitation of fabric and dye formulation in the cloth increases the dyeing rate and uniformity. It minimizes creasing as the fabric is not held in any one configuration for very long. The lower liquor ration allows shorter dye cycles and saves chemicals and energy.

2. In soft flow dyeing machines, the fabric is transported by a stream of dye liquor. However, the transport is assisted by a driven lifter reel.

3. These machines use a jet having lower velocity that used on conventional jet dyeing machines.

4. The soft flow machines are gentler on the fabric than conventional jet machines.

5.4.4.1   *Polyester dyeing curve*

Polyester can be very difficult to dye properly. The dyeing of hydrophobic fibres like polyester fibres with disperse dyes may be considered as a process of dye transfer from liquid solvent (water) to a solid organic solvent (fibre). Dyeing flow chart of 100% polyester is shown in Fig. 5.14.

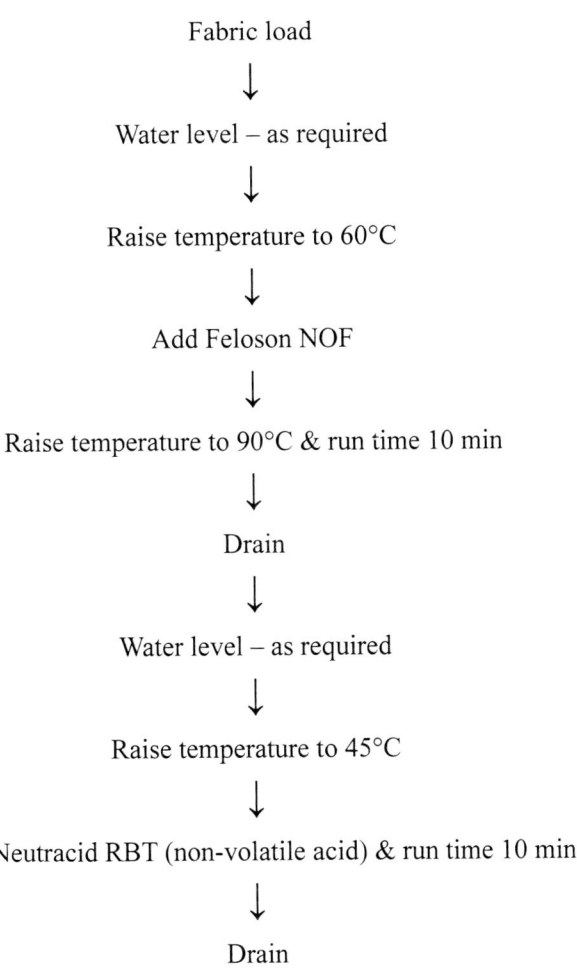

Fabric load
↓
Water level – as required
↓
Raise temperature to 60°C
↓
Add Feloson NOF
↓
Raise temperature to 90°C & run time 10 min
↓
Drain
↓
Water level – as required
↓
Raise temperature to 45°C
↓
Add Neutracid RBT (non-volatile acid) & run time 10 min
↓
Drain

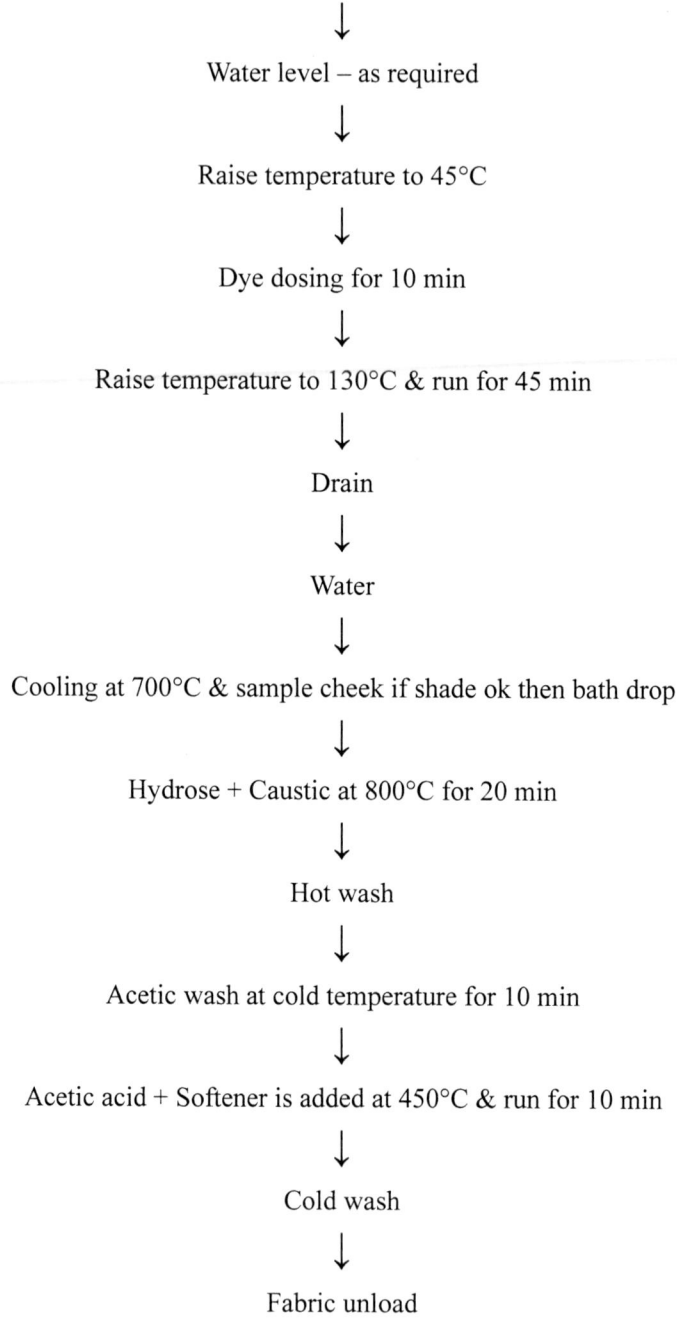

Figure 5.14. Process flow for automotive 100% polyester knitted fabric

Dyed texturized yarn/grey texturized yarn → Warping on Benniger (Germany) → Warp knitting machine → Knitted fabric (if grey) will go for dyeing in soft flow jet dyeing machines.

Dyed Knitted fabric → Shearing machine (Vollenweider, Switzerland) → The pile will be cut by blade cutters in the shearing machine → Then it will go for scouring in jigger/jet dyeing machine.

Upholstery piece dyeing is divided into two types wherein the headlining is dyed in beam dyeing and upholstery is dyed in jet dyeing using high light fastness dyes and UV absorberer and then finished on a stenter. Due to the inherent property of shrinkage of automotive fabrics, there should be attachment of a weft straightener in the stenter. Lastly, it will go for lamination. The lamination machine is made in Germany where the reverse side of the fabric is pasted with polyurethane foam with the help of a flame. Nylon screen is also used to cover it from the reverse side. In some places, they also do acrylic back coating. This foam/acrylic is fire retardant. The checking and packing into rolls is then done.

5.4.4.2   Process flow for automotive 100% polyester woven fabric

Warping → Warp beam prepared using air texturized/spun polyester yarn → Dornier looms → Woven fabric → Scouring → Drying on stener → Checking → Lamination → Checking and packing in rolls.

Owing to the high fastness requirements, only selected disperse dyes can be used for this application. Most of the dyes have to be imported and UV absorbers have to be added in the dyebath to further improve the light fastness.

5.4.4.3   Dyeing of fire retardant fabric dyeing

Trevira CS fabric is dyed in jet dyeing machines. The process is followed as mentioned earlier. After dyeing, it is squeezed and dried in stenter at a lower temperature of 140–150°C. Dispersion dyes take on fire retardant somewhat faster than the corresponding normal polyester types. Due to this, light shades can be dyed at boiling temperature without adding carrier. Medium to dark shades are dyed without carrier at a maximum temperature of 120°C. Disperse dyes used are high energy dyes from Dystar, Germany known as Dianix FG dyes or Foron RD dyes from Clariant, Basle. In raw white for piece dyeing or as multicoloured yarns, flame retardant textiles can be made in all desired qualities. In general, dyeing with coloured and patterned fabric results in adequate wash fastness. Preliminary tests are recommended; however, there is a possibility of staining adjacent material. Dyes on Trevira CS are fast to washing at boiling temperatures in bright shades and with selected dye-stuffs. A chlorine fast finish is possible with the appropriate dyestuffs. Boiling is not recommended for drapes and decorative fabrics.

5.4.4.4   Dyeing of polyester/lycra fabrics

Yarns containing elastane are continuing to make significant inroads into the knitted fabric market. The high stretch and rapid recovery properties of elastane provide unsurpassed elastic recovery properties which have been utilized by fabric and garment makers to provide elastic stretch and impart garment body clinging, shaping and shape retention, and silhouette shaping and slimming properties. Indeed, it is now estimated that 35–40% of all apparel in developed markets may contain some elastane in filament form.

However, fabrics containing polyester/elastane yarns present some formidable practical processing problems in dyeing and finishing, and dyers and finishers must adhere to the recommended guidelines from the fibre producers and from dyemakers to achieve the best performance. With single wrapped or double wrapped elastane with polyester filament, the dyeing process is designed to dye the polyester only, so careful dye selection is absolutely imperative.

Elastane filaments are well-known to be sensitive to high temperatures, to chlorine bleaching and to biological attack from microorganisms. Heat setting is a vital stage in the wet processing sequence. Polyester/elastane is generally heat set in the temperature range 182–188°C to minimize the possibility of yellowing occurring at higher temperatures, while the heat setting time of 20–35 s should not be greatly exceeded.

Elastane filaments are normally supplied with 2–5% of lubricant mainly based upon low molecular weight polydimethylsiloxane. Other lubricants are used on polyester so that the polyester/elastane yarn requires scouring prior to dyeing. Pre-heat setting can remove some of the fibre process lubricants through volatilization, and these oils can condense and sometimes give rise to oil spots when processing subsequent fabrics.

Aqueous or solvent scouring can be utilized for removal of the lubricants. Aqueous scouring in soft water using special surfactants to prevent re-deposition of the removed silicone lubricant is widely used, although this does not remove all of the lubricant. Solvent scouring is more thorough, using perchloroethylene, but this can remove not only more of the silicone, but also some of the additives in the elastane which are present to improve the resistance to heat and ultraviolet radiation.

An essential aspect of dyeing polyester/elastane weft knitted fabrics is to minimize the tension at high temperatures. If this is not ensured then the fabric will be stretched and the stretch/recovery properties will be adversely affected. For this reason, jet dyeing machines of the soft-flow type are normally used and the normal practical precautions during dyeing must be carried out to prevent rope marks and creases.

Dye selection is the key to successful dyeing of polyester/elastane fabrics. Dyers should aim for right-first-time dyeing, which may necessitate careful preparatory work in the dyehouse laboratory on the specific polyester/elastane fabric to be dyed. It is generally advisable to avoid making shading additions in dyeing because the extended dyeing time at the dyeing temperature can adversely affect the fabric elastic properties.

Disperse dyes have a high substantivity for elastane fibres, and this staining is increased with high dyeing temperatures. For this reason, polyester/elastane fabrics are dyed at temperatures below 130°C which is normally used for 100% polyester fabrics. Even when using dyeing temperatures of 115°C, there will be severe disperse dye staining on the elastane and the elasticity of the elastane can be decreased.

It is therefore essential to give an alkaline reduction clear after dyeing. Commonly this is carried out at pH 10–11 and 70–80°C using either sodium hydrosulphite (dithionite) or alternatively thiourea dioxide. The treated fabric must then be washed well to remove the dye decomposition products and any other impurities to ensure that the highest standards of colour fastness to wet treatments, light and rubbing (crocking) are attained.

When polyester microfibre/elastane fabrics are to be dyed to medium and heavy depths, the fineness of the polyester microfibres leads to a large increase in the surface area of the polyester. Consequently the dyed microfilaments appear lighter and much more disperse dyestuff is required, particularly for heavy depths of colour. As a result, the disperse dye staining on the elastane can be intensified, requiring intensive after treatment to remove staining and surface dyestuff.

An alternative approach to the use of disperse dyes with their attendant heavy staining problems on elastane is to use a more expensive polyester fibre, namely, cationic-dyeable polyester.

It is well known that high temperature process will affect the stability and elasticity of lycrafibre. It is also sensitive to very high amounts of carrier.

Similarly extreme pH variations within the processing sequence can impair the lycra fibre characteristics. It can absorb very high amounts of disperse dyes without forming a colourfast chemical bond. The silicone oil in the lycra fibre blend can cause the disperse dye to sweat out during subsequent treatments and storage. All disperse dyes have a high tendency to desorption and thermomigration out of the lycra fibre. Thermosetting and storage impair the wet fastness results.

Accordingly the processing can be done in two ways. The sequences are as follows:

1. Relaxation → Heat setting in hot air → Scouring → Dyeing → Finishing
2. Relaxation → Scouring (open width) → Heat setting in hot air → Dyeing → Finishing

Relaxation is required in the lycra fabric in the form of conditioning to provide residual shrinkage after the tension caused during knitting or weaving. It can be done by passing the fabric over a steam table or continuous steaming. Heat setting is done at 185–195°C for 30–45 s.

This process will impart dimensional stability and improve the fabric structure. Still then 5–10% shrinkage will remain in the fabric. The lycra blends will contain an excessive amount of oils and other impurities. It has to be scoured properly. Chemicals to be used are s andozin NIS liquor – 2 g/L and sirrix 2UDI liquor – 1 g/L. pH is 4.5–5 by use of acetic acid. Temperature is 90°C. Time is 30 min. Alternate scouring process is as follows: c hemicals used are sandoclean PCJ liquor – 2 g/L and sandopur RSK liquor – 1 g/L. Soda ash is 0.5 g/L. pH is 9. Temperature is 98°C. Time is 30 min. After scouring give a hot wash and cold wash. The dyeing of various types of lycra blends has to be taken great care. High dyeing temperatures and high amounts of carrier will affect the fibre stability and elasticity.

The dyeing temperature should be between 100°C and 120°C with a small amount of carrier. The carrier to be used should be odour free and ecofriendly. It must be like dilatin POE (clariant). The amount of carrier ranges from 1% to 5%. Normal disperse dyes are recommended for dyeing of lycra fabrics.Knitted fabric dyeing (soft flow)

Introduction
TEC series high temperature soft flow knit fabric dyeing machine (Fig. 5.15) is designed for superior requirement of the compact structure, crease sensitive fabrics and sensitive shade. TEC is the most energy saving equipment in the market. For example, the water consumption per kg for light, medium and dark shades are 28, 38 and 48, respectively. There are three different loading capacity machine available as jumbo, midi and mini which are manufactured by M/s Fong's National Engineering Co., Ltd. China (Fig. 5.16).

Figure 5.15. Knit fabric dyeing machine

Figure 5.16. Jumbo, midi and mini dyeing machines, M/s Fong's National Engineering Co., Ltd. China

Principal
Material and liquor both in circulation
Advantages
Suitable fabric weight 100–600 g/m²
Capacity
Max. capacity per chamber up 300 kg
Liquor ratio – 1:4

Table 5.4. Technical data knit fabric dyeing

Items	TEC Series		
	JUMBO TEC	MIDITEC	MINITEC
Capacity (per tube)	300kg	250kg	200kg
Design Pressure		3.0 bar	
Max operation temperature		140°C	
Min MLR (cotton)	1.4	1.4	1:4.5
Min running MLR (excluding fabric absorption		1:1.4	
Max lifter reel speed	400m/min	350m/min	350m/min
Nozzle		Standard size: 120mm dia.	

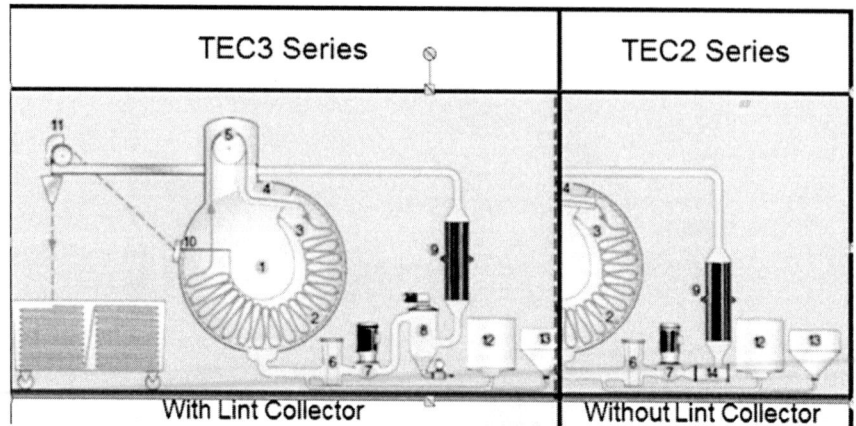

TEC3 Series	TEC2 Series
With Lint Collector	Without Lint Collector

Features
Vertical main pump
- High efficiency SS centrifugal pump
- Vertical and compact type, save installation area
- Lower down height of suction level for enhancement of low MLR operation
- High efficiency motor, more environment-friendly and economic

Figure 5.17. Vertical main pump

Vertical heat exchanger
- Heating and cooling gradient are guaranteed.
- Drain condensed water and dye liquor rapidly.
- Shorten medium exchange time.
- Save process time and energy.
- Prevent scaling formation, extend service life.

Figure 5.18. Vertical heat exchanger

Lint collector (patented)

Figure 5.19. Lint collector

- Collect the loose fibre and pile at the discharge area to avoid blockage of the filter
- No need to dismantle filter, no washing/cleaning required, save time and save labour when compared with traditional filter
- No stoppage of the machine due to blockage of the filter
- More efficient to remove lint when compared with traditional filter system

Vertical multi-function stock tank
- Interior heat coils for heating
- Circulating pump for mixing
- Discharge to service tank

Figure 5.20. Vertical multi-function stock tank

Internal cleaning device
- Cleans internal parts efficiently
- Save time in cleaning for next batch
- Reduces contamination

Figure 5.21. Internal cleaning device

Process sequence

Grey knit loading in soft flow – bleaching and dyeing in soft flow in tubular form – unloading – rope opening – hydro extraction– open width relax drying –compaction – cutting and garmenting.

6

Denim dyeing

Denim is a sturdy cotton warp-faced textile in which the weft passes under two or more warp threads. This twill weaving produces a diagonal ribbing that distinguishes it from cotton duck. The most common denim is indigo denim, in which the warp thread is dyed, while the weft thread is left white. Most denim fabric is yarn dyed; the warp yarns are dyed with indigo, and the filling yarns are left undyed. However, solid shades are becoming more popular and can be dyed by various methods.

There are a number of modifications or alternatives in the dyeing process that are routinely used to change the overall look or performance of the fabric. With the advent of denim garment washing techniques, the consistencies of the indigo dyeing process and its modifications have become crucially important in determining the quality and performance of indigo denim products.

Generally there are two most popular methods of dyeing denim fabric. They are as follows:

3. Rope dyeing
4. Sheet dyeing

6.1   Rope dyeing

Rope dyeing is considered a superior dyeing technology where the dyeing uniformity achieved is better than other indigo dyeing technologies like slasher dyeing/sheet dyeing. However, rope dyeing is also a more difficult dyeing technology. One needs to master its nitty-gritties to get the best out of the system.

Believed to be the best possible indigo dyeing method for yarn, the threads of denim yarn are initially twisted into a rope, then undergo a repetitive sequence of dipping and oxidization. The more frequent the dipping and oxidizing, the stronger the indigo shade.

Rope dyeing consists of twisting the yarns into a rope that is then quickly dipped into indigo baths. It is considered the best method for dyeing denim, as the short dyeing time does not allow the indigo to fully penetrate the fibres, thus creating ring-dyed yarn that fades better and faster than fully dyed yarn (Fig. 6.1).

However, rope dyeing is a also a more difficult dyeing technology. Flow chart of rope dyeing is given in Fig. 6.2.

Figure 6.1   Rope dyeing

Ball warping
↓
Rope dyeing
↓
Re-beaming
↓
Sizing
↓
Weaving
↓
Finishing

Figure 6.2   Flow chart of rope dyeing

The passage of yarn in rope dyeing is as follows:Pre-scouring → Hot wash → Cold wash → Dye baths → Hot wash → Cold wash → Application of softener

6.1.1   Pre-scouring

1. The objectives of pre-scouring are the removal of wax content from cotton, removal of trapped air from cotton yarn and making yarn wet.
2. This is done at 90°C.
3. We use the following ingredients at pre-scouring stage:
   (a) *Caustic soda.* Its quantity depends upon the quality of cotton fibres used in the mixing. Generally we take 2–4% of caustic soda. It removes the wax by the action of saponification.
   (b) *Wetting agent.* It is anionic in nature.
   (c) *Sequestering agent.* Even with the use of water softening, it is very difficult to find the desired softness in water (about 2–3 ppm). So we use the agent to make the water soft.
4. Why trapped air should be removed? The reason for this can be understood as follows: In 1 kg of yarn, there is approximately 2 L of air. One litre of air decomposes 1.8 L of sodium hydrosulphite. It will cause uneven dyeing and more consumption of sodium hydrosulphite (hydro).
5. Absorbency of yarn may be checked after scouring.

6.1.2   Hot wash

As some caustic is carried by the yarn after pre-scouring, hot water is given at 70–80°C. If this is not done, this yarn will go into the dye bath which will change the pH of the dye bath.

6.1.3   Cold wash

After hot wash, yarn temperature is more. To bring it back to its room temperature, cold wash is given to it.

6.1.4   Dye bath

6.1.4.1   Indigo dyeing

The properties of the indigo dye account for the wide variety of colour designs that are available on denim materials. Indigo is unique as a major textile dye, because it has a very low affinity for the cotton fibre. Because of the low substantively of the indigo, the ball warps dyeing process ring dyes cotton. Unlike almost all other commercially successful dye stuffs, the indigo dye concentrates in the outer layers of the cotton yarn and fibre during the dyeing process. This produces an intense ring of colour around a white core in the cotton yarn and the cotton fibre thus the name ring dyeing. When using most other dyes, if the ring-dyeing effect occurs, it would be considered a dyeing defect.

Indigo dye in its normal form is a vibrant blue, it is insoluble in water, and it will not dye cotton fibre. To dye cotton, the indigo must be converted to a water-soluble 'leuco' form and then applied to the cotton. This process is known as chemical reduction. Reducing agents such as sodium hydrosulphite with sodium hydroxide chemically convert the indigo dye to its soluble form. This also temporarily converts the dye from its blue colour to a very pale greenish yellow colour. The leuco form of indigo is readily absorbed by the outer layers of the cotton yarn. Once in the fibre/yarn, the indigo is made insoluble by oxidizing the yarn by passing the yarn through the air (skying). In fact, the dye will start to oxidize immediately when exposed to the air. The oxygen in air converts the dye back to its original blue and insoluble form. Thus the dye becomes trapped inside the outer layers of the cotton yarn. This results in a small amount of dye being deposited on the surface, resulting in only light blue dyed yarn. To obtain deep blue indigo dyed yarns, the colour must be built in layers.

The dye is layered by using multiple passes of the rope of yarn into the soluble dye and then exposing it to the air for oxidation. This multiple passing of yarn into dye is called dips. Normally, this process is repeated from 3 to 12 times to build up a deep indigo blue colour. The number of dips is limited to the number of dye boxes on the dye range. If the concentration of indigo dye in the dye boxes is doubled, this will result in slightly darker denim. This acts as a multiplier when labelling the denim. A double concentration of dye in nine dye boxes makes it 18-dip denim. Tripling the concentration makes it 27-dip denim. When even darker shades are desired, a sulphur black or blue dye can be applied to the yarn before indigo dyeing. This is known as a sulphur bottom. If the sulphur dye is applied after the yarn has been indigo dyed, it is known as sulphur top.

In rope dyeing, ball warps are continuously fed into the rope or chain-dyeing range for application of the indigo dyeing. Typically, 12–36 individual ropes of yarn are fed side-by-side simultaneously into the range. The ropes are kept separate from each other throughout the various parts of the dye range. For example, if the total number of ends on the loom beam is 3456, and each ball would have 288 ends, then the dye set would have a total of 12 ball warps. If there can only be a multiple of 10 balls on the dye range, then there would be 345 ends on 9 balls and 351 ends on the tenth ball.

The ropes are first fed into one or more scouring baths, which consist of wetting agents detergents and caustic. The purpose of these baths is to remove naturally occurring impurities found on the cotton fibre such as dirt, minerals, ash, pectin and naturally occurring waxes. It is very important to remove these materials to guarantee uniform wetting and uniform dyeing. The ropes are subsequently fed into one or more water rinsing baths.

Indigo is not a perfect vat colour. It may be called a trash vat colour. The constant of substantivity for other colours is 30, for indigo it is only 2.7. So there is a need of five to six dye baths and make the use of multi-dip and multi-nip facility to increase the penetration. The dyeing is done at room temperature as indigo belongs to IK class of vat dyes, where dyeing is done at room temperature and oxidation is done by air only and not by chemicals. If oxidizing agents are used, they will cause stripping of colours.

Indigo is not soluble in water. So it is reduced with sodium hydrosulphite. Then caustic soda is added to make sodium salt of vat colours to make it soluble. To reduce 1 kg of indigo, 700 g of sodium hydrosulphite is required. However, some extra hydro needs to be taken to avoid some decomposition of hydro. Practically it is prepared in the following sequence:

(a) Take indigo
(b) Add caustic
(c) Then reducing agent

When caustic is added to indigo, it is an exothermic reaction. It is allowed to cool down, then before sending it to feeder, sodium hydrosulphite is added. Reducing agent is not added first as it will be decomposed first, so consumption of it will increase. It is also not advisable to take solubilized vat, as offered by some companies due to the following reason: If it is used after 6 months, there will be a decomposition of sodium hydrosulphite. It will become partially soluble. Then to make it soluble again, more hydro has to be added.

6.1.4.2 Core and ring dyeing effect

This effect is obtained by multidip–multinip facility. pH of the dye bath should be kept in between 10.5 and 11.5. At this pH, sodium salt of indigo is monophenolic form. At this form, the strike rate of dye is very high. So after washing, there will be a better dye effect. At pH 11.5–11.7, this affinity is less, so dye effect will be less prominent. pH is controlled by the addition of caustic soda.

6.1.5 Washing

Rubbing fastness of indigo is very important. On a scale of 1–4, it is 2. Washing is done to improve the rubbing fastness. Wash at 60°C → Wash at 60°C → Wash at room temperature → Wash with softener

6.1.6   Why softener

The rope is going to be opened at long chain beamer. If the softener is not used, opening will be hampered. It is generally 1.2% of the weight of the yarn. It is a cationic softener. It is always having pH in the range of 4–55. Softening is done at room temperature. If high temperature is used, there is always some chance of tendering of yarn.

6.1.7   Importance of high concentration of free hydrosulphite

The clearest shades with minimum reddish streaks are observed at by relatively high concentration of hydrosulphite. On the other side, with lack of hydrosulphite, the leuco indigo is less dissolved and thereby adheres to a greater extent to the fibres. With lack of hydrosulphite, furthermore, the amount of unreduced dyestuff by oxidation at the upper level of the liquor and through activation of unfixed dyestuff, gets separated from the fibrous material would constantly rise as the reducing agent for creating leucoform would be missing. Under these circumstances, a reddish-bronze-like shade results due to dispersion of not reduced dyestuff in the yarn. The minimum proportion of hydrosulphite should be around 1.3–1.5 g/L in case of rope dyeing and 3–4 g/L in case of sheet dyeing. Also to avoid the lack of hydrosulphite or indigo at certain places in the immersion, vat, the whole quantity of the liquor should be circulated two to three times every hour.

6.1.8   Indigo dyeing process control

1. *Concentration of hydrosulphite:* It is measured by vatometer. It should be from 1.5 g/L to 2.5 g/L, or by redox potential of dye bath which should be from −730 mV to −860 mV.
2. *Caustic soda or pH value*: Should be from 11.5 to 12.5.
3. *Dye concentration in dye bath*: It is measured by spectrophotometer. It should be in g/L

6.1.9   Guidelines

High indigo concentration = Shade is greener and lighterLow indigo concentration = Shade is dull and redHigh pH or caustic concentration = Redder and lighterLow pH or caustic concentration = Greener and darker

6.1.10   Dipping time

Longer the dipping time, better will be the penetration and lesser will be the ring dyeing effect. It varies from 15 to 22 s.

6.1.11   Squeeze pressure

High pressure will lead to lower wet pick up and result in lesser colour and better penetration. At rope dyeing, squeeze pressure is 5–10 ton, that is, wet pick up is as low as 60%. Hardness of squeeze roller is about 70–75° shores. If sqeeze rolls are too hard then there are chances of slippage and uneven yarn tension. If squeeze rollers are too soft then shading will occur. Surface of the squeeze rolls should be ground twice a year.

6.1.12   Airing time

It should be 60–75 s. Longer airing time results in high tension on the yarn and subsequent processes will become difficult.

6.1.13   Drying

Insufficient or unevenly dried yarns will result in poor rebeaming.

6.1.14   Effect of pH

At pH of 10.5–11.5, there will be formation of more monophenolate ions, which lead to higher colour yield, as strike rate of the dye to the yarn bundle is very high, and wash down activities will be very good. At pH higher than this, dye penetration will be less and wash down characteristics are also poor.

6.2   Sheet dyeing

Slasher, or sheet, dyeing combines dyeing and sizing into a single process. Warp yarns are repeatedly passed in warp beam form through several baths of indigo dye before being sized and wound for weaving.

Slasher dyeing is considered to be lower quality than rope dyeing – the dye does not penetrate well and the colour tends to be uneven. However, recent mechanical improvements have helped make it a more viable option (Fig. 6.3).

Figure 6.3   Slasher dyeing machine

6.2.1   Slasher/sheet dyeing

In continuous slasher/sheet dyeing and sizing machine, direct warping beams are used, instead of ball warping logs in case of indigo rope dyeing system. The slasher dyeing machine is capable of handling Ne count form 9/s to 30/s (OE and slub both). At the back end of the slasher/sheet dyeing range, the direct warping beams are creeled. The yarn sheet from each beam is pulled over and combined with the yarns from the other beams so that multiple sheets of yarns can be made (Fig. 6.4).

Figure 6.4   Creeling of direct warping beam in sheet dyeing machine

In sheet dyeing range, the total number of required ends for a weaver beam are dyed, dried, sized and dried simultaneously. The back direct warping beam contains 380–420 ends, similar to rope, but the ends are distributed evenly over the width of the flanges and the end lay parallel to each other.This continuous slasher dyeing range eliminates a few intermediate processes of the rope dyeing, such as rebeaming and sizing. The yarn sheet from the back beam passes through wash boxes, where it is treated with caustic and subsequently washed with normal water. After squeezing the excess water, the yarn sheet passes through dye baths and skied for oxidation as in the case of rope dyeing. This develops the indigo coating on the yarn. After dyeing, the dyed yarn is washed by passing through three to four wash boxes and finally squeezed before allowing it to pass through drying cylinders. The dyed yarn then enters into the sow box, where it is sized. Subsequently the yarn sheet is dried.The yarn sheets then passes through a set of stainless steel split rods, which separate them into individual sheets, equivalent to the number of section beams in the creel. After passing through the split rods, the yarn sheets are collected into single sheet and passed through an expansion comb at the head stock, which separate individual yarns. The expansion comb can be adjusted to the desired loom beam width. Slasher dyeing range typically consists of 1–2 wetting vats, 4–8 dye baths and 3–4 rinsing troughs. The immersion and oxidation times lie between 10–20s and 45–60s.

6.2.2   Process sequence in indigo slasher/sheet dyeing

The passage of flow of yarns in slasher dyeing is shown in Fig. 6.5.

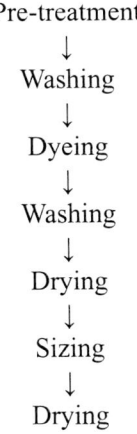

Pre-treatment
↓
Washing
↓
Dyeing
↓
Washing
↓
Drying
↓
Sizing
↓
Drying

Figure 6.5   Flow chart processes involved in slasher dyeing

6.2.2.1   Pre-treatment

Pre-treatment process in sheet dyeing consists of treatment of the cotton yarn sheet with caustic and wetting agent. Pre-wetting is carried out to get proper dyeing of the sheet. Pre-wetting is carried out with a wetting agent, at room temperature. In some cases, if well penetration of the dye is required, the yarns are treated with strong caustic soda solution followed by hot wash and cold wash treatment prior to dyeing. The pH of the bath is 11.8–12.

6.2.2.2   Washing

Cold washing is carried out at room temperature.

6.2.2.3   Dyeing (indigo blue dye)

The dyeing is carried out in four to eight dye boxes. The dyeing is carried out withindigo powder,

- sodium hydrosulphite,
- caustic soda and
- dispersing agent.

Typical dipping time of sheet in each tank is 15 s and oxidation time is about 90 s.

6.2.2.4   Washing

The dyed yarn sheet is subsequently to remove the unfix dye from the yarn surface. However, the number of washing tank may vary. Typical wash type is cold wash at room temperature and hot wash at temperature of 70°C.

6.2.2.5   Drying

The dyed yarn sheet is dried by passing it through drying hot cylinders.

6.2.2.6   Accumulator

The function of accumulator is to store the extra yarn sheet when the machine is stopped or at the time of size beam doffing, so the dyeing cannot be stop.

6.2.2.7   Sizing

The yarns are sized to achieve the required strength.

6.2.2.8   Drying

The dyed sheet is dried by passing through drying cylinders.

6.2.3   Advantages of slasher/sheet dyeing

Slasher dyeing ranges have a number of advantages. Slasher dyeing range produce sized beam directly which is ready to use in weaving. Sheet dyeing method has the following advantages:

1. Slasher dyeing is more comfortable for producing lightweight denims.
2. These machines require less floor space.
3. Enable smaller production runs.
4. Have a quicker turn over time.
5. The technology is less capital intensive and the machinery cost is less.
6. The cost of production is less.
7. Other type of dye can be use to dye the cotton in this range. Hence the slasher dyeing technique can produce a wide variety of colours other than indigo blue.
8. Rope opening is avoided, as in the case of rope dyeing.
9. The immersion and oxidation times are much shorter than rope dyeing.

6.2.4   Disadvantages

1. In sheet dyeing, there is a problem of centre to selvedge shade variation.
2. The hydrosulphite consumption is much higher owing to the greater surface.
3. *Slasher dyeing* is considered to be lower quality than rope dyeing – the dye does not penetrate well and the colour tends to be uneven. However, recent mechanical improvements have helped make it a more viable option.

6.3   Rope dyeing and sheet dyeing

The difference between rope dyeing and sheet dyeing is summarized in Table 6.1.

Table 6.1. Difference between rope dyeing and sheet dyeing

	Rope Dyeing	Sheet Dyeing
1	In rope dyeing there is an opportunity at rebeaming to repair broken ends	No Such Opportunity
2	More than one slasher sets can be dyed at one time	Only one slasher set may be dyed at one time

3	Possible to mix yarns of different colors- one can get denim stripes at rebeaming	not possible
4	No need to start and stop the machine at each set, so shade matches perfectly	Need to start/stop the machine. Difficult to achieve the target shade until hundreds of meters of yarns have been run
		when the slasher dyeing machine slows down at the end of each yarn set, the wash down shade will be altered
5	Large number of yarns are difficult to open at rebeaming- not very suitable for lighter weight yarn	Advantageous for lighter weight fabrics > 16s
6	No extra ends	Extra ends

7

Continuous dyeing

Batch dyeing is one of the widely used technique for semi-continuous dyeing process. It is mainly used in the dyeing of cellulosic fibre like cotton or viscose (knit and woven fabric) with reactive dyes. Pad-batch dyeing is a textile dyeing process that offers some unique advantages in the form of versatility, simplicity, and flexibility and a substantial reduction in capital investment for equipment. It is primarily a cold method that is the reason why it is sometimes referred to as the cold pad-batch (CPB) dyeing.

In the process of semi-continuous dyeing that consists of pad batch, pad jig and pad roll, the fabric is first impregnated with the dye liquor in, what is called a padding machine. Then it is subjected to batch wise treatment in a jigger. It could also be stored with a slow rotation for many hours. In the pad batch, this treatment is done at room temperature, while in pad roll, it is done at increased temperature by employing a heating chamber. This helps in fixation of the dyes onto the fibre. After this fixation process, the material in full width is thoroughly cleansed and rinsed in continuous washing machines. The only difference between continuous and semi-continuous dyeing process is that in semi-continuous dyeing, the dye is applied continuously by padding and the fixation and washing remain discontinuous. Liquor ratio in semi-continuous dyeing is not of much importance and is not taken as a parameter. One of the widely used techniques for semi-continuous dyeing process is the pad-batch dyeing. Figure 7.1 shows a schematic diagram for the semi-continuous dyeing process.

In older days, cotton fabric (both woven and knitted) was dyed using jigger, winch, etc., then synthetics and cotton synthetics were dyed in jet dyeing machines, beam dyeing machines, etc., one limitation in these machines were small batch sizes. Then batch to batch colour difference was not accepted by the garment manufacturer. Today's market drivers focus on reduced lead times to support speed to market, higher productivity and elimination of non-conformance for cost effectiveness. To overcome this, continuous dyeing ranges (CDRs) were started which includes enhanced productivity (as high processing speeds of 50–80 m/min are possible), better process control leads to better reproducibility and production flexibility because the lot size could be from 200 to 10,000 m and more. Besides these following are the advantages of continuous dyeing.

4. Cost-efficient process
5. Uniformity of dyeing
6. Enhanced fabric brilliancy and lustre
7. Improved levelness and solidity of shade
8. Short through time

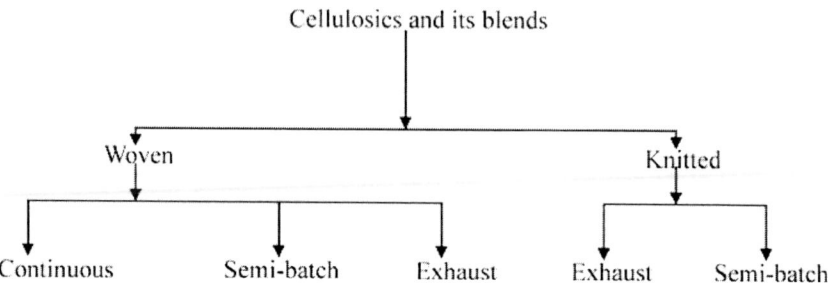

Figure 7.1 A schematic diagram for the semi-continuous dyeing process

Continuous dyeing processes are suitable for lightweight to heavyweight fabrics. Semi-batch for knitted fabric is pad batch, whereas semi-batch for woven fabric is pad batch and pad jigger. Semi-batch and continuous dyeing processes are used to produce uniform shades with quality and economy for longer lengths of production. The CPB is the most effective method of dyeing both woven and knitted fabrics.

The continuous dyeing is a two-stage operation which involves continuous preparation and continuous dye application and fixation which means instantaneous physical application of dye by padding. The dye application padding mangle is the most important component for shade uniformity without giving selvedge-to-selvedge or centre-to-selvedge shade variation. In India, knit fabrics are commonly dyed in rope form on winches, jet or air/soft flow machines. This process while being relatively simple and scalable poses several disadvantages and problems which are familiar to any dyer experienced in his field. They are as follows:

a. Formation of crease marks
b. Uneven dyeing (especially on viscose)
c. Hairiness of surface 'pilling' from mechanical friction
d. Tension in fabric causing high rest-shrinkage values
e. High consumption of water/steam/chemicals/dyestuffs
f. Usage of salt causing problems in effluent treatment

An alternative to the dyeing in rope is the process of dyeing in open width, also called CPB process. It has been successfully operated on knits in Europe for more than 20 years and in India for more than 10 years. This method allows a low-cost and high quality dyeing while avoiding some of the problems experienced in rope dyeing. The global market demands are supported by the use of CPB processing for woven fabrics. The key factors are continued pressure on cost, improved quality (both appearance and fastness) and reduced lead times in the global textile supply chain. These requirements combined with shorter production runs – due to reduced lead time or multi-phased seasons – means the dye house must be capable of excellent lab to bulk transferability.

There are five main methods of piece dyeing:

1. Beam dyeing
2. Beck dyeing
3. Jet dyeing
4. Jig dyeing
5. Pad dyeing

Pad dyeing is nowadays known as continuous dyeing. The fabric is passed through large pad rollers which squeeze the dye onto the fabric. The material passed through the liquor and squeeze rollers would be expected to absorb 50–100% of their weight in the dye liquor. The lower the percentage pick-up , the higher must be the concentration of dye liquor. Pad dyeing is usually done in a CDR and is useful for dyeing a large yardage of fabric. The method is particularly suitable for the dyeing of 100% cotton and polyester/cotton-blended fabrics.

Benninger, Kusters and Harish are the main manufacturers of CDR. Presently CDRs working in the world are as follows:

1. Asia – 50%
2. Europe – 20%
3. Africa – 10%
4. South America – 10%
5. North America – 20%

Dyes used for continuous dyeing of cellulosics are as follows:

1. Reactive dyes – 45%
2. Vat dyes – 25%
3. Sulphur dyes – 15%
4. Direct dyes – 10%
5. Naphthol dyes – 5%

A major contribution to reactive dye application procedure has been the development of pad-batch method. Today the short-time pad batch is one of the most important of all padding process. With the right choice of dyes, this process becomes economically more significant since it saves energy, labour and time. Dyestuff with relatively lower affinity and high reactivity make them most suited for CPB technique. The reactive dyes suitable for CPB method are

1. Colron RGB series (vinyl sulphone dyes)
2. Colron SF dyes (meta base vinyl sulphone dyes)

Different methods of reactive dye application are as follows:

1. Pad-batch method.
   a. Pad (alkali)-batch (cold) process
   b. Pad (alkali)-batch (warm or hot) process
2. Pad-dry method
3. Pad-steam method

7.1   Pad-batch method

7.1.1   Pad (alkali)-batch (cold) process

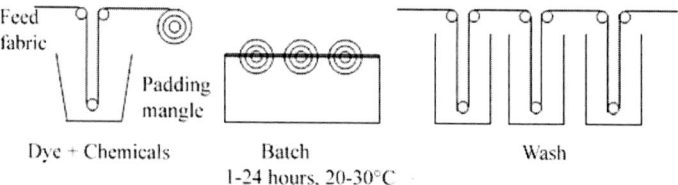

Figure 7.2 Pad-batch method

Steps:

1. The fabric is first padded in a padding mangle with reactive dye in the presence of an alkali.
2. The padded fabric is rolled in a batch and the batches are wrapped by polyethylene sheets and stored in wet condition for 1–24 h at 200–300°C in a room.
3. During the storage period, the rolls may be kept slowly rotating to prevent seepage of the dye liquor.
4. After storing time is finished, fabric is washed in a rope washing machine to remove the unfixed dye from fabric surface (Fig. 7.2).

NIP dyeing (popularly known as pad-batch dyeing) is perhaps the most sustainable method for textile dyeing. The CPB dyeing is known since many years to the dyers and they have been practicing this method as well. CPB dyeing is a more environmentally sound and higher quality dyeing method for woven and knitted cotton/viscose fibres. The process removes salt from the effluent, reduces the use of water, energy reduces the volume of effluent and occupies less space on the production floor (Fig. 7.3).

Pad   Batch   Washing off

Padding condition
- Batching time = 16 hours
- Batching temperature = 25°C
- Padding solution temperature = 25°C
- Liquor Pick up = 50–60%

Alkali solution (Sodium silicate (38°Be) 50mℓ/ℓ + Caustic Soda (38°B) X mℓ/ℓ

Colorant : CPB Alkali solution - Sodium Silicate Method

Dye (Total g/ℓ)	< 10	20	30	40	50	>60
Caustic Soda (38°B) X mℓ/ℓ	12	14	16	18	20	22
Sodium Silicate(38°Be)	50mℓ					

Alkali solution (Soda ash 20g/ℓ + Caustic Soda (38°B) X mℓ/ℓ

Colorant : CPB Alkali solution - Sodium Silicate Free Method

Dye (Total g/ℓ)	< 10	20	30	40	50	>60
Caustic Soda (38°B) X mℓ/ℓ	4	5	7	9	11	13
Soda ash	20g/ℓ					

Figure 7.3 Cold-pad batch dyeing process for cellulose

The CPB dyeing method gives you the required fastness standards with ease and minimum costs and less energy; however, dyers have been reluctant to use this method.

1. Salient feature of CPB dyeing can be summarized as follows:
   - Less surface area used in plant
   - Low amount of effluent
   - No salt residue in effluent water, other specialty chemicals such as anti-migrant, levelling and fixing agents not used
   - Conventional exhaust dyeing system emits up to 1 kg salt/kg of fabric

- Low steam consumption, almost 50% lesser
- Low electricity consumption
- Improved fabric quality, added lustre, reduced labour cost
- This process works at low temperature and low liquor ratio, thus saving energy and water considerably
- Ease of running small batches
- A high reliability and reproducibility as compared to other continuous methods.
- A soft hand feel and far better penetration of dyes
- Very good build-up for medium and heavy shades, reduced tailing and listing

After considering the above-mentioned facts CPB method is most economical and workable.

2. Selection of dyes

For any dyeing method, selection of dyes and process are very important. The dyer must consider following factors before selecting the dyes.

- Dyes chemistry
- High solubility at room temperature
- Similar substantively (low to medium) to prevent tailing. Very good compatibility
- Similar reactivity to obtain excellent reproducibility and high fixation
- Diffusion coefficient. High degree fixation value
- Higher alkali stability to prevent hydrolyzation
- Unaffected by fixation time variation

Solubility: In dye chemistry, there are two main factors which govern the dye chemistry of reactive dyes chromospheres and reactive system.The chromospheres are responsible for the colour gamut light fastness, chlorine fastness, solubility, affinity and coefficient of diffusion. The reactivity is responsible for the application properties such as dye fibre bond stability, efficiency of the reaction with the fibre, affinity and alkali requirements of dyeing.

Reactivity: The reactivity of the dye can be responsible for low and high dyeing time, temperature and pH alteration can modify the reactivity and two types of different reactivity can be dyed at the same conditions by adjusting pH and temperature.

Substantively: Mainly dependent on the nature of chromospheres. A higher substantively will have low solubility of the dyes high fixation rate, lower diffusion of coefficient, less sensitive towards temperature, pH and dyeing conditions, high risk of unlevel dyeing. Difficult to remove unfixed dye.

Dyes of better solubility can penetrate quickly in the fibre and gives better fixation and levelling, the dye solubility can be increased by increasing temperature, adding urea and reducing electrolytes. This knowledge of factors controlling the reactive dyes helps in producing correct right first time dyeing specially the CPB dyeing. Colourant has wide range of reactive dyes which are engineered looking in to the suitability for CPB application.

3. Machine (padding mangle) parameters
    • Padding cylinders should be horizontally mounted
    • Padding through should have small volume of 20–40 L for cooling front and backside of entering fabric two water cylinder should be available
    • Padding through should be clad in a cooling jackets to maintain constant liquor temperature ideal temperature is around 22–23°C
    • Lab padder should be horizontal and liquor should be place between the nips of cylinder
    • Dyed lab samples should be kept at the same temperature of production dwelling time area
    • Dwelling area should be air conditioned and kept 3–5°C above padder temperature
    • Lab padder and batching area should be away from sunshine, air current, chemicals fumes
    • Dyed and alkali supplying pump should run always 4:1 ratio and checked regularly
    • During dwell time batch rotation should be around 10–30 rpm.
    • Trough turn over time of the padding liquor should be less than the dye hydrolysis rate
    • Fabric should be wrapped properly after the padding of the dye so that no air come in contact with the fabric, also prevents the drying of selvedge.
    • Longer batching time and longer time required to asses correct shade are weak points of this process but manageable

Batching (wrapped in plastic): The fabric is wrapped in plastic shading to prevent drying out and exposure to air. Drying out causes migration of dyes resulting in edge marks.

Carbonation: Reaction between carbon dioxide (from air) and caustic soda.

$$2NaOH + Co_2 + H_2O \rightarrow Na_2CO_3 + H_2O$$

The quality of commercial soda ash is also very important, as it can contain high amounts of sodium bi carbonate. The presence of sodium bi carbonate results in a second reaction.

$$NaHCo_3 + NaOH \rightarrow Na_2Co_3 + H_2O$$

4. Colourant recommended robust trichromy for CPB

Table 7.9 Recommended colourant trichromy for CPB

Light and earth (niche) shades up to 16 GPL

    Colron golden yellow SF

    Colron red SF

    Colron blue SFGR

Medium shades 16–35 GPL

    Colron golden yellow CES

    Colron ruby S3B

    Colron deep red GLX

    Colron dark blue CS

Dark and extra dark shades above 35 GPL

    Colron golden yellow CES

    Colron deep red GLX

    Colron ruby S3B

    Colron navy SGI

    Colron navy GLX

For highlight fastness

    Colron yellow CNRG

    Colron red CN2BL

    Colron blue CNR

For black's

    Colron black W2N

    Colron black DN

    Colron black GLX

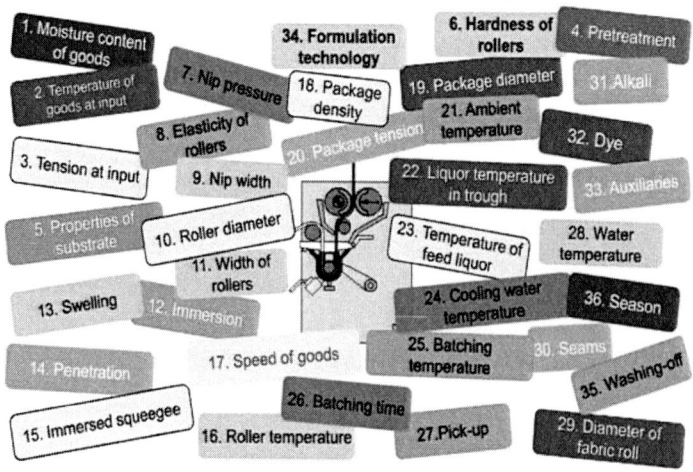

Figure 7.4 Parameters affecting CPB dyeing

7.1.2   Pad (alkali)-batch (hot) process

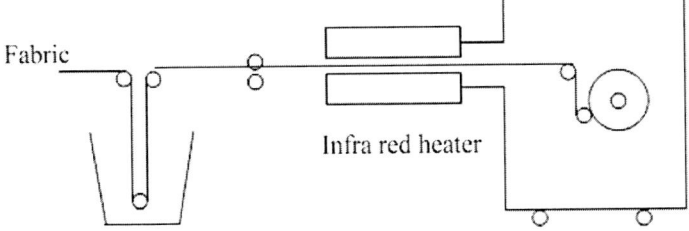

Fabric

Infra red heater

Padding mangle   Hot batch chamber

Figure 7.5 Pad-batch (hot) process

Steps:

1. The fabric is first padded in a padding mangle with reactive dye in the presence of an alkali.
2. The fabric is then passed in between infrared heater to pre-heat the padded fabric to 50–90°C.
3. The fabric is then batched on a large diameter roller in a hot chamber. The batching is done under controlled conditions of temperature and humidity for a sufficient time to ensure diffusion and fixation of the dye in the fibre. During this period, the batch is kept slowly rotating to avoid the seepage of dye liquor.
4. The cloth is then washed in a rope washing machine to remove the unfixed dyes (Fig. 7.5).

7.1.3   Pad-dry method

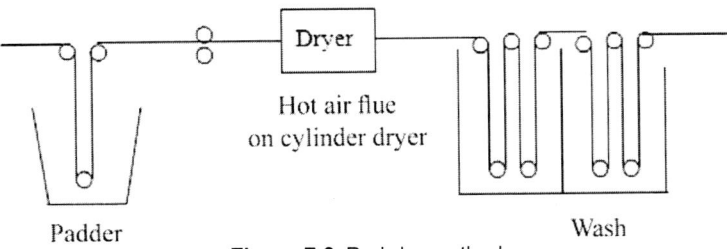

Dryer

Hot air flue
on cylinder dryer

Padder   Wash

Figure 7.6 Pad-dry method

Steps:

1. Fabric is first padded in a padder with reactive dye in presence of an alkali.
2. Padded fabric is then passed through a squeezing roller into a dryer. As a dryer cylinder, stenter may be used. During drying due to higher

temperature, fixation of dye in fibre increases and at the same time water is removed by evaporation.
3. After drying, fabric is washed in a washing machine to remove un-fixed dye (Fig. 7.6).

7.1.4   Pad-steam method

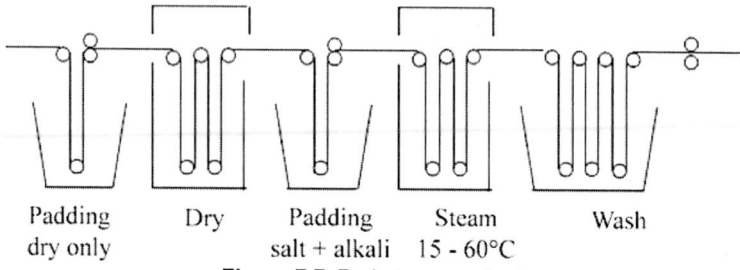

| Padding | Dry | Padding | Steam | Wash |
| dry only | | salt + alkali | 15 - 60°C | |

Figure 7.7 Pad-steam method

Steps:
1. Fabric is first padded in a padder with the dye.
2. It is then passed through between two squeezing rollers in a dryer. Drying should be done slowly; otherwise precipitation of dye due to quick removal of water may take place leading to lower colour value.
3. After coming out from dryer, fabric is padded in a padder containing salt and alkali. Due to salt, exhaustion of dye takes place and due to alkali, fixation occurs.
4. Fabric then passed through a steamer where it is kept for 15–19 s. Due to high temperature here, fixation rate increases.
5. In this step, fabric is washed in a washing machine to remove the unfixed dye (Fig. 7.7).

7.1.5   Factors involved in the dyeing process

The factors that need to be considered are as follows:
1. Machinery and production area, laboratory
2. Padding cylinders should be horizontally mounted.
3. Padding trough should have small volume of 15–25 L.
4. For the cooling, two water cooled cylinders should be available at the front and back sides of entering fabric.
5. Padding trough should be clad in a cooling jacket to maintain a constant liquor temperature. Ideal temperature is around 22–23°C.
6. Laboratory padder should be also horizontal and liquor should be placed between the nips of cylinders.

7. Dyed laboratory samples should be kept at the same temperature of production dwelling area.
8. Dwelling area should be air conditioned and kept 3–5°C above padder temperature.
9. Laboratory padder and batching area should be away from sun shine, air current and chemical fumes.
10. Dye/alkali supplying pump should run always at 4/1 ratio and be checked regularly.
11. Batch rotation should be around 5–10 rpm.

7.1.5.1   Fabric

Fabrics should be uniformly bleached, and if possible mercerized. Absorbency should be high and be same along width of fabric. Fabric should be uniformly dried and cooled down prior to padding. Fabric stitching should not be overlapped.

7.1.5.2   Dye

Dyes should have high solubility at room temperature water. It should have similar substantivity to prevent tailing and should have similar reactivity to obtain excellent reproducibility. Dyes should have higher alkali stability to prevent hydrolyzation. Since alkaline stability of vinyl sulphon dyes is rather low, bifunctional reactive dyes are more preferred for CPB dyeing because the reproducibility ratio is much higher and washing off property is superior.

7.1.5.3   Washing

For efficient washing of CPB dyed fabrics, 7- to 8-chamber washing tank is sufficient. In the first 2–3 tank, excess amount water is used, to remove silicate and drop the pH to 8–8.5. Temperature in these tanks should not be above 50°C. If pH does not drop or washing machine is short, it is wiser to wash rest of the chambers at 50°C and in the second pass at boiling temperature.

1. Chamber 50°C
2. Chamber 50°C
3. Chamber 50°C, pH 8–8.5
4. Chamber 98°C
5. Chamber 95°C
6. Chamber 98°C
7. Chamber 70°C
8. Chamber 40°C

7.1.6   Special features of pad-batch dyeing process

The CPB dyeing is a more environmentally sound and higher quality dyeing method for woven and knitted cotton/viscose fibres. The process removes salt from the effluent, reduces the use of water, energy, reduces the volume of effluent and occupies less space on the production floor.

- Significant cost and waste reduction as compared to other conventional dyeing processes.
- Total elimination of the need for salt and other speciality chemicals. For example, there is no need for anti-migrants, levelling agents and fixatives that are necessary in conventional dyebaths.
- Optimum utilization of dyes that eliminates speciality chemicals, cuts down chemical costs and waste loads in the effluent. All this results in a formidable reduction in wastewater treatment costs.
- Excellent wet fastness properties.
- Pad-batch dyeing cuts energy and water consumption owing to low bath ratio (dye:water) required for the process. This is because unlike other dyeing processes, it does not function at high temperatures.
- A uniform dye quality is achieved with even colour absorbency and colour fastness.
- As compared to rope dyeing, pad-batch dyeing produces much lower defect levels.
- In pad-batch dyeing, qualities like high shade reliability and repeatability are common. This is because of high reactivity dyes with rapid fixation rate and stability.
- Lastly pad-batch dyeing can also improve product quality. The fabric undergoing the CPB dyeing process is able to retain a uniformly coloured appearance. It shows added lustre and gives a gentle feel. The fabric gives a brighter look in shades.
- Reactive dyes are vastly used in dyeing and printing of cotton fibre. These dyes have a distinctive reactive nature due to active groups which form covalent bonds with –OH groups of cotton through substitution and/or addition mechanism. Among many methods used for dyeing cotton with reactive dyes, the CPB method is relatively more environment friendly due to high dye fixation and non-requirement of thermal energy. The dyed fabric production rate is low due to requirement of at least 12 h batching time for dye fixation. The proposed CPB method for dyeing cotton involves ultrasonic energy resulting into a one-third decrease in batching time.

Benefits of CPB dyeing are as follows:

- Relatively low cost of equipment
- Less surface area
- Low amount of effluent
- No salt residue in effluent water
- Conventional exhaust dyeing system emits up to 1 kg salt/kg of fabric
- Low steam consumption, 50% less
- Low electricity requirement, 30–40% less
- Improved fabric quality
- Reduced labour cost

7.1.7   Special features and applications

Pad-batch dyeing is a process change that offers several significant advantages, primarily cost and waste reduction, over other conventional dyeing processes. This method eliminates the need for salt and speciality chemicals such as anti-migrants, levelling agents and fixatives required in conventional dyebaths. In turn, efficient use of dye and the elimination of speciality chemicals reduce chemical costs and waste loads in the effluent resulting in reduced wastewater treatment cost.

Pad-batch dyeing reduces water and energy consumption due to the low bath ratio (dye:water) required for the process and since it does not operate at high temperatures as required in other dyeing processes.

Consistent dye quality is achieved resulting to even colour absorbency and colour fastness. This method also produces much lower defect levels than rope dyeing, wherein the fabric is transported through the dyeing machine in a loosely collapsed form resembling a rope. When used in this process, high reactivity dyes have rapid fixation and stability resulting in shade reliability and repeatability.

Pad-batch dyeing also improves product quality. The fabric that undergoes the CPB dyeing process retains a smooth, uniformly coloured appearance with added lustre and a soft touch and drape. It is also brighter in shade, has superior wet fastness properties and has a better look and feel.

7.1.8   Thermosol dyeing method

Recipe
- Dyeing depth: X g/L
- Migration inhibitor: 2 g/L
- Intermediate drying: 110°C × 60 s

- Pick-up: 60–80%
- Thermosoling: 200°C × 90 s

The process involved in thermosol dyeing method is shown in Fig. 7.8.

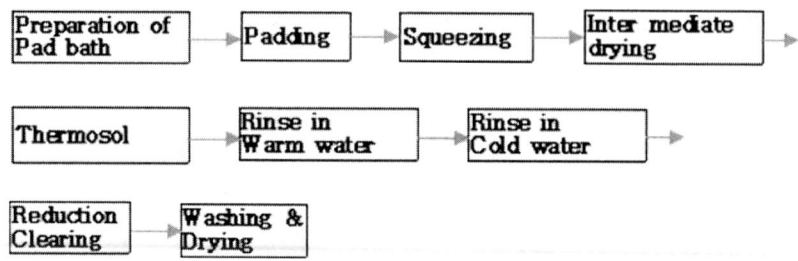

Figure 7.8 Thermosol dyeing method

7.1.9   Paper heat transfer dyeing method

Recipe
- I lkaron JP Hi conc dyes: X
- Methylalcohol: 10–30
- Sodiumalginate: 10–20
- Water: 50–80
- Total: 100

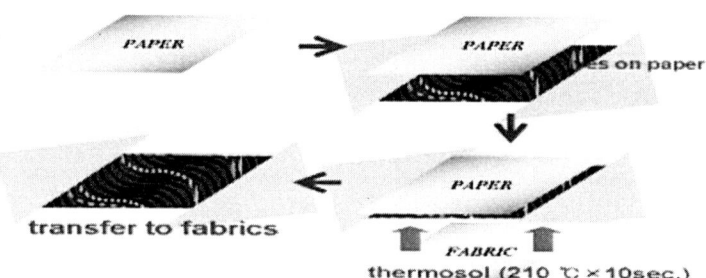

Figure 7.9 Paper heat transfer method

Fastness properties
- Light: ISO 105-B02 xenon
- Sublimation: ISO 105-P01, 180°C × 30 s
- Rubbing: ISO 105-X12, wet/dry
- Washing: ISO 105-C06, 60°C × 30 min
- Perspiration: ISO 105-E04 acidic and alkaline

A great variety of fully continuous dyeing processes has been developed, which is described in the following sections.

7.2   Pad-dry-pad steam process

It is a double-pad process classic procedure for dyeing of heavy weight woven fabrics, primarily for cotton goods and long yardages (Fig. 7.10). Main advantages of the pad-dry-pad steam process are as follows:

a. High productivity
b. Good appearance
c. High colour yield

Pad with dye liquor $\rightarrow$ Dry $\rightarrow$ Chemical pad $\rightarrow$ Steam $\rightarrow$ Wash-off

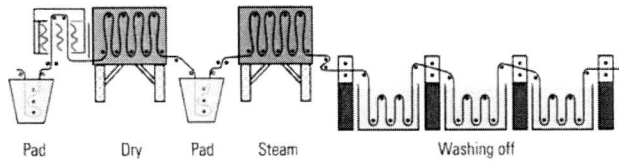

Pad   Dry   Pad   Steam   Washing off

	Chemical	Amount
Padding	Synozol Dyes	Xg/ℓ
	Migration Inhibitor	10g/ℓ
	Wetting Agent	1g/ℓ
	Reduction Inhibitor	1g/ℓ
	Liquor Temperature	25°C
	Liquor Pick up	60%
Drying	Temperature	120°C
	Time	1min
Alkali Padding	Soda ash	20g/ℓ
	Caustic soda(38°Be)	10mℓ/ℓ
	Glauber's salt	200g/ℓ
	Liquor Temperature	25°C
	Liquor Pick up	80%
Steaming (Fixation)	Temperature	102°C
	Time	1min

Figure 7.10 Pad-dry-pad steam dyeing process for cellulose

1. Pad liquor temperature is 20–30°C. Modified padding mangles are used. Both cold brand and hot brand reactive dyes can be used in this process. Liquor pick-up, 50–80%.
2. Drying at 110–130°C in hot flues.
   Chemical pad (salt, alkali) on a padder. Pad liquor temperature is 20–25°C. Liquor pick up, 70–80%.
3. Steaming for 45–90 s in saturated steam. It is done in a dye house steamer. Dye fixation takes place during steaming.
4. The goods are then washed-off and scoured as they move down the range.

Pad-dry steam process: This is a one bath continuous process without any addition of salt. It is mainly used for cotton goods and long yardages. Following are the advantages of the pad-dry steam process:

a. High productivity
b. High colour yield
c. Salt-free dyeing
d. Lower migration of dyes
e. Good washing off

Pad with dye and alkali → Drying → Steaming → Wash-off

In this process, reactive dye solution containing urea (50 g/L), ludigol/resist salt (10 g/L; to protect dye from reducing fumes) and padded at room temperature, and the padded goods are then dried in a hot flue at 100–120°C, after which the goods are steamed for 6–8 min at 100–105°C. The steamed goods are then washed-off.

Following are the precautionary measures need to be checked:

a. Soda ash should not be added to the dye liquor until immediately before dyeing starts.
b. The dyes have less tendency to migrate because of intermediate drying operation.
c. Steaming should be adequately carried out with saturated steam to ensure uniform moisture absorption by the dry goods.

7.3   Pad steam process

It is a one-bath continuous process without intermediate drying. It is particularly suitable for heavy woven fabrics for which intermediate drying would not be economical and there would be a risk of dyestuff migration. Following are the advantages of the pad steam process:

a. High productivity
b. Suitable for mainly light to medium shades

c. Suitable for terry towels and cord

Padding with alkaline dye solution → Steaming → Washing off

This is also called wet-steam process, where the fabric is padded with dye solution and alkali, steamed without intermediate drying and then washed-off. The dyes are completely fixed by steaming for a minute at 102–105°C to avoid dye hydrolysis. The main requirements are cooled fabric, cold pad liquor, small trough, rapid liquor turn over, etc. Sulphur dyes can also be used in this process (Fig. 7.11).

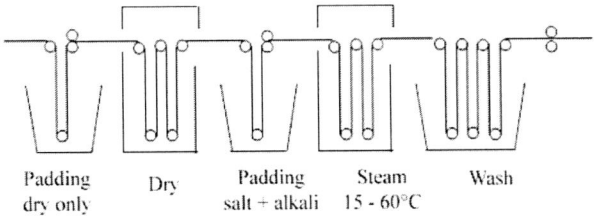

Padding   Dry   Padding   Steam   Wash
dry only   salt + alkali   15 - 60°C

Figure 7.11 Pad steam method

Steps:

1. Fabric is first padded in a padder with the dye.
2. It is then passed through between two squeezing roller in a dryer. Drying should be done slowly; otherwise precipitation of dye due to quick removal of water may take place leading to lower colour value.
3. After coming out from dryer fabric is padded in a padder containing salt and alkali. Due to salt exhaustion of dye takes place and due to alkali fixation occurs.
4. Fabric then passed through a steamer, where it is kept for 15–19 s. Due to high temperature here, fixation rate increases.
5. In this step, fabric is washed in a washing machine to remove the unfixed dye.

7.3.1   Pad-steam dyeing machine/process

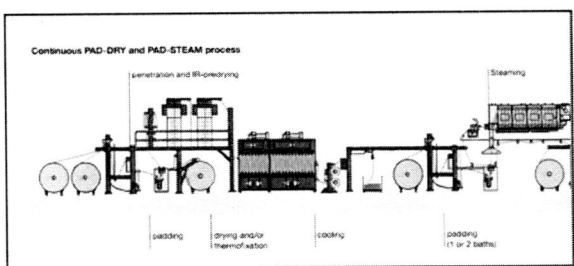

Figure 7.12 Pad-steam dyeing machine/process

Pad-steam dyeing is a process of continuous dyeing in which the fabric in open width is padded with dyestuff and is then steamed. Pad steam is an ideal machine for reactive dyeing of cotton and its blended fabrics. Light, pale and medium shades can be dyed in this machine. Continuous roller steamer is used for diffusion of reactive, vat, sulphur and direct dyes into cellulosic fibres in an atmosphere of heat and moisture that is created by saturated steam injected into the steamer.

7.3.2   Purposes of the machine

It can be used as a pad batch for reactive dyeing in which batch is left for 12–18 h for the completion of the reaction. For time saving, the fabric passes through the steamer for 1 min and the reaction is completed. It can also be used for reduction clearance in which we treat polyester/cotton fabric with caustic and sodium hydrosulphide to remove the disperse dye from cotton. Stripping of the fabric can also be done on this machine, that is, colour can be removed completely by adding higher amount of caustic and sodium hydrosulphide. It can be used for the development of vat dyes. The dyed fabric can be washed in this machine.

7.3.3   Process that can be done on this machine

- Reactive dyeing
- Pad-batch dyeing
- Reduction dyeing
- Stripping
- Vat development/vat dyeing
- Hot and cold washing
- Pad steam
- Wet chemical pad

7.3.4   Main sections of the machine

7.3.4.1   Inlet Section

Inlet section consists of following parts:
- Plaiter/batcher
- Tensioner rollers
- Free guide roller
- Stationary rollers

7.3.4.2   Padding section

Padders use for padding. The pressure of the padders is 1.5–2 bar. Two types of pressure used in kuster padders: hydraulic and pneumatic. The central pressure is hydraulic and side pressure is pneumatic. We can adjust the pressure of the padders, to prevent the listing problem. Liquor is picked in the fabric; afterwards the excessive liquor is squeezed out by means of padders at pre-determined pick-up% set by applying pressure on the padders.

7.3.4.3   Kuster padder

Uniform squeezing pressure over the entire working width independent of fabric width (no selvedge pressure). No side-to-centre shading, due to smooth treatment of the selvedges.

Advantages:
- Uniform liquor application over the whole fabric width
- Different liquor application in the range of side-centre-side zone is possible
- Attractive price
- Easily operation
- Reliable and economical padder

7.3.4.4   Steamer

Here in steamer temperature required for the fixation is given to the fabric. This temperature is achieved by saturated steam. The purpose of using saturated steam is that the chemicals used for developing should not dry on the surface of fabric preventing fabric from stains. Here roof temperature is given to avoid water dropping that causes spotty dyeing. Here water is not given at the entry of steamer because to prevent developing chemicals that just applied before going into steamer so water lock is given at the end of steamer.

7.3.4.5   Washer

Washing is carried out to remove unfixed dyes. After steamer, fabric flows from seven to eight washers. Most commonly first to four washers are used for washing of salt or chemicals which are being applied in trough of pad steamer. In 5th, 6th washer, oxidation is done if required. If oxidation is not required then soaping is done in 5th, 6th washers. Neutralization is done in 7th washer by using acetic acid. There are eight chambers counter flow system is used. The chambers are used for number of processes according to the requirement of dyes by showering of different type of chemicals.

7.3.4.6   Dryers

In the last of the pad steam machine, there are three groups of drying cylinder for dry the fabric. Each group has 12 hot cylinders, but last one has 10 hot and 2 cool cylinders. All the cylinders are Teflon coated. Their purpose is to remove water molecules from fabric.

7.3.4.7   Batcher

The fabric from drying cylinders passes through some tension rolls and also from anti-static rods which absorbs the charge from the fabric. This rod has 5 kV voltages and 1800–18,000 Å of current after passing through this fabric is winded on the batcher.

7.3.5   Pad-pad-steam process

This process is only for specific heavy fabric qualities and for very absorbent fabric – like towels. Good colour yields are obtained even when dyeing dark shades. A special pad liquor applicator is required. This is a 'wet-on-wet' process without intermediate drying. In this process, the cloth is first padded with dye solution and consequently padded with alkali (wet-on-wet) without intermediate drying. The alkali padded material is then steamed. Reactive dyes having high reactivity and substantivity are suitable for this process with limitation of dyeing pale shades only.

7.3.6   Pad-air/steam process

The pad-air/steam process is one-bath continuous dyeing method for dyeing light-to-dark shades. It is mainly used for cord, terry towels and regenerated cellulose (Figs. 7.13 and 7.14). The advantages of pad-air/steam process include:

a.   high colour yields
b.   salt and urea not necessarily required

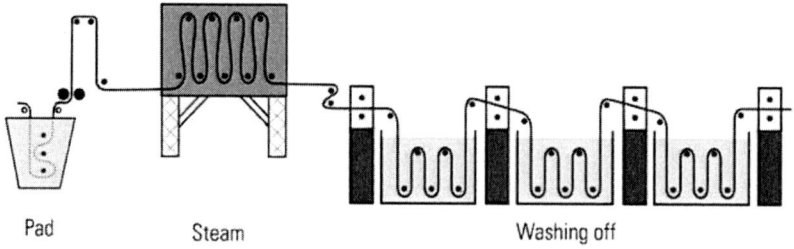

Pad   Steam   Washing off

	Chemical	Amount
Padding	Synozol Dyes	Xg/ℓ
	Soda ash	20g/ℓ
	Caustic Soda (38°B)	7g/ℓ*
	Wetting agent	1g/ℓ
	Liquor Temperature	
	Liquor Pick up	65%
Steaming (Fixation)	Temperature	102°C
	Time	1min

Dye (Total g/ℓ)	< 10	20	30	40	50	>60
Caustic Soda (38°B) Xmℓ/ℓ	4	5	7	9	11	13
Soda ash	20g/ℓ					

Figure 7.13 Pad-steam dyeing process for cellulose

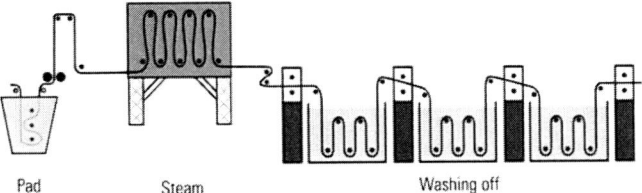

Pad   Steam   Washing off

	Chemical	Amount
Padding	Synozol Dyes	Xg/ℓ
	Soda ash	10g/ℓ*
	Glauber's salt	60g/ℓ*
	Wetting agent	1g/ℓ
	Liquor Temperature	25°C
	Liquor Pick up	100%
Steaming (Fixation)	Temperature	102°C
	Time	1min

Dye (Total g/ℓ)	< 10	20	30
Glauber's salt (g/ℓ)	60	90	90
Soda ash (g/ℓ)	10	20	30

Colorant Washing Off Conditions   *Glauber's Salt & Soda ash table

Condition	Chemical	Temperature	Time
Cold Rinsing	–	25°C	1min
Hot Rinsing	–	70°C	1min
Soaping	1g/ℓ Soaping agent	98°C	1min
Hot Rinsing	–	70°C	1min
Cold Rinsing	–	25°C	1min

* For dark shades a second soaping bath at 98°C is recommended

Figure 7.14 Pad-steam alternative process for heavy pile fabrics
(e.g., towels, corduroy)

7.3.7   Pad-dry-thermofix process

This is a salt-free, continuous one-bath method that ensures satisfactory coverage of dead cotton. Drawbacks include lower fastness properties, pollution of exhaust air and effluent and soiling of equipment by urea. This process is usually recommended for dischargeable dyeing and shades of turquoise and brilliant green, where no other process is suitable (Fig. 7.15).

In this process, reactive dye is padded on to cellulose in the presence of sodium bicarbonate and urea (150–200 g/L). After drying, dye fixation is carried out in a curing chamber at 125–160°C for 2–5 min depending on the depth of the shade. Then washing-off is carried out.

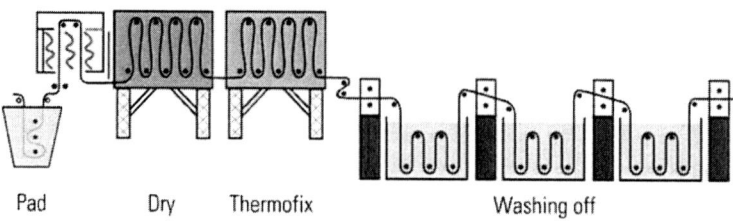

Pad   Dry   Thermofix   Washing off

	Chemical	Amount
Padding	Synozol Dyes	Xg/ℓ
	Urea	100g/ℓ*
	Migration Inhibitor	10g/ℓ
	Reduction Inhibitor	2g/ℓ
	Soda ash	20g/ℓ
	Wetting Agent	1g/ℓ
	Liquor Temperature	25°C
	Liquor Pick up	60%
Drying	Temperature	120°C
	Time	1min
Thermofix (Fixation)	Temperature	160°C
	Time	1min

Dye (Total g/ℓ)	< 5	10	20	> 40
Urea(g/ℓ)	25	50	75	100

Bath stability up to 3 hours   *Urea table

Figure 7.15 Pad-dry thermofix dyeing process for cellulose

7.4   Pad-thermosol-pad steam process

It is the process for continuous dyeing of polyester cotton blends. It may be either disperse reactive combination or disperse vat combination.

Pad with disperse & reactive dyes → Infrared drying → Thermosol-pad steam → Washing-off → Drying

The temperature in infrared drying is 120°C, time 45–60 s, residual moisture 30%. The temperature in thermosol is 210–220°C and time is 15–30 s. The second padding is done by using glauber salt – 250 g/L, caustic soda – 15 g/L, soda ash – 20 g/L. The temperature in the steaming is 102–104°C and time is 45–75 s.

Table 7.1 Colron dyes for high light and light fastness: dyestuff selection for cold-pad batch/pad-dry-pad steam/e-control/pad steam/pad thermofixa

Colron dyes for high light fastness

0 to 16 GPL

    Colron yellow CNRG

    Colron red CN2BL

    Colron blue CNR

    Colron blue 2BGLX (up to 5 GPL)

16 GPL & above

    Colron golden yellow CES

    Colron orange GLX

    Colron ruby S3B

    Colron deep red GLX

    Colron blue SFGR

    Colron navy blue SGI/marine GLX

    Colron navy GLX (above 40 GPL)

For navy shades

    Colron navy blue SGI

    Colron marine GLX

    Colron navy CR

Support dyes

    Colron yellow CNRG

    Colron golden yellow CES

    Colron orange GLX

    Colron deep red GLX

    Colron ruby S3B

For black shades

    Colron black W2N

    Colron black GLX

    Colron super black G

    Colron super black R

Support dyes

    Colron orange GLX

    Colron golden yellow CES

For brilliant yellow, green & royal, violet

Colron yellow CN5G (CPB)

Colron yellow CNNP (non-phototropic)

Colron turq. blue GC (CPB)

Colron ocean blue SR

Colron blue R special

Colron dyes for light fastness (ISO) economy

0 to 16 GPL

Colron golden yellow SF

Colron red SF

Colron blue SFGR

Colron blue CS

16 GPL & above

Colron golden yellow CES

Colron orange GLX

Colron ruby S3B

Colron deep red GLX

Colron marine GLX

Colron navy GLX

For black shades

Colron black SDN

Colron black WNN

Support dyes

Colron orange GLX

Colron golden yellow CES

For brilliant yellow, green & royal, violet

Colron turq. blue GC (CPB)

Colron ocean blue SR

Colron blue R special

[a] Suitable for continuous and semi-continuous dyeing. Excellent reproducibility, reliability, productivity and flexibility. Fastness with cost-effectiveness.

7.5   E-control: a new process for dyeing

In recent years, only minor technical application developments have been made to the continuous processes used to apply reactive dyes. There have not been any fundamental changes or real innovations, or where there have been, they have not become established in the industry. That was true, at any rate, until ITMA 1995 in Milan when Monforts and Zeneca (now part of DyStar) unveiled a joint development. This process delivers customer controlled colouration. Benefits combined with the innovative machine design from Monforts.

Economy and ecology are among the most used catchwords of our time, and their significance is of the greatest possible importance for the survival of our industry. Where continuous dyeing with reactive dyes is concerned, economic efficiency means employing dyeing process and techniques, which help to withstand international price cut and stay competitive. Simplified or reliable processes with greater colour yields yet with a smaller usage of auxiliaries and energy must be the aim.

In the coming century, the forecast of cotton fabrics persistence in the world market is evident from the rapid growth in the popularity of woven leisurewear. The restrictions imposed by environmental legislation, growing competition in the world market, increased costs of resources and rapidly changing fashion demands, have put the pressure on the continuous dyes to compete with the present and future requirements. To survive in the market, necessary modifications in the machinery are essential along with the optimization in the process. For this purpose, an evaluation was made of E-control method, a new dyeing technique with some modifications. The E-control process was jointly developed by BASF (GB) (previously Zeneca) and Monforts. In contrast to the process already described, E-control is characterized by the fact that no ancillary substances such as urea, water glass, soda, sodium hydroxide solution or salt are required. Reactive dyestuffs with high reactivity can be used for the E-control process. Dichlorotriazine (Procion MX) dyestuffs were employed in the development of this process and are used in practice.

The necessary economical and environmental benefits, which could be achieved with respect to the other conventional processes for continuous dyeing, were assessed. The main priority was to assess the colour yield achieved by different processes and to point out fastness requirement level. Since then 65 units have been installed worldwide, allowing more economical and more ecologically friendly dyeing by continuous application.

Why we need E-control process? Following are the reasons:
1. The number of shades to be dyed has risen steadily
2. At the same time, textile finishers are expected to meet higher fastness specifications
3. Environmental regulations have become far more stringent in many parts of the world
4. Batch sizes have declined dramatically
5. Price pressure has risen enormously
6. Delivery times are far shorter than in the past

One of the major challenges facing textile finishers is the reduction in batch sizes. After all, capital investments tend to be geared to a defined production capacity. Output levels thus have a lasting impact on the capacity utilisation of a unit. Such changes evidently have a direct impact on textile mills. To maintain output levels, far more batches have to be dyed within the same time. Naturally, that also increases the number of lab matchings required. Let us look at capacity utilisation in more detail. Many mills aim for 75% capacity utilization. Assuming an average batch size of 11,000 m, that means about 60 min are available to change shades. By contrast, if the average batch size were 1000 m, a 60 min changeover time would reduce capacity utilization to under 20%, making economical operation impossible. To achieve 75% capacity utilization in these conditions, the changeover time would have to be cut to 5 min. Dyeing processes and products therefore have to be adjusted to reflect these conditions. Production personnel also need to adapt. That was the trigger for the development of the E-control process.

7.5.1   The basic principle of the E-control process

The E-control process comprises just three steps:

a. Application of the dye
b. Drying
c. Washing-off

The dyeing system must be carefully balanced to ensure complete fixation of the Colron reactive dyes within the drying time. However, complete fixation of reactive dyes is not possible in normal drying conditions because of the wet bulb temperature, which is a characteristic of all convection-based drying systems. It means that during drying the temperature of the moist goods is far lower than the surrounding temperature. In normal drying processes, the temperature on the goods can drop to 50–55°C. The exact temperature depends on the humidity of the air used to dry the goods. If humidity is 25% and the air temperature is 120°C, the temperature on the surface of the goods is around 68°C. If humidity is 30%, the temperature on the goods is around 71°C. The E-control process uses this physical fact to fix the reactive dye during drying.

7.5.2   The key variables

There are three key variables like time, temperature and humidity and are used to ensure accurate control of the dyeing process in the Thermex and hot flue. The goods should spend about 2–3 min in the hot flue. However, it should be noted that these are average values: the actual drying time depends principally on the material to be dried. Dyes, auxiliaries and alkali systems for the E-control process recommended dyes. Evidently, the dyes used in the E-control process have to be suitable for the process conditions. Since dyeing is completed in approximately 2–3 min, reactive dyes with high to medium reactivity are particularly suitable. In practice, modified hetero bi-functional reactive dyes are used and poly-functional dyes are also used. Dyes with double-MCT anchors such as Colron HE and H Conc are not recommended for the E-control process. Alkali and auxiliary soda ash or soda ash/NaOH are normally used as the alkali system. To ensure optimum reliability, it is essential to dose the alkali and dye solutions separately. For example, the following amounts of alkali are recommended for Colron dyes: amount of dye up to 10 g/L > 10–30 g/L > 30 g/L soda ash 20:20:20 NaOH 50% – 2 mL/L 4 mL/L. Common auxiliaries used in continuous dyeing processes can be used, e.g., wetting agents, antimigration agents and, if necessary, sequestering agents. A mild oxidant can be added to prevent reduction of the dyes. Urea is not essential for the E-control process but up to 50 g/L may be advantageous in some circumstances.

Recipe
- Colron golden yellow RNL: 2.15 g/L
- Colron brilliant red RB: 1.59 g/L
- Colron black B: 6.03 g/L
- Wetting agent: 2–4 g/L
- Antimigrating agent: 10 g/L
- Mild oxidant:2 g/L
- Soda ash: 20 g/L

7.5.3   Features of the E-control process

The demands made on future-oriented dyeing systems include the following criteria:

- Economical, even when dyeing small batches
- Rapid colour changes
- Simple to use
- Suitable for brilliant, very pale and very heavy shades
- High reliability due to excellent reproducibility
- A simple lab dyeing method must be available
- Environment friendly
- Process should not restrict the obtainable fastness properties
- Short processing times. In other words, the aim is to achieve optimum results in the shortest possible time. The fewer the variables affecting the dyeing process, the higher the probability of achieving

this goal. According to this definition, 25 variables mean 25 parameters that can adversely affect the outcome of the dyeing. This process contains 40% fewer variables, which means the possibility of making a mistake is reduced by 40%. That is only possible because the E-control process does not use a separate fixation step. By contrast, all other continuous dyeing processes require separate fixation of some sort. There are also a number of other variables that are not directly related to the dyeing process. As a rule, far too little attention is paid to these parameters, especially the quality of the dyes and chemicals used.

7.5.4   Product quality

It is amazing how many mills invest in very expensive machines yet use them to apply poor-quality dyes, auxiliaries and chemicals. However, it is a well-known fact that a process is only as good as the weakest link in the chain. In practice, using inferior products prevents mills getting the best out of their machinery. No-one would dream of running a racing car on low-grade petrol and oil because they know they would have no chance of winning the race. The same applies to textile dyeing with the E-control process. The advantages of this process, which include excellent reproducibility, can only be fully achieved by using high-quality dyes and chemicals.

Lab processes can maximize the efficiency of a dyeing process. A good lab process is needed to ensure that reproducible dyeing can be produced quickly and easily. The more reliable a dyeing process can be transferred from lab to production conditions, the more effective use is made of the money invested in the machinery. The more accurately the lab method simulates bulk dyeing, the better the results. In collaboration with a lab machine manufacturer, Colourant has therefore developed a method that allows accurate reproduction of the E-control dyeing conditions in the lab. This comprises a type DH lab steamer especially adapted for the E-control process. This allows regulation of all process parameters that influence the drying and dyeing results on the Monforts thermex hot flue → Temperature → Steam content → Air flow → Time. As an optional extra, a radiation pyrometer can be used to monitor the temperature of the goods. This allows optimum adjustment of the dyeing parameters to meet the requirements of different types of goods.

Using the E-control process, we can greatly improve efficiency in the lab. This process compares the time required for an E-control dyeing and a pad-dry pad-steam dyeing. In this comparison, the E-control process requires 35% less time. The process also has a logistical benefit: since the goods are dry at the end of the E-control dyeing process, they do not have to be washed-off immediately and can be stored without problem. If the comparison is extended to the cold pad-batch process, an additional batching time of at least 8 h per dyeing has to be taken into account. This extremely simple yet accurate lab process therefore helps meet the need to speed-up processing times and injects a new meaning into the over-used cliche quick response. As an alternative to this laboratory steamer, all lab processes commonly used for cold pad-batch processes can be used. However, the performance is in no way comparable and these processes greatly increase the overall time required for the process. A quick lab process is no use if it cannot be transferred reliably to production conditions. Special attention was therefore paid to this in the development of the process. This compares lab and bulk dyeings on various materials. All lab dyeings were produced on the Mathis unit. The illustrations show very clearly that there is a very high standard of lab-to-bulk reproducibility on a wide variety of articles. Values below DE 1.0 (measured in accordance with CMC) can be achieved without difficulty and can be optimised on production machinery.

E-control can be used to dye all shades that can be obtained with reactive dyes. That includes brilliant turquoise, deep bordeaux, deep navy and, of course, black. So far this process has been used for fabric weights from 70 to 500 g/m. All cellulosic fibres can be dyed. Colourant has extensive experience of this process on cotton, Tencel, viscose and linen. The E-control process needs far less energy than, e.g., the PDPS process because it cuts out the extra fixation step (steaming). Unlike the PDPS process, the E-control process does not require any salt. The amount of alkali is similar. A total of 61,000 kg salt is needed in the PDPS process, to dye 1,000,000 m of fabric assuming a fabric weight of 350 gram per running metre, a liquor pick-up of 70% and 250 g/L salt. The environmental profile of industrial production processes is becoming an increasingly important factor. It therefore has to be considered when selecting the optimum dyeing process. The key criteria are energy consumption, chemical requirements, effluent contamination and pollution of exhaust air. One factor that is often overlooked is that the most environmentally friendly process is often also the most economical. E-control is a case in point. By reducing energy requirements, it cuts the amount of energy that has to be generated and thus paid for. Similarly, dyehouses do not have to purchase chemicals they do not need. Moreover, they do not have to wash out so they cannot contaminate effluent or exhaust air. E-control is therefore both economical and eco-friendly. A comparison of production cost with the pad-dry-pad-steam-process.

Nevertheless, like all dyeing processes, E-control does have its limitations. The first is that Indanthren dyes can be dried by using the Thermex hot flue but not fixed as an air-free steamer is required for this. Pale to medium shades can be dyed on PES/CO blends using a modified one-bath TTN process but a separate thermosol step is required after the E-control step. Consequently, at least two Thermex chambers are required. More experience is needed to obtain reproducible dyeings on very lightweight qualities (<60 g/m) than on heavier fabrics. The E-control process is not entirely suitable for heavyweight pile and loop pile goods such as terry towelling because there is a risk of frosting. The appearance of some goods dyed under E-control conditions differs from their appearance in the PDPS process. This means the two processes are not always interchangeable.

E-control is the reliable and economical new continuous dyeing process. E-control is a rapid and reliable method of dyeing a wide range of substrates. It does not require steaming or curing and ensures high reproducibility and excellent fixation. It was developed in conjunction with Monforts and is based on the Thermex hot flue technology. E-control is environment-friendly and dyeing can be carried out without silicate, urea and salt, thus reducing recipe costs and effluent load. E-control is easy to use in the lab and lab-to-bulk reproducibility is very good. Excellent reproducibility and high fastness levels can be obtained on both lightweight and heavyweight cotton, viscose and Lyocell qualities with selected Colourant reactive dyes.

With the trend towards smaller lot sizes per colour in continuous dyeing and the requirement for simple processes, the E-control process for reactive dyestuffs was developed in the mid-nineties. It has now established itself with Colourant technology worldwide.

The E-control process is a simple, quick and economical one-pass pad–dry wash-off continuous dyeing process with drying in a hot flue at 120–130°C and controlled humidity (25–30% by volume) to obtain fixation in 2–3 min.

7.5.5   Process

- One Step, Simple & Economical Continuous process
- No unproductive batching sequence
- Ideal for short lot dyeing
- Effective Wash-off
- A wide variety of fabric can be dyed
- Rapid shade matching in laboratory
- Energy efficient

7.5.6   Performance

- High productivity due to short process
- Environment friendly as no salt, urea and silicate are used
- Soft handle of fabric due to mild fixation conditions

- Migration is minimized by rapid fixation and humidity control
- No crushing of pile fabrics
- Improved penetration of different fabrics
- Presence of humidity at high temperature
- Very good lab-to-bulk and bulk-to-bulk reproducibility

7.5.7   Benefits

- Ideal choice for short or long lot dyeing
- Full colour range
- Excellent reproducibility of colours
- No risk of browning of cellulose fibres, thanks to low fixation temperature bright colours possible
- No urea, no fumes, no machine contamination
- No salt (easy wash-off of unfixed dyestuff)
- Instant colour viewing
- Accurate, easy and rapid laboratory check method
- No unproductive batching sequences, no moiré effects
- Migration minimised by rapid fixation and humidity control
- No crushing of pile fabrics (e.g., corduroy)
- Improved penetration of difficult fabrics due to presence of humidity during fixation
- Energy-efficient thanks to optimum humidity control
- Ideal process for fast change technology

E-control knit is suitable for all types of reactive dyestuffs. Although the process requires no salt or urea, it achieves up to 5–8% higher yields from dyestuffs than with other processes. Production costs can be up to 45% lower than for discontinuous methods of dyeing.

E-control knit also offers more certainty: If the dwell time was not exactly observed in the past, differences in dyeing were the result. After the discontinuous process, the fabric had to be immediately further processed. E-control knit is a dyeing process with no uncertainties: the fabric is dry after the dyeing process and does not have to be immediately washed out. The dyeing result is absolutely reproducible, even with large batches.

There is no moiré effect. The fabric surface is not – as with other processes – subjected to mechanical strain. All types of reactive dyestuffs can be used without the need for salt or urea. The dyestuff yield is 5–8% higher than with other processes. The production costs are significantly lower (up to 45% lower than those of the discontinuous processes). When dyeing cotton/

elastane fabrics, the fabric no longer has to be sewn into a tubular form. Thermex ranges can continue to be universally employed.

The ideal reactive dyes used for E-control in industries are as follows:

a. Light shades: Colron CN series
b. Medium: Colron SF series
c. Dark: Colron RGB series

Table 7.2 The ideal reactive dyes used for E-control in industries

Suitable in E-controll	Depth in use
Colron golden yellow RGB [P]	Medium/dark
Colron ultra yellow RGB	Medium/dark
Colron red RGB	Medium/dark
Colron ultra red RGB	Medium/dark
Colron deep red RGB	Medium/dark
Colron ultra carmine RGB	Medium/dark/extra dark
Colron blue RGB	Medium/dark
Colron navy blue RGB	Medium/dark/extra dark
Colron ruby S3B	Medium/dark/extra dark good washing fastness
Very light and medium	
Colron yellow CNRG	Light
Colron yellow CNNP	Light/medium
Colron yellow CN2R	Light
Colron scarlet CN3G	Light
Colron red CNR	Light
Colron red CN2BL	Light
Colron red CN4B	Light
Colron blue CNR	Light
Colron navy CR	Medium/dark
Colron deep red GLX	Medium/dark
For black shade	
Colron black CNGR	
Colron black WNN	
Colron black W2N	

Colron super black G

Colron super black R

Colron black SD

Process controll parameter

Dyes: X g/L

Primasol AMK: 10.0 g/L (anti-migrating agent)

Primasol NF: 2 g/L

Resist salt : 5 g/L   Must require as a mild oxidative agent

Acetic acid : 0.1 g/L

Alkali: soda and caustic lye as depth shade

Metering pump must required for maintain liquor ratio

Colour preparation: The colour was prepared by the automatic colour dispenser.

Checking the DLC: The liquor prepared in the bulk was checked in the lab. The shade was slightly lighter thus 7% depth was added and sampling was carried out.

The parameters in the laboratory are as follows:

1. Mangle parameters: Pressure: 2 bar, speed: 2 m/min
2. E-control lab parameters: Humidity: 27%, air temperature: 115°C
3. Machine parameters – bulk:
   (a) E-control: Incoming fabric temperature: 34°C, trough temperature: 31.9°C, pH of trough solution: 11.2, IR temperature: 800°C
   (b) Mangle pressure: LMR: 16 17 16, pick-up: 60%, pick up after IR: 37.5%
   (c) Chamber 1: Humidity : 27%, temperature: 120°C, fan: 60
   (d) Chamber 2: Humidity: 27%, temperature: 120°C, fan: 70
   (e) Chamber 3: Humidity : 27%, temperature : 120°C, fan: 70

With the above conditions, 300-m sampling was carried out to check out the shade and centre to selvedge. The shade was found to be OK, but slight centre to selvedge difference with respect to depth of shade was observed. Thus the pressure was adjusted and the final bulk was run.

The fabric was later washed-off.

a. Washer 1: Flow: 3 L/kg, temperature : 48°C
b. Washer 2: Flow: 3 L/kg, temperature : 60°C
c. Washer 3: Temperature: 95°C
d. Washer 4: Temperature: 95°C, soap
e. Washer 5: Flow: 3 L/kg, temperature: 85°C
f. Washer 6: Temperature: 55°C, acetic acid
g. Washer 7: Temperature: 45°C

The final shade was checked with DLC. The only difference was in strength (bulk approx 12% deeper. Tone was ok)

Observations and suggestions: There was humidity fluctuation in chamber no: 3 (as low as 23%). As humidity plays important role in E-control dyeing, humidity fluctuation should not be compromise. The temperature of the colour padding trough was high, i.e., 31.9°C. Thus the cooling jacket is strongly recommended at this stage.

7.5.8   E-control technology

Figure 7.16 E-control technology

7.5.8.1   *Applicator: Thermexhotflue*

Thermex 5500 is the modern system for the continuous dye shop and finisher. It is characterized by outstanding cost-effectiveness when dyeing both long and short batches. The latest future-oriented techniques and process technology ensure that the machine can be expanded into a continuous range. All in all, a universal hot flue for continuous dyeing, condensing and thermosoling with unrivalled reliability at the highest fabric speeds. Exclusively for E-control® processes for continuous dyeing of woven and knitted fabrics, exclusively designed and manufactured by A. Monforts Textile Machinen GmBH, Germany (Figs. 7.16 and 7.17).

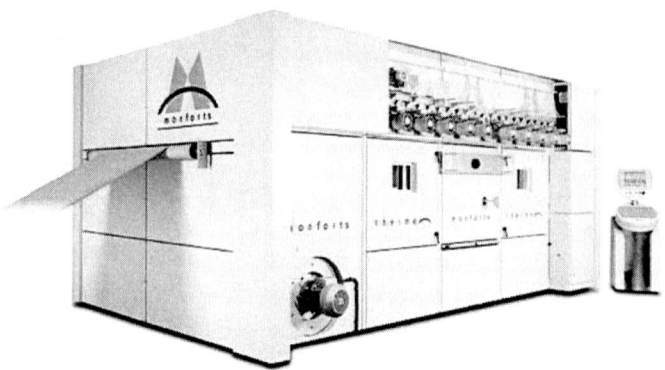

Figure 7.17 E-control® process by Monforts Textile Machinen GmBH, Germany

The E-control® process is a continuous dyeing process fixing reactive dyestuff onto cellulose fibres in one step by drying with controlled steam/air mixtures (pad-dry process) without common salt, silicate and urea in the dye liquor with only mild alkali and wetting agent. The fabric is kept 2–3 min in 125°C hot air and 25% steam out of the total volume of the chamber to attain 68°C wet pulp temperature on fabric, which is the fixation temperature. All kinds of reactive dyes can be dyed with E-control technology (Fig. 7.18).

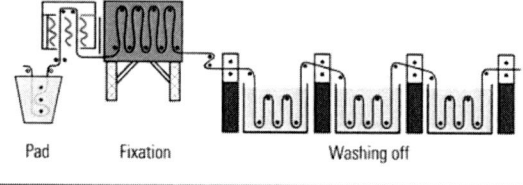

Pad   Fixation   Washing off

	Chemical	**Amount**
	Synozol Dyes	Xg/ℓ
	Migration Inhibitor	10g/ℓ
	Wetting Agent	1g/ℓ
Padding	Soda ash }**	20g/ℓ
	Caustic soda(38°Be)*	5g/ℓ*
	Liquor Temperature	25°C
	Liquor Pick up	60%
I.R. Drying	-	-
Fixation	Temperature	128°C
	Time	1min

Dye (Total g/ℓ)	< 10	20	40	50 ·	> 60
Caustic Soda (38°B) Xmℓ/ℓ	-	2.5	5	10	15

Caustic Soda (38°B) Table

** Mixing pump required with 4:1 (dye : chemical) ratio

Figure 7.18 Pad-humidity fix process (E-control, Monforts)

7.5.8.2   Process flow in E-control followed with post-treatment

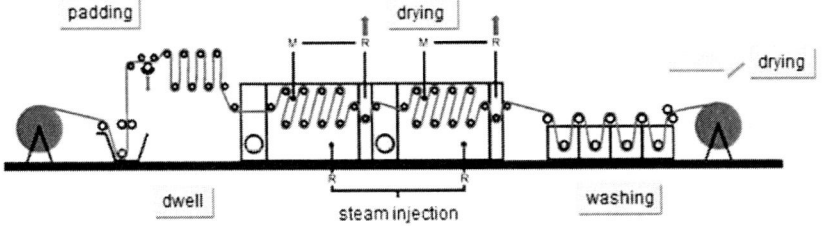

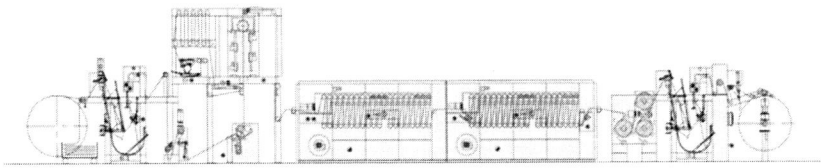

Figure 7.19 E-control dyeing machine

1. Chemical quantities for different processes

Used in the recipe:

    a. Pad-dry-pad-steam = 250 g/L, salt
    b. Pad-dry-thermofix = 100–150 g/L urea
    c. Pad-batch = 100 g/L urea

2. Chemical consumption/year for different processes

    a. 110 g/L silicate
    b. 30–35 m³/NaOH 38°C

3. Chemical consumption/year for different processes

Table 7.3 Chemical consumption/year for different processes

	Conventional	New	Saving
Remazol-dyestuff	62,66 g/L	57, 59 g/L	8.1%
Urea	100 g/L	—	100%
Wetting agent	3 g/L	3 g/L	0%
Silicate 42°C	110 g/L	—	100%
Caustic soda 38°C	33.5 cm³/L	20 cm³/L	40.3%

$G_{Fl./year}$	$=V$		g_0	B	60	22.5	220	f_1	[l/year]
	$=40$		0.32	1.52	60	22.5	220	0.75	[l/year]
	$=4.333.824$ L	of liquor per year							

	Conventional	New	Saving
Dyestuff/year	271,557 to	249,585 to	21,972 to
Urea/year	433,382 to	—	433,382 to
Silicate 42°C/year	476,721 to	—	476,721 to
Caustic soda	145,183 m³	86,676 m³	58,507 m³
	47.2 to firm	28.17 to firm	19.03 to firm

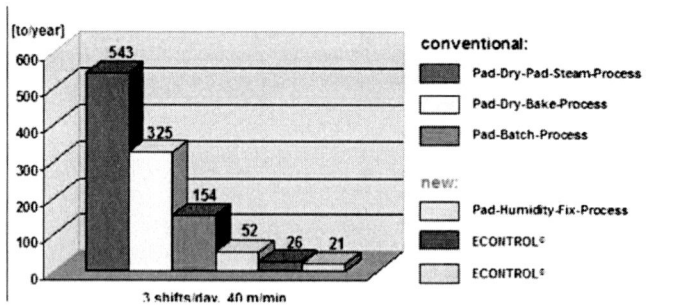

Figure 7.20 Chemical consumption/year – practical examples

4. Wash-off

Due to good fixation rate in E-control, post-treatment is quiet easy after E-control (Fig. 7.21).

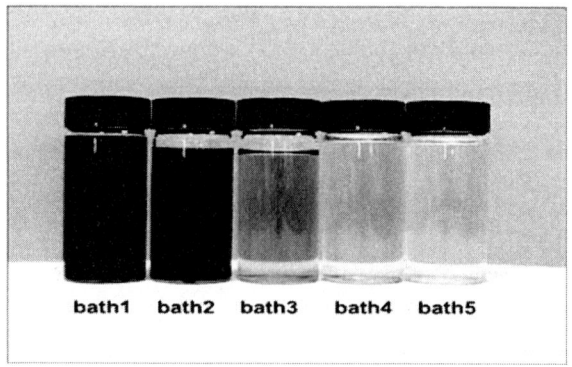

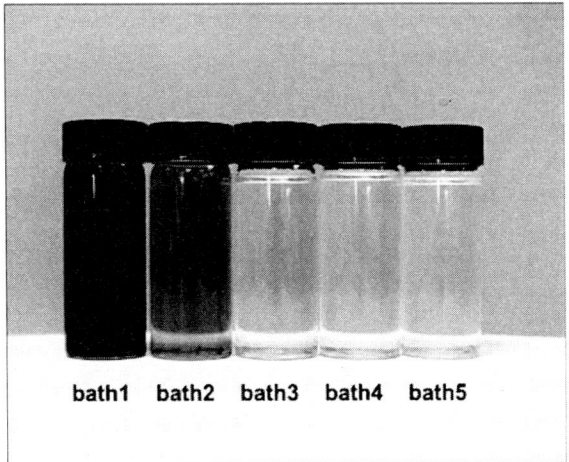

Figure 7.22 E-control: wash-off

7.5.8.3   Benefits of E-control

1. No infrared predrier required (apart from heavy fabrics)
2. No steamer required
3. Versatile equipment (can be used for drying)
4. Short down times with rapid cleaning equipment
5. No batching frames or rotating equipment required
6. Reduced chemical usage
7. No urea in the effluent or urea fumes discharged to the atmosphere (if urea is not required for dye dissolving)
8. No salt required
9. No sodium silicate
10. Soft handle
11. Baking for fixation (in pad dry thermofix) can produce a harsh handle on sensitive fabrics
12. Migration effects minimized
13. Dye is fixed before complete moisture is removed. Useful for pile and lightweight fabrics
14. Pile fabrics suffer no crushing
15. Corduroy quality improved
16. Better diffusion on viscose/tencel fabrics than other pad-dry/bake methods
17. No batching sequences – no moiré effects

7.5.9   Thermex E-control T-CA

Figure 7.23 Thermex E-control T-CA

The new single-bath continuous dyeing process for fabrics of polyester/cotton with selected reactive and dispersion dyestuffs. Downline of the pad-E-control thermosol process without reductive intermediate cleaning (Fig. 7.23).

8
Garment dyeing

Garment dyeing is the process of dyeing fully fashioned garments (such as pants, pullovers, t-shirts, jeans, sweaters, dresses, bathrobes, casual jackets, shirts, skirts and hosieries) subsequent to manufacturing, as opposed to the conventional method of manufacturing garments from pre-dyed fabrics. Most garments are made of cotton knit goods and/or cotton woven fabrics, although several other fabrics can be found in the whole or in part such as wool, nylon, silk, acrylic, polyester and others. Due to cost savings and fashion trends, garment dyeing has been gaining importance and popularity in the past years and will continue to do so in the future. Traditionally, garments are constructed from fabrics that are pre-dyed (piece dyed) before the actual cutting and sewing. The advantage of this process is the cost effectiveness of mass producing identical garments of particular colours. A major drawback with this approach is the risk associated with carrying a large inventory of a particular style or colour in today's dynamic market. Normally, it is used for lingerie, socks, sweater dyeing, etc.

8.1   Garment dyeing machineries

During the last decade, the market has been forcing textile companies to supply sports and leisurewear in extremely reduced times and in the trendiest colours of the moment. Obviously the standard textile production cycle, contemplating a dyeing process followed by make-up and distribution, which does not run with very short times, leads to a considerable loss of sales. The piece dyeing process ensures very short times from the customers' demand to the fulfilment of market needs for cool colours and stylish finishing processes. To meet large or small demands for a given colour as fashion requires, machine manufacturers now offer many garment dyeing and finishing machines (Fig. 8.1). These are generally rotating machines, similar to large-size industrial washing machines; the garments are loaded in special baskets for finishing operations. Paddle machines and rotary drums are the two types of equipments regularly used for garment dyeing. The machine size and equipment allow the maximum flexibility to meet the most different needs as described in the following sections.

Figure 8.1. Garment dyeing machine

8.2   Paddle dyeing machines

A process of dyeing textiles in a machine that gently move the goods using paddles similar to a paddle wheel on a boat. This is a slow process, but there is extremely little abrasion on the goods. Horizontal paddle machines (over head paddle machine) consist of a curved beck-like lower suction to contain the materials and the dye liquor. The goods are moved by a rotating paddle, which extends across the width of the machine. Half immersed paddles cause the material to move upwards and downwards throughout the liquor. The temperature can be raised to 98°C in such system. In lateral/oval paddle machines consist of oval tank to enhance the fluid flow and the processing the goods. In the middle of this tank is a closed oval island. The paddle moves in a lateral direction and is not half submerged in the liquor and the temperature can be increased up to 98°C.

The HT paddle machines work according to the principle of horizontal paddle machine; however, the temperature can be raised up to 140°C. PES articles are preferably dyed on HT paddles. In paddle machines, the dyeing can be carried out with 30:1 to 40:1, lower ratios reduces optimum movement of the goods, lead to unlevel dyeing, crease formation. For gentleness, the blades of the paddle are either curved or have rounded edges and the rotating speed of the paddle can be regulated from 1.5 to 40 rpm. Circulation of the liquor should be strong enough to prevent goods from sinking to the bottom. Paddle machines are suitable for dyeing articles of all substrates in all forms of make ups. The goods are normally dyed using PP/PET bags.

Paddle dyeing machines are generally used to dye many forms of textiles but the method best suits to dye garments. Heat is generated through steam injection directly into the dyebath. The machine works like this, the paddle circulates both the bath and garments in a perforated central island. It is here only that the chemicals, water and steam for heat are added. The overhead paddle machine is nothing but a vat with a paddle that has blades of full width. The blades generally take a dip of few centimetres into the vat. This action stirs the bath and pushes the garments down, thus keeping them totally submerged in the dye liquor. This machine is made of high quality stainless steel (SS) for durability, most suitable for dyeing, bleaching and relaxing of all sorts of knitted piece goods of wool, polyester, nylon, cotton, rayon, cashmeres and blends. Direct and indirect heating systems are installed at the bottom of the machine for heating, heat preservation. Precise forward and reverse controller for paddle wheel offers uniform turning to the fabrics and ensures even dyeing result. Paddle wheel with fine polishing avoids material damage, advanced control system with reasonable design and easy operation with minimum space requirement. Maximum working temperature is 98°C, and heating speed (according to steam pressure of 6 kg/cm^2) is approximately 5°C/min.

The process steps are

- Chemical mixing
- Load preparation
- Pre-dyeing treatments
- Dyeing cycle
- Post-dyeing treatments
- Rinsing
- Unloading
- Liquor discharge
- Disposal

Advantages of paddle dyeing machine are

- Steam heated
- Very efficient liquor flow
- No harm to garment structure
- An uniform patchless dyeing
- Low liquor ratio
- Rapid heating and cooling
- Quick drop and fill

Application of paddle dyeing machine: Paddle dyeing machine offers itself as a suitable dyeing platform for all the types of piece goods. This typically includes rugs, socks, bed spreads and other types of garments and fabrics.

8.3   Rotary drum dyeing machines

Rotary drum machines are preferred for garments, which require gentle handling, such as sweaters. A high liquor ratio is required for paddle machines, which is less economical and may limit shade reproducibility. Many machinery companies have developed sophisticated rotary dyeing machines, which incorporate state-of-the-art technology. Following are the characteristics of rotary drum machines used for garment dyeing.

These machines work on the principle of 'movement of textile material and a stationary liquor'. The rotary drum dyeing machine consists of rotating perforated cylindrical drum, which rotates slowly inside a vessel of slightly bigger in size. The internal drum is divided into compartments to ensure rotation of goods with the drum rotation, and the outer vessel holds the required quantity of dye liquor. High temperature drum machines are capable of processing the garments up to 140°C.

Features of modern rotary-dyeing equipment include the following:

- Lower liquor ratio
- Gentle movement of goods and liquor (minimizes surface abrasion)
- Rapid heating and cooling
- Centrifugal extraction
- Variable drum speed with reversal capability (adaptable to a wide variety of goods)
- Continuous circulation of goods (improves migration control)
- Easy of sampling
- Variable water levels with overflow rinsing capabilities
- Large diameter feed and discharge lines (minimizes filling and draining time)
- Microprocessor controls
- Lint filters
- Pressure dyeing
- Auto-balancing drums

One feature that can be used to reduce abrasion on delicate garments or to minimize tangling is a compartmental chamber, sometimes referred to as a 'Y' pocket. The rotary drum machines are very simple to operate and are quite compact in size. The cost of unit is also not high.

Drum dyeing-centrifuging machines are also called 'multipurpose drum machines' or 'multi-rapid dyeing centrifuging machines' since these machines can perform scouring, dyeing, centrifuging and conditioning successively with automated controls. The goods are treated in a perforated inner drum housed within an outer drum (dyeing tank). Inner drums without dividing walls are provided with ribs that carry the goods along for a certain time, partially lifting them up out of the liquor. These machines can operate at very low liquor ratios and can dye the goods up to 98–140°C. This is suitable for knits as well as other garments. Liquor circulation can be intensified using additional jets. Drums can be rotated in both the directions.

8.3.1   Tumbler dyeing machines

These machines are being used for small garments either in loose form or in open mesh bags. Design wise the tumbler dyeing machines are similar to the commercial laundering machines.

The principle of operation is to load the material into perforated inner SS tanks, which rotates round a horizontal shaft fixed at the back of the drum. The drum is divided into compartments for moving the goods with rotation of drum. A variety of tumbling machines have higher rotation speeds and can spin dry at the end of the cycle. These are similar to dry-cleaning machines.

Rotating drum machines are more efficient and cleaner to operate than paddle machines. The more vigorous mechanical action often promotes more shrinkage and bulking, which may be desirable for some articles. To handle higher quantities and large production of similar pieces the latest machines are provided with several automatic features and sophistication.

8.3.2   Toroid dyeing machines

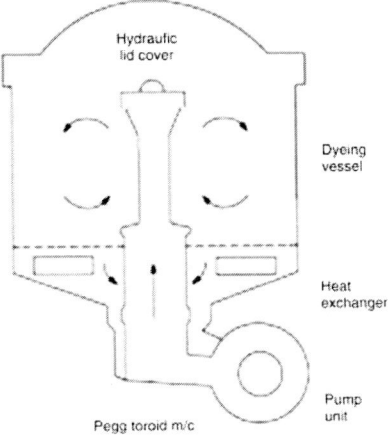

Figure 8.2. Toroid dyeing machine

In these machines, the garments circulate in the liquor in a toroidal path with the aid of an impeller situated below the perforated false bottom of the vessel. Movement of the goods depends completely on the pumped action of the liquor. High-temperature versions of this machine operating at 120–130°C were developed in the 1970s for dyeing fully fashioned polyester or triacetate garments. The liquor ratio of such machines is about 30:1.

8.3.3   The gyrobox

The machine has support in the form of a large wheel, which is divided into 12 independent non-radial compartments. The goods are placed in these compartments .The wheel runs at a moderate speed of 2–6 rpm. The main advantages of this machine are

- Reduced M:L
- Different types of garments can be dyed simultaneously
- Flexible loading
- Fully automatic operation

8.3.4   MCS readymade garment dyeing machine

The rotodye machines are suitable for dyeing pure cotton, wool, polyester, cotton blends in the form of T shirts, sweaters, bath rugs and accessories, socks and stockings.

8.3.5   Modified Pegg Toroid Whiteley garment dyeing machines

This is an improved version of Toroid machines. The additional features are

- The machine is suitable for both atmospheric and pressure dyeing
- Full automation up to hydroextraction
- The design features, speed and performance are simplified to make the machine more versatile and free from operating problems

8.3.6   Advantages of garment dyeing

- Handling of smaller lots economically
- Enables various special effects to achieve
- Distressed look can be effectively imparted
- Unsold light shades can be converted into medium and deep shades
- By the time the garment has been in a boiling dyebath and then tumble dried, it will have adopted its lowest energy state and will not suffer further shrinkage under consumer washing conditions
- Latest fashion trends can be effectively incorporated through garment wet processing by immediate feedback from the customer

8.3.7   Disadvantages of garment dyeing

- High cost of processing
- A little complicated dyeing
- Garment accessories like zips, buttons, etc., impose restrictions
- The garments produced from woven fabrics create many problems and it has been found that the existing textile treatment styles as developed for piece dyed fabric cannot be just assembled for garment

wet processing operation such as garment dyeing, unless they have been engineered from the original design stage for garment dyeing

8.4   Dip dyeing

Garment dyeing is an opposition of the conventional method of making garments from previously dyed fabrics. Knitted fabrics, twill weave fabrics and/ or other woven fabrics made from cotton yarns are subjected to garment dyeing. Garments made of cotton material are easy to dye up but besides cotton several other fabrics can be found in whole or partially made of wool, silk, nylon, polyester or acrylic are subjected to garment dyeing. Dip dye (also known as 'tip dyeing') is a hair colouring style that involves dipping the ends of the hair into either a naturally coloured dye or a bright coloured dye – which is generally a more popular choice nowadays used in clothing. There are various techniques used for garment dyeing; recently some latest technology have developed for garment dyeing in our industry: (1) Tie dyeing, (2) Dip dyeing, (3) Spray dyeing, (4) Over dyeing, (5) Cold dyeing, (6) High white dyeing, (7) Washable dyeing, (8) Reverse dyeing, (9) Top dyeing.

What are the differences between the spray dyeing and dip dyeing? In spray dyeing, the colour effect is visible only one side it may be the face side or back of a garment depend on what I want. In side of the garment is not coloured. Pigment dyes are suitable for spray dyeing. We may get same effect by dip dyeing but not for all designs. We may spray back side of the garment. But in dip dyeing garment is dip into the dye solution so the colour effect is visible both sides, that is, the face side and back of a garment. When the garment dry, slightly it will lighter. The top of the dyed band is gradually lighter than the bottom part. In garment dip dyeing, dip dyeing machine is used but in spray dyeing spray gun is used (Fig. 8.3).

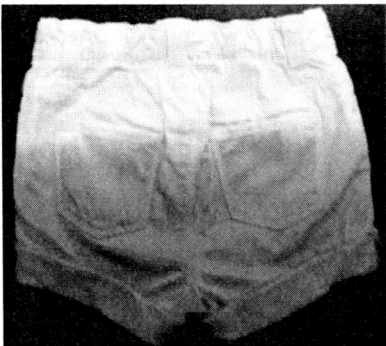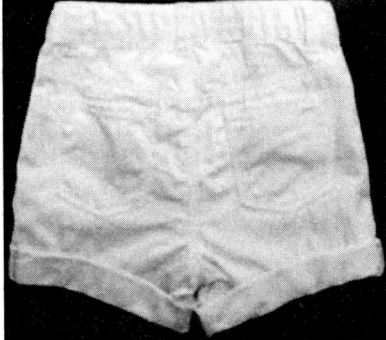

Figure 8.3. Dip dyed garments

In garments dyeing, dip dyeing is becoming more popular day by day as the fashion trend and customer demand. Due to cheap and easy process, it is very popular among manufacturers. It is also as called gradient dyeing. It is the process, the fully fashioned garments are arranged to dip into a dye bath (vessel, bucket, bowl, etc.) accordingly a pre-determined area or height. This process runs again and again. According to demand, garments are to be soaked into dye bath up to a certain limit for enough time. To dye up the garment as deeper shade to light shade from bottom to top, the gradient time will be maintained for certain area of the garment.

8.4.1   Application of dip dyeing machine

There are two types of dip dyeing machines: garment and rope dyeing machines. The rope dip dyeing machine is applicable for loose type rope form of the acrylic and knitted cotton fabric, and the processes involved are scouring, dyeing, bleaching, washing and soaping. Technical parameter and main specifications are

18. Dyeing tank working volume: 2800 L.
19. Capacity of dyeing cotton and knitted fabric: 105 L
20. Cloth guiding folder block: 10 blocks; cloth guiding folder distance: 180 mm
21. Working width: 1800 mm
22. Dyeing tank: 2100 mm; hydroextractor (remove excess water); dryer (for drying the garments); garment may be wet/dry state; supply binder 1–1.5 g/L (Catanizer 2–2.5 g/L if any); Dip a sample into the dye bath (as per shade); keep garment in dyebath and shake until it turns the desired colour; shaking 15–20 times for 10–15 min; hanging the garment on the dye bath (5–6 h); curing (80-90°Flow chart of dip dyeing shorting garment scouring (detergent 1–1.5% at 70°C)
23. Machine size: L × W × H 3390 × 3060 × 2510 mm
24. Motor power: 1.5 kW

Step 1: Wash and dry the fabric or garment to remove any finishes that resist dye. Fill plastic bin with hot water to a depth of at least 6 inches. Stir in dye, starting with a small amount and adding more as desired. Dip a sample strip of fabric or paper towel to test the colour. If too dark, add water; if too light, add dye. Let solution settle, so there are no bubbles at the surface.

Step 2: Determine where you want the top of the dyed band to be; use pins to mark that line on fabric. Holding your fabric as straight as possible, dip into dye solution to just below the pin line (make sure you do not dip the pins; the fabric will wick the dye higher). Keep fabric in dye until it turns the desired colour (it will dry slightly lighter).

Step 3: Rinse fabric in cold water to remove excess dye until the water runs clear. Hang to dry. If desired, wash item with synthrapol detergent. Dip dyeing hoody shirt/jacket for 30–40 mins. When the garment dry slightly, it will lighter. The top of the dyed band is gradually lighter than the bottom part. In dip dyeing face and back side of garment dyed but in spray dyeing only one side is dyed.

Dyeing can be done in normal temperature, Which is suitable for hank yarn and garment of cotton, silk, acrylic fabric, artificial wool, etc.

8.4.2   Process flowchart for dip dyeing

- Scouring (80°C)
- Neutralization (with acetic acid)
- Enzyme (at 45°C temperature and pH 4.5–5.5)
- Dyeing (90°C temperature for 3–5 min)
- After treatment (fixing agent)
- Softener (commercial softening agent)
- Bath drop

Normally in garment dyeing, two types of dyestuffs are used.

1. Direct dyes
2. Reactive dyes
   (a) Hot brand
   (b) Cold brand

For dip dyeing, direct dye is used though it has low colour fastness property; moreover, it takes less time for colour fixation. Necessary items used in the dyeing process are

- Dye bath (vessel, bucket, bowl or any other container)
- Hanger
- Washing Machine
- Stirrer
- Rubber gloves

8.4.3   Pre-treatment

As the fabric is not ready for dyeing after making cloths, it is treated with caustic soda and peroxide solution to perform pre-treatment process which helps to make it more absorbent and free from natural colour as well as other impurities.

Recipe
- Caustic soda (NaOH): 2 g/L
- Per-oxide (H_2O_2): 4 g/L

- Detergent: 1 g/L
- Temperature: 80°C
- Time: 15 min

Working procedure

A lot of garments are now taken into the dyeing machine for easy performing the pre-treatment process. (Good to know that, scouring, dyeing or any normal wash can also be done into a washing machine.) Now add caustic soda, hydrogen peroxide and detergents according to the above recipe. Now maintain the liquor ratio at 1:10 at 80°C. Now perform this treatment for 15 min. Well scoured garments are treated as more absorbent, where colour permanency will be high. In this process, the garments are to be more absorbent, removing the natural or unwanted colours to make genuine white, oils as well as other impurities. Good scouring plays an important role during dyeing, colour fastness and appearance. The whole materials are now subjected to rinse wash to remove the caustic and other chemicals. Now the garments are neutralized with acetic acid so that any of the caustic soda does not exist with the material. The pH of acetic acid should be kept 4.5–5.

Note

If the garments are made with zipper of cotton tape frequently be torn during the scouring process, so nylon or polyester tape is recommended to use here.

8.4.4   Enzyme

After scouring process, the full quantity of garment needs to treat with enzyme which will help to remove the hairiness and projectile fibres as well as make the garments ready for dyeing. You can use acid enzyme as 2 g/L at 45–55°C for 15 min. After enzymatic treatment one or two rinse wash to be done for washing it properly. (pH during enzymatic treatment should be kept 4.5–5.)

Recipe

We have used this for dyeing 380 pieces of baby girls' shorts.
- Direct dye: 355 g
- Salt (glubar salt): 10 g/L
- Labelling and wetting agent: 1 g/L
- Water: 2500 L
- Temperature: 95°C
- Time: 5 min

Working procedure

Make the dye bath ready with direct dye as normal dyeing procedure. At first, we will be sure the garment is made from scoured and bleached fabric, otherwise scouring or pre-treatment will have to complete. After adding dyes, salt and wetting agents, it needs to stir to make the dye bath properly. For getting more colour fastness and reduce fixation hours, we can use some binding agents as we need. The temperature of dye bath remains around 90–95°C.

The garment which is used to dyeing need to attach with hanger or something else to hang it from, then water uses to wet the garment properly before taking it underneath the dye bath. Now start to place the garment into the dye bath as desired height. You may require help in doing this so as not to spill the dye everywhere. Once the garment has been soaked in the dye bath for long enough time, then remove it carefully so that colour spot does not transfer to the non-dyed portion. If so, run the garment, from non-dyed section to dyed section, under warm water until water runs clear. This will remove any excess dye from the garment. Continue the process till your desired shade is developed and maintain the shade at different areas of the garment. To get as deeper to lighter shade at bottom to top, you can soak the garment into dye bath as long enough time. The process of dip dyeing is not so crucial but it takes extra precautions, care and many more times. During dyeing, it will take spraying some water to the garment to wet it and helps to absorb colours by the material.

Once your desired shade produced, it needs to take it into a washing machine.

8.4.5   After treatment

It is the process usually doing after dyeing for colour fixation and colour permanency as well as improving the colour fastness. This is a chemical treatment and the recipe is as follows.

Recipe
- Commercial fixer: 0.5 g/L
- Softener (cationic): 0.5 g/L
- Temperature: 40°C
- Time: 3–5 min
- M:L ratio: 1:10

Working procedure
After completing the dyeing, it will be required some process to have sufficient dye molecule penetration to the most inner part of the fibres. In the dye bath, now adds the fixing agent (commercial fixer) according to the recipe. Temperature in this bath will keep at 40°C and treatment time is 3–5 min. By this treatment, the material will be treated as finally coloured. Most of the dye particles will finally enter into the core of fibres and the fibre will swell at some extent so that the pores of fibres will be reduced and dye materials cannot come out from the innermost part. It will increase the colour fastness

of the materials. Before this treatment is started, a rinse wash will be done to remove extra dyes which adhere with the materials and after this process also another rinse will be done to remove unfixable dyes or sediment dyes at the bottom of the dye bath. Now softener is added to the bath to make the material softer, improving hand feel and brightness.

Now just hang up the garments for 20 min to squeeze extra water. (Do not use hydro extractor to reduce staining or uneven colour adhering.) You can us e oven to dry the garments as well, you can get prominence colour performance without staining or shade variation. Now the garments are cured at curing chamber at 150–200°C to dry finally and properly. You can use dryer for a while at the final stage.

Index

judgments is higher. Individuals ultimately act from their perceived reality as opposed to an objective reality, and this perception gap leads to significant managerial problems.[7] It is why the authors support the position that evidence-based, scientifically grounded approaches like managerial forensics heighten the probability of operational success and corporate restructuring.

Quest for the Tools of Managerial Forensics

In this book, the authors expand the understanding of managerial forensics, offer tools for its corporate application, and identify strategies for corporate turnarounds and revival. The book is organized into three sections and covers these three topics. In the concluding chapter, findings and conclusions are presented and the equivalent of a "surgical tool kit" for managerial forensics is presented.

The book is divided into four sections. Part I defines the pioneering approach of managerial forensics and the existing research on the causes of organizational failure. Part II introduces a compilation of insights and experiences from leading business professionals and academics around the world who support the organizational-approach managerial forensics advocates. These chapters offer specific examples and tools that managers and consultants can use in analyzing organizational malaise or morbidity. Part III focuses on ways managerial forensics can be used to restore organizational health and contribute to business turnarounds. Part IV contains the authors' concluding thoughts.

Part I: Understanding Managerial Forensics

- Introduction (*J. Mark Munoz and Diana Heeb Bivona*)
- From Problem to Cause (*Diana Heeb Bivona*)

---

[7] Ferris, G.R., G. Adams, R.W. Kolodinsky, W.A. Hochwarter, and A.P. Ammeter. 2002. "Perceptions of Organizational Politics: Theory and Research Directions." In *Research in Multi-level Issues Volume 1: The Many Faces of Multi-Level Issues*, eds. F.J. Yammarino and F. Dansereau, 179–254. Kidlington, OX: Elsevier Science.

The authors hope that, through the expanded knowledge of managerial forensics and the application of its approach, many organizations can be helped and revived. The application and tools of managerial forensics can be useful for:

- Entrepreneurs or corporate executives—as they seek ways to revive their companies or enhance corporate performance;
- Consultants—when they endeavor to help their clients find solutions to problems;
- Legal entities—in cases where ethical violations, fraud, negligence, legal violations, or corporate irresponsibility exist;
- Government institutions—on occasions in which local, national, and international laws are broken, tax evasion occurs, money laundering takes place, and in matters of national security; and
- Academic institutions—as business schools seek to identify innovative and practical approaches to organizational improvement.

Much has been written in the literature on management on the need for taking an evidence-based, scientifically grounded approach to business decision making. This book expands on the need for this approach. As a pioneering work on the subject of managerial forensics, the editors, authors, and contributors of this book were confronted with the challenges and opportunities of embarking into a new frontier. While there is significant literature on the practice of consulting and business turnaround, the notion of taking an investigative approach to assessing the practice of management or "managerial forensics" was coined by the authors and used in this book for the first time. With limited academic resources to build this new body of knowledge, the editors and contributing authors drew upon their combined theoretical knowledge and business consulting expertise to shape a novel paradigm.

Not known to many, the practice of managerial forensics has been applied in many business organizations. The process typically commences when the board of directors or the management team asks a question such as "Why are we on the brink of bankruptcy?"; "Why are we no longer profitable?"; or "Why are we in poor operational health?" These types of questions trigger an organizational response and put the firm in an investigative mode. It often leads to the creation of task forces or the recruitment of external management consultants to help find the answers and the right solutions. Consequently, there are different solutions for diverse problems found. In many cases, the problems are never completely uncovered or are influenced by the often myopic view of those involved. There are all too many examples in recent history of failed companies that may have benefited from the managerial forensic approach, including Blockbuster, Borders, Radio Shack, and Fresh and Easy.

The practice of management consulting is one where standardized solutions are often hard to find and execute. Companies are unique and problems can be varied and complex. As a point of clarification, the practice of managerial forensics is not about finding standard solutions, but rather the application of a well-defined approach that increases the probability of finding the right solution.

In this book, we drew upon the successful track record of management experts and experienced consultants to help assemble this "toolkit" of forensic approaches to best examine managerial practices in organizations. It is an exploratory effort with much room for further growth and improvement. It is our hope that this endeavor stimulates interest in the subject and paves the way for discussion on dynamic ideas on management in the future.

CHAPTER 2

From Problem to Cause

Diana Heeb Bivona

Managerial forensics is the rigorous application of managerial processes to solving business problems. It is about critically evaluating the framework, system, and dynamics of a business with the right set of tools that will allow managers to better understand and design policies and guide decisions to bring about effective changes. Managerial forensics is grounded in fact and the methodical pursuit of answers. It is not a postoperative critique or an autopsy as to what led to the death of a business. Instead, it is a diagnostic approach that can be used to develop a plan of treatment for an ailing business. In short, it is about the recovery and renewal of a business and not an obituary.

Managerial forensics starts with the identification and diagnosis of the problem. If you fail to diagnose accurately the real issue, you end up treating the symptoms rather than the cause. Subsequently, the problem not only gets worse, but the failure of the business also accelerates. It sounds simple enough; so why then do we often fail to recognize the problem before it reaches critical mass?

Failure to Recognize

For most managers, day-to-day operations consist of putting out one proverbial fire after another. Frantically rushing from one issue to the next, our time is spent buried in what seems like mounds of minutia. Time is no ally in the ever-changing dynamic business environment. As problems arise, we often fail to recognize their significance for a number of reasons. Denial, an inaccurate perception of just how bad the problem is; little experience in managing; or a lack of skills in decision making are common reasons. These are all common explanations cited in the literature as to why a problem is typically not caught earlier.

When problems reach critical mass and grab our attention, the pressure to find a solution mounts. Action is needed and needed quickly. Pressure mounts and managers scramble. Hastily crafted solutions are generated from limited information and decisions are then made and implemented haphazardly. Such decisions are not typically the best ones, failing to address adequately the underlying problem. Failure is not a surprising outcome, when the proposed solutions offered, were not thoroughly vetted. The result is a trip back to square one with time and resources wasted.

Logic Should Trump Intuition

Any serious problem that has plagued a business for some time is likely to be complex and not easy to fix. The greater the complexity and number of causes, the more challenging it is to diagnose the problem. Another issue is the variation in how problems present themselves. Problems can be intermittent, change the way they present, on occasion briefly resolve themselves (only to return later), and even defy analysis because the diagnostic tools and processes to detect do not exist. Then, there is the human variable.

Correctly diagnosing problems can be particularly difficult when human emotions, feelings, and behavior are thrown into the equation. Behavioral decision researchers have determined that humans apply two modes to processing information and making decision called System 1 and System 2 thinking. We perform System 1 thinking automatically, instinctively, and often emotionally. We rely on our intuition and "rules of thumb" to reach our conclusions.

Intuition, also known as a "gut feeling" or a "hunch", is an unconscious form of thinking that results in decisions tied to feelings rather than analytical thinking. Scientists, psychologists, economists, and managers have been fascinated by the role intuition plays in their respective fields. Business icons Steve Jobs, Bill Gates, and Richard Branson have tipped their hats to intuition and the role it has played in their success. Steve Jobs viewed intuition as a "very powerful thing," telling his biographer Walter Isaacson that it was "more powerful than intellect." Even Bill Gates, who acknowledged his gut instinct, had been wrong on several occasions, confirmed in a CNN interview that he still preferred to trust his intuition when it came to new product developments.

They are not the exception by any means. A PR Week—Burson-Marsteller CEO survey found that 62 percent of CEOs confirmed that their gut feelings were highly influential in guiding their business strategies. Only 44 percent indicated that they based business strategies on internal metrics and financial information or competitor analysis. Our reliance on our gut feeling cannot be ignored in the decision-making process. Instead, we need to remain alert to intuition in decision making as it often creates bias.

It is estimated that we have 170 interactions with others every business day, and the likelihood that we are called upon to make decisions daily with limited information is high. In those circumstances, we come to rely on gut feeling. However, are we becoming too reliant on our intuition instead of the facts when solving problems and making decisions?

Because intuition is a subconscious undertaking, we often under-estimate the degree of dependence we have on it. As a result, we come to rely more heavily on intuition and less on systematic analysis. No one is suggesting that intuition should be discounted. It is an important component in both problem solving and decision making, and when limited information is available, and a split-second decision is needed, we lean on it heavily to see us through.

Because our intuition in System 1 thinking can be flawed, the need for System 2 thinking to guide our decision making is great. System 2 thinking, popularized in Noble prize winner Daniel Kahneman's book, *Thinking, Fast and Slow*, relies more on logic and deliberation. In theory, System 2 thinking is supposed to balance our reliance on System 1 thinking when making decisions. The reality though is that we seem to rely on System 1 thinking in times of crisis. Why?

The primary reason is that System 1 thinking generates an immediate payoff that we find gratifying. We enjoy what we believe is a concrete solution to a problem found in the here and now. This immediate resolution is far more satisfying and rewarding than expending time and energy to exploring, examining, and potentially resolving what may be a more abstract or complex problem.

Turning to System 2 thinking would require using far more cognitive abilities than we may elect to use. This is not to suggest that we are not interested or capable of using these higher cognitive abilities. It is more an issue of time for many managers. As management guru Peter Drucker once pointed out, managers do not spend a lot of time making decisions. Time is a precious commodity and our time is monopolized by the multitude of other tasks we are expected to perform on a daily basis. To sit down and devote a block of time in critically examining a problem and finding a solution are often viewed as a luxury instead of a necessity with far-reaching consequences. More System 2 thinking, that is, logical and deliberate thinking, is definitely needed in the problem-solving and decision-making processes.

No matter how experienced we are as managers, intuition cannot serve as a substitute for logic and deliberation. This is not to suggest that intuition should be completely ignored as it does have its place. It should, however, take a backseat role in decision making.

Starting with the Problem

The first step in problem solving is to find and define the problem. This is not an easy task for reasons discussed, but it is paramount to success. Having worked in a variety of industries over the last 30 years, I have been intrigued by the problem-solving process. In theory, the accepted broad framework for managerial problem solving has been to find and define the problem, generate alternative solutions, evaluate alternatives and select the solution, implement the solution, and evaluate the results.

One would expect that managers spend a significant amount of time identifying and defining the problem. After all, how can you generate possible solutions if you do not know exactly what the problem is? The reality though is that many managers spend very little time on this important initial step, trusting their instincts and a handful of visible symptoms to make a diagnosis. Thoughtful deliberation and valid reasoning appear to be absent and the idea that speed is of the essence the driver. However, imagine, as eluded to by Albert Einstein that if greater time and energy were devoted to defining the problem, how much easier the solution would be?

Where do we then start in our quest to find and define the problem? For this answer, we turn to science and in particular medicine for inspiration. Differential diagnosis is a systematic diagnostic methodology. In medicine, it is used to weigh the probability of one disease versus that of other possibilities.

Performing a differential diagnosis for business will require an extensive knowledge base, skills in gathering data, assessing performance, and the ability to synthesize relevant information efficiently in a logical and deliberate manner. It does take some time and effort, but by identifying the problem correctly, we can then prescribe the appropriate method of effective resolution. Therefore, we propose the adoption of a similar differential diagnostic approach in managerial forensics. These steps are explained next.

Step 1: Collect Information

Collect all information about the company and the current situation and begin by creating a list of *symptoms*. For instance, poor sales, declining profits, poor cash flow, increasing debt load, inadequate inventory, and a reduction in assets are examples of symptoms indicating a deeper problem.

It is challenging and time consuming to sift through the data and identify symptoms, but there are notable benefits. Through the collection and analysis of data, new approaches to solving a problem often emerge. Solutions also emerge more readily from performing such a thorough analysis. In short, an investment of time on the front end of proper problem identification provides rewards on the back end.

Step 2: Identification of Potential Causes

The next step is to list all possible causes for the symptoms you have identified. Start with the obvious causes of the symptoms first. This is where we once more turn to our medical colleagues. Medical students are instructed to follow Occam's razor when faced with finding an explanation of symptoms in a diagnosis. Students are told "when you hear hoof beats, look for horses, not zebras" meaning look for the simplest explanation first.

For example, if a patient presents to a physician with a runny nose, chills, and fever, it is likely that it is flu and not malaria. Thus, by examining and performing a simple test for the flu first saves valuable resources trying to rule out a diagnosis of malaria. The same approach can be applied in business by remembering to examine analytically the most logical and simplest explanation first before delving into the more complex one.

In situations where the primary cause is not visible or readily known, a deeper analysis of systems and processes is needed. A thorough review of these business components may expose several issues that merge into a pattern that may point the way to the identification of a specific cause. Another more specific reason may also emerge during the evaluation.

In medicine, physicians use a mnemonic based on etiology to ensure that a thorough, systematic, and organized differential of a patient occurs. The mnemonic used in medicine is VINDICATES:

- V—Vascular
- I—Inflammatory
- N—Neoplastic
- D—Degenerative or Deficiency
- I—Idiopathic, Intoxication
- C—Congenital
- A—Autoimmune or Allergic
- T—Traumatic
- E—Endocrine
- S—Psychosocial or Something else

We propose the adoption of a similarly styled mnemonic to ensure a thorough and systematic review of business systems in identifying potential causes. That mnemonic is SOAP. A fitting analogy as we view this process of eliminating problems as a *cleansing*:

- S—Sales and marketing
- O—Organizational structure and leadership

- A—Accounting and finance
- P—Production and operations

Managers may struggle if unfamiliar with the association of certain symptoms with its related cause. If that is the case, focusing on specific business systems (sales and marketing, operations, etc.) instead of seeking to identify one specific cause will prove helpful. Focus on the system first, and then think of the probable causes unique to that system. This should ultimately aid in reducing the number of potential causes.

Finding and identifying the underlying cause will require in-depth knowledge of the critical systems and processes that a company relies on to be successful. It requires a manager to have a well-rounded view of the business to start and then the ability to focus in on the functional areas of the organizational processes to determine where the problem exists. Successfully identifying the cause dictates the need for this systematic approach.

A note of caution as you begin to develop a theory as to what the underlying cause may be: It is not uncommon that as we start to formulate the cause we receive additional information or data that may conflict with the initial hypothesis. Our inclination may be to ignore that conflicting data or attempt to interpret it in such a way that it supports our existing theory when in fact it may support dismissing our hypothesis regarding the cause and searching for a new one. Be cognizant of that possibility and rigorously question your motivations and reasoning for rejecting any new information received.

By using the SOAP mnemonic, we can ensure that each area of business is scrutinized, and a comprehensive list of potential causes developed. Due diligence during this phase is vital in not only capturing the information required to identify possible causes, but in making a final decision.

Step 3: Prioritize

Next, we prioritize the list of potential causes by focusing on the most urgent or serious problems first. Not all causes are equally pressing or grave, but in order to select the correct tools and corrective measures, it is essential to identify and select those that are. Problems that threaten the immediate continuation of business operations are the priority.

As Dwight Eisenhower is attributed with saying, "What is important is seldom urgent and what is urgent is seldom important." Prioritizing can be challenging in terms of trying to determine what is urgent and serious. The determination of causes that are classified as urgent and serious are those that can quickly bring operations to a halt if they continue much longer. Prioritizing a list of potential causes lets us take a more targeted approach instead of a shotgun approach to diagnosing the problem. It also allows us to determine what needs to be addressed immediately and what can be deferred.

The Eisenhower Matrix is used in time management to prioritize tasks by urgency and importance. The use of this type of matrix was popularized by Stephen Covey's book, *The 7 Habits of Highly Effective People*. It can prove helpful in developing an approach to prioritizing problems and decreasing the need for running potentially unnecessary diagnostics on a business, diverting attention from the most urgent and serious problem at hand. A similar matrix has been modified for use in the prioritization of possible causes, and as you create your list of possible causes, place those causes in the appropriate matrix box as shown in the following:

An example of the Eisenhower decision matrix

Urgent and serious	Not urgent but serious
Urgent but not serious	Not urgent or serious

Step 4: Process of Elimination

The fourth step in the process focuses on ruling out potential reasons. The best place to start is with the most urgent and serious potential causes. Ruling out the potential sources of the symptoms can occur by using various established business diagnostic tools and other methodologies to eliminate that probable cause or to confirm a possible cause in another quadrant. If you have a possible urgent and serious cause that you cannot eliminate, focus your attention on it until you have explained and defined a corrective course of action. Once you have addressed the most urgent and serious causes, you can then address causes in the remaining three quadrants, as these are not as urgent.

Throughout this differential diagnosis process, time should be spent focusing on the problems and questions essential to the company's continued operation. Is the company stable? Are we still bleeding cash? What might be causing it? How will we pinpoint the cause? How will we address it? Spend your time investigating and deliberating these answers.

If you continue to struggle with pinpointing the cause, you may need to go back to the beginning and ask more probing questions, obtain collateral information from employees or stakeholders familiar with the problem, and pull additional supporting data. A fresh set of eyes may also be needed. Senior colleagues and trusted mentors may be able to lend insight and direction to your analysis. Do not hesitate to utilize their knowledge. The important takeaway in using the differential diagnosis approach is to remain focused, methodical, and systematic in your approach to identifying the actual cause of the problem.

Using a differential diagnosis approach to identifying business problems is beneficial for several reasons. It allows us to decide whether the business is in a critical and unstable position. It assists us in identifying which diagnostic tests are appropriate. The differential diagnostic guides us in conducting an investigation to confirm or refute the possible causes of business problems reducing our reliance on intuition. Finally, it gives us the ability to plan an effective, more focused turnaround strategy.

The benefits of investing our time and energy in accurately identifying the underlying causes of business problems are many. By doing so systematically and methodically, we improve our ability to resolve those issues and turn a failing business around. Granted it is only the first step, but it is an important one.

Bibliography

Beshears, J., and F. Gino. 2015. Leaders as Decision Architects (cover story). *Harvard Business Review* 93, no. 5, pp. 51–52.

Bryant, R.J. 2013. "Identifying Single Points of Failure in Your Organisation." *Journal of Business Continuity & Emergency Planning* 7, no. 1, pp. 26–32.

Caruth, D.L., G.D. Caruth, and J.H. Humphreys. 2009. "Towards an Experiential Model of Problem Initiated Decision Making." *Journal of Management Research* 9, no. 3, pp. 123–32.

Covey, S.R. 1989. *The 7 Habits of Highly Effective People*. New York: Simon and Schuster.

Drucker, P.F. 1967. "The Effective Decision." *Harvard Business Review* 45, no. 1, pp. 92–98.

Hood, J. 2006. "CEO Survey 2006: All Things Considered." *PR Week,* November 6.

Humphreys, J., and H. Langford. 2008. "Managing a Corporate Culture 'Slide'." *MIT Sloan Management Review* 49, no. 3, pp. 25–27.

Markle, R., and A. Feibelman. 2015. "Building an Effective Data-Driven Business." *Electric Perspectives* 40, no. 3, pp. 62–65.

Schermerhorn, J., Jr. 2011. *Management.* 11th ed. Hoboken, NJ: John Wiley and Sons.

Wanless, J. 2002. "Using Intuition at Work." *Women in Business* 54, no. 3, pp. 44–47.

CHAPTER 3

Featherston Resources: A New, Clean, Green Fertilizer Business That Failed

Sergio Biggemann and Alan Collier

Featherston Resources Limited (FRL) had access to a natural asset, a diatomite mine, with a potential market value as high as $1 billion.[1] This prospect captured the imagination—maybe the greed—of each set of directors who took control of the company during its 20-year life. But today, this asset still sits in the ground with FRL no longer in control of it—with all the assets of value in FRL having been sold to an investor with larger pockets who may be able to realize the potential. During its life, FRL scarcely extracted any of the material from its mine, and never made a profit. In order to explain how this came about let us start from 2011 when FRL still controlled these assets.

… Yes Mr Chin, I understand you need an answer urgently … Yes, I know … I give you my assurance that we want to work with your company … Yes we know how big this business could be … I'll explain this to the board and call you back … No, don't worry, you will hear from me soon … Good-bye Mr Chin.

Emma Weston hung up the telephone and looked around her small office; the day was sunny but there was not much to see out the window. She was sitting at the processing plant that she had helped commission just one year before. While everything had started off running smoothly,

---

[1] Unless otherwise noted, values are expressed in U.S.$.

some problems with the drying equipment arose a few months ago and had not yet been resolved. This created a situation that risked putting all her efforts in jeopardy. While Emma walked toward the kitchenette to grab her cup of coffee, she reflected on the comfortable position with a big company she had left to accept the challenge that Tim Goodacre, her former boss, put in front of her: to lead FRL to success in manufacturing and marketing Adveco, a natural fertilizer with the potential to significantly change the agricultural fertilizer industry while generating immense profits. She wondered whether she had made the right decision by accepting Tim's offer.

Mr. Chin represented AgroFertz,[2] a Chinese company that wished to buy significant quantities of Adveco's main component: Black Diatomite™. He had been pressing Emma for some months to sign a long-term deal to supply Black Diatomite to his company to take advantage of the increasing demand for fertilizers in China and to increase its business presence in New Zealand. Emma knew that if she were to accept AgroFertz's proposition she could generate enough cash flow to support FRL's long-term plans, but that the offer involved some risks that may affect FRL's ability to realize its full potential. This left Emma in a quandary and she had been delaying making a final decision. However, time was running out for the company and she needed to make a decision on the AgroFertz offer soon.

Rapid economic development in China had resulted in huge increases in food consumption and the consequent increased demand for fertilizers. FRL was well placed to satisfy some of the demands of the Chinese market. However, Mr. Chin was not interested in buying the processed product that offered FRL higher margins and profits; Mr. Chin wanted the unprocessed precursor, Black Diatomite. As Mr. Chin explained to Emma, he wanted the unprocessed diatomite because Chinese soils are different from those in New Zealand, and AgroFertz was in a better position to customize the product for local use. In addition, all the other inputs needed to manufacture finished fertilizers could be procured at a lower cost in China than in New Zealand, making sense for AgroFertz to

---

[2] A real person and company, although the names of both have been changed here.

process the diatomite in China. Mr. Chin was offering a tough bargain—he was aware of FRL's capital squeeze and had declined to offer FRL a partnership in the China venture. The most he was prepared to offer FRL was a royalty for the intellectual property required to convert Black Diatomite into processed fertilizer.

How It Started—The First Managers

FRL was founded in 1997, but the story began over a decade before. In 1983, a large international company involved in manufacturing construction materials, Holcim, purchased land containing a diatomite deposit located in Foulden Hills, Central Otago, in the South Island of New Zealand. Holcim obtained a permit from the government to extract minerals such as diatomite located within the land. However, when the date for renewal of the extraction permit fell due in 1983, Holcim failed to renew the permit, allowing Dave, a speculator, to take advantage of this legal oversight. Dave incorporated a company, Envirosil, and obtained for Envirosil an extraction permit from the government for the diatomite deposit from under the nose of Holcim. However, neither Dave nor Envirosil had the funds to develop the deposit, and so attempted over some years to negotiate with Holcim either to have Holcim buy the rights that it had lost through its oversight or to negotiate access to the diatomite deposit across Holcim's land so that Dave and Envirosil could access the deposit. Not surprisingly, Holcim was greatly annoyed by Dave's actions and refused to negotiate, leading to a standoff.

In 1997, Dave invited a friend and business colleague, Alan Walker, an experienced geologist and businessman, to assist by assessing the financial potential of the diatomite deposit and to negotiate with Holcim on behalf of Envirosil. Alan was successful in negotiating with Holcim to obtain for FRL (as successor to Envirosil) all the rights needed to exploit the diatomite deposit including access across Holcim's land. Alan was also key to raising the capital of $8 million used during his period as a director. In this way, FRL became the holder of all the relevant rights while Dave, along with his other investors, became shareholders in FRL.

The founding board of FRL comprised Dave (the speculator), Alan Walker (the geologist), Alan Ralph (a chemical engineer), with David

Kember (a lawyer) as the chairman. Because of his key role in the formation of FRL and his prior experience as a geologist and businessman, and
because he was trusted by the other board members, Alan became, on
the death of Dave only two years later, the only executive member of the
board. Alan was especially valuable because he had become well aware of
the potential commercial uses of diatomite.

One of the outcomes from Alan's negotiations to reach an accord
with Holcim was an obligation placed on FRL to pay Holcim $100,000
each year for access to the mine, adjusted according to the international
commodities index, payable whether or not FRL was extracting minerals.
Of course, because FRL saw itself as a mining company with a rich resource,
this sum was a mere bagatelle. As time progressed, however, and little revenue was being earned, this cost became a significant burden on FRL.

However, before we explain the problems faced by this management
team, we should explain a little about the asset that FRL had acquired.

The Asset

Diatomite, also known as diatomaceous earth, is the naturally occurring
fossilized remains of diatoms, which are single-celled aquatic algae. They
belong to the class of golden brown algae known as Bacillariophyceae.
Diatomite in its habitual form is a near-pure sedimentary deposit
consisting almost entirely of silica. The Greeks first used diatomite over
2,000 years ago in pottery and bricks.[3]

The diatomite at Foulden Hills was notable because it was formed from
an inland freshwater lake[4] unlike most diatomite deposits that are marine
based. The Foulden Hills deposit comprised two parts: the majority of the
mineralization was Black Diatomite, a carbonaceous form of diatomite,
while the remainder was oxygenized and therefore white diatomite. The
Black Diatomite in the deposit contained a high level of organic material

---

[3] Industrial Minerals Association—North America (IMA-NA). http://www.
ima-na.org/?page=what_is_diatomite
[4] A so-called lacustrine environment; that is, it is formed at the bottom or along
the shore of lakes, as geological strata, unlike most diatomite deposits that have
a marine origin.

that required only a relatively small amount of processing to convert it into a useable soil conditioner and fertilizer product.

There are many diatomite deposits throughout the world, but deposits of high purity that are commercially viable are reasonably rare.[5] Globally, large reserves of diatomite are found in China, Europe, Japan, Mexico, the United States, and parts of South America. Black diatomite is particularly rare because most diatomite deposits undergo oxidization and have their key nutrients leached over time from the raw product.

The white diatomite in the FRL deposit also had an unusual and valuable characteristic: it exhibited a very high absorption capacity for liquids, up to 350 percent of its own weight, in contrast to most diatomites that have a much lower absorption capacity of about 100 to 120 percent. This characteristic was very useful for the application of the FRL deposit in a wide range of industrial products, in particular for use in waste product cleanup, as an insulation material, and as a desiccant.[6]

The FRL diatomite deposit at Foulden Hills had an estimated total resource of 20 million tons, of which 6 million wet tons, converting to at least 2 to 3 million dry tons, was suitable for commercial exploitation. FRL planned further tests, which were expected to prove that the deposit was even larger.

As a raw resource, the Foulden Hills diatomite had a value as unprocessed diatomite exceeding $700 million in 2007 prices once extracted.[7] However, adding value to the raw material by processing it into an end product offered much higher potential profits.

The First Managers Went Fishing for Opportunities

Alan knew from the beginning that the Foulden Hills diatomite had numerous industrial applications including its potential use in agriculture.

---

[5] http://www.ima-na.org/?page=what_is_diatomite

[6] A desiccant protects products by absorbing moisture that would otherwise cause the deterioration of the product.

[7] "U.S. Geological Survey, 2012, Diatomite Statistics." 2013. In *Historical Statistics for Mineral and Material Commodities in the United States (2013 Version): U.S. Geological Survey Data Series 140*, T.D. Kelly and G.R. Matos, comps. http://pubs.usgs.gov/ds/2005/140/ (accessed June 28, 2013). Calculated as US$237 per ton for 3 million dry tons.

He and other members of the board were also aware of other potential applications of diatomite. These multiple uses of diatomite made Alan and his fellow directors and shareholders enthusiastic about the potential economic value of the resource. David Kember, former chairman of the founding FRL board and later one of the larger shareholders, explained that they were in the business because of its innovative potential, its potential to improve crop yields, its benefits for the environment, and its export potential for New Zealand.

Of course, none of this could be done without cash to extract the diatomite, conduct R&D on its potential uses, and set the company on a business footing. Shares in FRL were offered on a restricted basis to family, friends, and sophisticated investors. In this way, the requirements for extensive disclosure required by the law would not be activated. Between 2000 and 2007, FRL raised about $8 million from 130 investors who then became shareholders in FRL.

During 2000, the FRL board, on Alan's advice, decided that it would use the diatomite in whatever market it could enter, such as the waste product cleanup market, for which diatomite was eminently suited,[8] as an insulating product capable of offering light, cost-effective fire protection and building insulation, or as a desiccation agent for shipping containers. At that stage, it would be true to say that the company was attempting to identify a product capable of using its rich resource. FRL engaged in research using diatomite in several fields. In one joint venture begun in 2000, FRL acquired an interest in a company in Australia that seemed to have the promise of offering FRL an opportunity to use its raw material as a high-technology building material. The idea appeared sound, but was going to require a lot of R&D to develop a product and generate the cash flow FRL needed to finance its further research, while the company still had to operate and pay costs such as the access fee to Holcim. After some three years of making losses from its Australian joint venture, FRL

---

[8] "[FRL] wants to have an open-cast mine and processing plant at Foulden Hills station, about 10 kilometres east of Middlemarch, so the absorbent material can be used nationally and internationally in the rapidly growing environmental clean-up market" (from *The Dominion* newspaper, June 28, 2000).

decided to sell its interest, which it did at a financial loss, with FRL still far from realizing its objective of creating a value-added product.

While the waste product cleanup market was still a potential market for the diatomite, as a result of experiments the company had conducted in agriculture, the FRL board began to see its future in processing the raw diatomite into an environmentally friendly fertilizer. To exploit this new direction, the company needed more capital to undertake systematic R&D, to develop full extraction capability, to build production facilities, and to develop this new market.

FRL managed to set up a small processing facility and began experimentation and trials of fertilizers in different locations in New Zealand. Nevertheless, by about 2007, FRL's 130 shareholders still had nothing more than some failed opportunities and the promise of a product as a result of $8 million having been spent. More capital would be needed if progress toward a commercially viable fertilizer product were to occur. Not surprisingly, after 10 years, the original investors were becoming weary of unfulfilled predictions and lack of progress and, if nothing changed, were unlikely to contribute any more capital, meaning that FRL risked being wound up.

The Fertilizer Market

The Black Diatomite and the white diatomite at Foulden Hills offered particular advantages over most diatomite. First, they were natural alternatives to the hard chemical fertilizers that had been used in agriculture traditionally; they could be impregnated with any combination of nutrients or conditioners to improve soil quality and can suppress some plant diseases naturally owing to the presence of silicon. Second, because of the synergy between silicon and the added NPK,[9] the product was a slow-release fertilizer, meaning that less fertilizer was required for the

---

[9] NPK is an abbreviation that refers to the ratio of important elements in a fertilizer or soil amendment. N stands for nitrogen, which is responsible for strong stem and foliage growth. P stands for phosphorus, which aids in healthy root growth and flower and seed production. K stands for potassium, which is responsible for improving overall health and disease resistance.

same effect, resulting in lower costs for farmers and less environmentally damaging nutrient runoff. As a result, the Foulden Hills deposit had significant advantages in the market because it could be differentiated from other fertilizers by virtue of its being a clean and green alternative to chemically derived fertilizers.

The most popular fertilizers, by use, were anhydrous ammonia (around 82 percent nitrogen), urea (a solid compound that contains around 42 percent nitrogen), superphosphate, monoammonium phosphate (MAP), and diammonium phosphate (DAP) (the latter two containing a mix of nitrogen and phosphate). The escalating use of these synthetic fertilizers was causing widespread environmental concerns. Some of these concerns included a contribution to depletion of world fossil fuel reserves; leaching of nitrates and phosphates (in particular) from farms into waterways and river systems, promoting algal bloom and other aquatic vegetative growth, which in turn degrades water quality and can cause eutrophication;[10] release of the greenhouse gas nitrous oxide in some storage or application conditions; increasing soil acidity owing to excessive use of ammonia and contribution to soil degradation; an increase in some pests owing to excessive nitrogen use, resulting in the increased birth rate, longevity, and fitness of pests; introducing contaminants and heavy metals into soils owing to the use of chemical fertilizers some of which contain mercury, arsenic, cadmium, uranium, and polonium-210; contributing to anthropogenic climate change by the production, application, storage, and use of chemical fertilizers; and the cost of chemical fertilizers, which affects poorer communities and developing nations by causing socioeconomic stratification.

The Second Managers Change Business Direction

Although Alan felt that he had done his best in identifying opportunities for the company and guiding its early life, the shareholders were of the view that the company needed a change of management if the company

---

[10] The process by which a body of water becomes rich in dissolved nutrients from fertilizers or sewage, thereby encouraging the growth and decomposition of oxygen-depleting plant life and resulting in harm to other organisms.

was to produce and sell fertilizers. Because the company needed to raise further capital, the board and key shareholders made it a condition for contributing capital that Alan step aside.

Enter Tim, who was appointed managing director of FRL in November 2007 and insisted that all members of the founding board had to be replaced as soon as practical. He was of the view that, after 10 years, there was too much baggage to be carried by continuing with the founding board, while the presence on the board of the original entrepreneurs was inconsistent with the direction he saw for the company. In order to satisfy this demand, members of the founding board progressively retired, to be replaced with others believed to possess the skills needed to take FRL forward in its new direction. All members of the founding board (except the chairman) resigned between January 2007 and April 2008, with the founding chairman of directors, David, who resigned on September 24, 2008, being the last to depart.

As often happens in the case of small emerging companies, there was no international executive search done for a new managing director; rather, Tim had been approached to become managing director because he was known to a horticulturalist who was, in turn, known to a member of the founding board. From his assessment of the company and its product, Tim quickly concluded that FRL should concentrate on developing, manufacturing, and marketing processed diatomite as a new form of fertilizer. Even if done somewhat informally, the appointment appeared, at first, to be a propitious one.

Tim had the experience to see the potential because he came from an ideal background to understand and deal with farmers: He had been a senior executive concerned with international markets and sales at the Australian Wheat Board, and was later CEO of the monopoly New Zealand marketing agency for kiwifruit, Zespri. He had excellent connections in Australian agriculture as well as New Zealand agriculture and horticulture, and he had substantial experience in business. At the same time, his experience was principally in large companies, so the needs of a nascent small company were something he would have to learn along the way.

Tim believed that Adveco, as the processed diatomite became known, was a revolutionary product capable of customization to meet the requirements of individual soils and crops and, potentially, the needs of every farmer and horticulturalist. Black Diatomite was an especially valuable resource for agriculture, Tim said

> ... because of its natural attributes and also the opportunity to deliver [growers] a business which really did relate to their own business and work in a partnership with their own business. So I saw this as an opportunity to work with growers and give them a product which was not only a good product but also a business which provided them with a good basis for a much more expanded business.

The product, together with a close relationship with individual growers, would represent the main advantage of FRL over the longer-established suppliers. Tim summarized his vision of the business: "... a good relationship, working with growers, getting their confidence in the product, giving them good service, a whole thing."

After only a few months as managing director at FRL, Tim made the decision to employ a former business associate, Emma Weston. Emma was a lawyer with an MBA from the Australian Graduate School of Management; she had a similar background to Tim and had worked with him at the Australian Wheat Board, but was younger and in a better position to become a full-time executive. She also had extensive experience in the management and operation of both large and small agricultural companies and possessed significant experience in grower sales and marketing—ingredients that would be essential if FRL was going to succeed.

In April 2008, Tim appointed Emma as the chief operating officer (COO) and, in November 2008, she became CEO and managing director in his stead. Upon Emma's appointment, Tim relinquished his role as CEO and managing director and became nonexecutive chairman of the restructured board.

With Tim as nonexecutive chairman of an entirely new board, and with Emma assuming the role of CEO and managing director, they had management control of a small company with a large resource, some R&D activity, a partially developed product, and very limited capital. In order to grow, additional capital was needed urgently.

While the founding directors had relinquished control entirely to a new board containing no members of the founding board, these former board members, including Alan, continued to hold a substantial equity interest in the company.

The transition between the old and new management did not occur without significant friction. The new managing director had been skeptical about Alan's managerial credentials, while, in Alan's view, neither Emma nor Tim had experience in either the mining or industrial minerals industry or in manufacturing—two vital skills needed if FRL was to be the successful vertically integrated company capable of returning large dividends to the shareholders as planned. The parties remained estranged with the result that the knowledge Alan possessed was not passed on to the new management for the benefit of FRL. On top of everything, Alan was claiming a substantial sum for unpaid fees relating to his work for FRL while on the board.

Both Tim and Emma brought cash to the company and received equity in the company for the cash they contributed and as compensation for working in a start-up company. Tim purchased several hundred thousand shares in FRL at $0.50 per share. In addition, Tim and Emma both bought shares, along with others, in a 2009 capital placement at 10 cents per share. Some existing shareholders believed that these allocations were done on improperly favorable terms to the new management, and seriously discounted the real value of the company. FRL would eventually have 217 shareholders.

The Company

By now, FRL had plans to become a vertically integrated company that extracted the raw material, processed and manufactured the raw material into a value-added fertilizer product, and then marketed the carbon-rich, high-silica organic and blended fertilizers and soil conditioners in the Australian and New Zealand agricultural and horticultural markets.[11] Or, at least, that was the plan. FRL's headquarters was located in Sydney, Australia, while the diatomite resource and the manufacturing plant was

---

[11] See www.frlgroup.com

located in New Zealand. By this time, FRL had two subsidiaries, one in Dunedin, New Zealand (Envirofocus Limited), and one in Australia (Adveco Pty Limited).

Envirofocus held the key intellectual property rights, operated the New Zealand processing plant, conducted New Zealand-based research and product trials, and managed New Zealand distribution and sales. Adveco undertook key R&D functions, Australian-based research and product trials, and managed Australian distribution and sales.

FRL developed some impressive technology and, through Envirofocus, held New Zealand, Australian, and Chinese patents that covered the production method and formulation of the soil conditioning and fertilizer product. It was also pursuing a worldwide claim for patent protection under the Patent Cooperation Treaty in a number of jurisdictions including Canada, Europe, Japan, Korea, and the United States. FRL also claimed ownership of intellectual property rights in product formulas, processing methods, trademarks, and proprietary information relating to a range of soil conditioning fertilizers and plant treatment agents it had developed. However, as it existed, FRL was largely a one-product company that had to be capable of producing a quality product, with consistent attributes, in the volumes needed, at the time demanded by customers, in order to be viable. This was only possible if it had a reliable processing capability.

FRL's Products

The product developed by Adveco was a formulation of NPK based on Black Diatomite, which provided a rich source of carbon and plant-available silica. Four formulations were offered: Advedo NPK, Adveco MAX, Eco-Gro and Home-Gro.[12] Key benefits of Adveco products included:

---

[12] Adveco NPK was a controlled release fertilizer with superior moisture retention properties.
Adveco MAX was designed for use in the broadacre grains sector and is currently part of an ongoing wheat yield trial program.
Eco-Gro was a fully organic soil conditioning fertilizer product similar in formulation to Adveco NPK but with a lower analysis.
Home-Gro was developed for home garden and small plot use.

- A balance of macro- and micronutrients essential to plant health in a slow- or controlled-release pellet;
- Efficient nutrient delivery and decreased nutrient loss through leaching and volatilization (as compared with conventional high-analysis fertilizers);
- Disease suppression in treated plants, reducing the need for and costs of disease control;
- Improvement in fruit quality without compromising yield; and
- Moisture retentive properties.

A major benefit of the Adveco formulations was improved soil structure and a positive interaction with the soil food web (microorganisms, bacteria, worm count), which is the nutrient foundation for healthy soils.

Research had identified many beneficial effects of silicon in agriculture, in particular on plant growth and health. As described earlier, silicon is the basis of diatomite, and hence Adveco. Research suggests that these beneficial effects may be due to factors that include the modification of the physical properties of the soil, the facilitation of beneficial mineral elements, suppression of the uptake of toxic compounds, or by stimulation of the uptake and incorporation of silicon by the plant. It has been shown that the addition of silicon can help to reinforce a plant's cell walls by providing a physical incrustation that makes disease penetration more difficult.

Known beneficial effects of silicon in plant health include disease control, pest control, reduction of salt stress, reduction of freezing stress, improvement in plant growth, increased yield, reduced soil toxicities, and the reduced effects of drought. In soils that have high levels of soluble silicon, repeated cropping has reduced plant-available silicon to a point where the addition of silicon is often required to achieve maximum production. The evidence was overwhelming that the presence of silicon improves plant growth, and the Adveco formulations were ideally suited to meet this need.

FRL products were suitable for all major crop types from intensive horticulture to broad-acre cereal and fiber crops and even home gardeners. The main target markets were New Zealand and Australia for finished retail products, and developing countries for exports of Black Diatomite as a raw or part-processed product with or without a technology license to manufacture fertilizers. Farmers are notoriously averse to change, but Adveco fertilizers had the advantage that they did not require growers to adapt their methods of use because the product was similar in appearance and application requirements to existing fertilizers. In New Zealand, FRL had focused on the horticultural market, especially on crops of significance and scale such as kiwifruit, cherries (and other stone fruits), and apples (and other pip fruits). In Australia, growing wheat was likely to be the most lucrative market segment not only because of the size of the industry, but also because the market was dominated by two major suppliers who had a reputation of keeping prices high and services poor.

Emma believed that Australian grain growers, in particular, were amenable to using alternative fertilizer supply options given their general disenchantment with the traditional suppliers owing to their fluctuating prices and poor service levels. In all cases trialed, growers welcomed FRL's Adveco fertilizers on the basis that they wanted to try alternative products capable of showing positive trial data from a company with a known and trusted management. Owing to the indifferent quality of much of the soil, fertilizers account for approximately 13 percent of total farm cost inputs for Australian farmers, making the Australian market particularly attractive for FRL but, at the same time, challenging because growers are cost conscious and slow to embrace change.

While pursuing markets in Australia and New Zealand, FRL was also in advanced discussions with customers overseas, mainly in China (with Mr. Chin) and India, with the objective of selling raw or part-processed Black Diatomite as feedstock for fertilizers and soil conditioners. Discussions with potential Asian customers were focused on agreeing on multiyear offtake or similar arrangements with agreed or minimum annual price and tonnage targets or through take-or-pay arrangements with established price and volume. It was expected that all exports (to countries other than Australia) would be on a free-on-board basis.

Exports were also likely to include licensing fees from FRL patents and proprietary technology and in some cases product formulations as part of a value-added sale price. It was possible that export sales combining unprocessed or part-processed diatomite and intellectual property could expedite and drive sales in key markets.

Competition

Ravensdown and Ballance were the two leading manufacturers and distributors of fertilizers in New Zealand.

Ravensdown was the largest supplier of fertilizers in New Zealand, directly supplying more than half of all fertilizers used in New Zealand agriculture. Ravensdown is a grower cooperative and over the 10 years up to 2012 had expanded its traditional NPKS (nitrogen, phosphorus, potassium, and sulfur) fertilizer products to include liquid fertilizers plus a range of value-added products and services—from soil testing to agrichemicals. In 2011, Ravensdown entered the Australian market by acquiring United Farmers' Co-operative in Western Australia and through a joint venture with Queensland Cane Growers Association in Queensland, Australia.

Ballance Agri-Nutrients Limited was the other leading fertilizer manufacturer in New Zealand. The company also owned an ammonia–urea manufacturing plant, Summit-Quinphos, New Zealand's third-largest fertilizer company, and Super Air, one of the country's largest agricultural aviation companies. Ballance was a 100 percent farmer-owned cooperative originating from a series of company amalgamations and alliances that saw regional fertilizer cooperatives amalgamate under the umbrella of what was then Bay of Plenty Fertiliser.

The Australasian[13] fertilizer industry was made up of manufacturers, importers, blenders, distributors, agents, dealers, contract applicators, and associated services such as agronomic advisory, soil testing, storage, handling, and transport. The fertilizer industry was largely self-regulated, although there were clear-labeling laws in place and import and export certification processes. There was increasing pressure to review the environmental impact of conventional fertilizers and, in Queensland, Australia, legislation had been introduced into the State parliament to control the use of nitrogenous and phosphate fertilizers in areas with waterways under threat.

The Fertilizer Industry Federation of Australia comprised 21 members including companies such as OCP (national Moroccan phosphate

---

[13] Australasia comprises Australia, New Zealand, and Papua New Guinea.

company, the world's largest exporter of phosphates and derivatives); Mosaic (global leader in concentrated phosphate- and potash-based fertilizers); Foskor (phosphates and phosphoric acid); Wengfu and Sinofert (Chinese suppliers); Agrium (global producer and supplier of agri-inputs and services); PCS (the world's largest fertilizer enterprise by capacity); and CANPOTEX (international fertilizer marketing and distribution company jointly owned by Agrium, Mosaic, and PCS), as well as a large number of players based in and out of the Middle East.

Approximately 5 to 6 million tons of fertilizer was manufactured in Australia each year with the balance being imported. Superphosphate, which was usually manufactured locally from imported phosphate rock, made up 50 percent of domestic production. Around 62.5 percent of nitrogenous-based fertilizers, 33 percent of phosphate fertilizers, and 100 percent of potassium fertilizers were imported each year. Three major companies dominated the production of fertilizers in Australia: Incitec Pivot, Wesfarmers CSBP, and Impact Fertilisers.

Incitec Pivot Limited was Australia's largest fertilizer manufacturer with over 50 percent of the total market share. With historical production of over 3 million tons of fertilizer products per year, Incitec Pivot had divested its distribution and marketing businesses with the aim of becoming a low-cost, high-volume producer of urea, ammonia, ammonia sulfate, superphosphate, diammonium phosphate (DAP), and monoammonium phosphate (MAP).

Wesfarmers CSBP was established in 1914 as a grower cooperative; Wesfarmers was a highly diversified public company with CSBP being its chemicals and fertilizer division. CSBP had around 13 percent of the total market share but a dominant position in Western Australia. CSBP produced over 1 million tons of fertilizers annually, including super-phosphates, compound fertilizers, and home garden varieties. CSBP's dominant market share in the west of Australia had declined from about 90 percent in 1995 to 96 to 60 percent in 2012 with other suppliers such as Summit Fertilizers and United Farmers' Co-operative entering the market.

Impact Fertilisers, producing an average of 200,000 tons per year (tpy), was a Tasmanian manufacturer of single superphosphate. In July 2006, Impact Fertilisers merged with Swiss company Ameropa A/G, through which it imported and marketed urea, DAP, MAP, and muriate of potash to Tasmania and Eastern Australia.

Business Prospects

As shown in Figure 3.1, fertilizer prices had increased significantly over the past 40 years up to 2012; however, it was only in the last decade that prices had skyrocketed because of the increasing price of crude oil. Indicative prices of competitors' fertilizers are summarized in Table 3.1.

FRL had prepared projections of costs and revenues based on its plans to increase production by building a plant in Australia and expanding processing capacity to 120,000 tpy by 2019. Table 3.2 summarizes these figures.

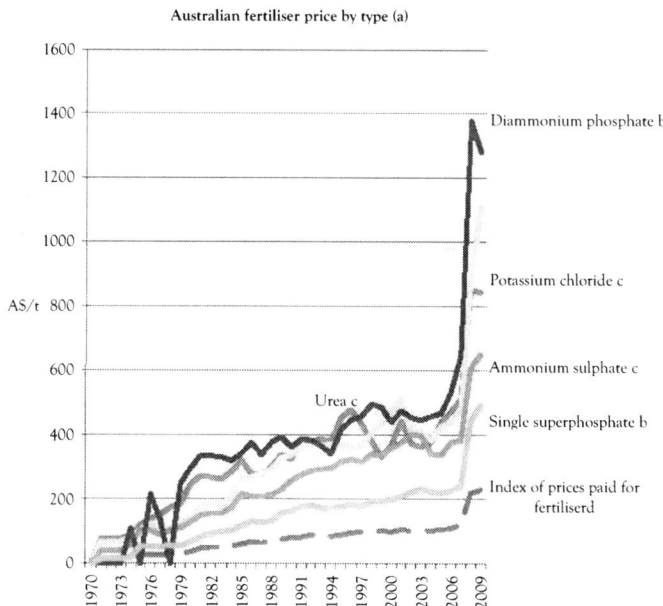

Figure 3.1 Prices of fertilizers in Australian market[1]

---

[1] *Source*: ABARE data, *Australian Commodity Statistics 2009, Rural Commodities, Farm Inputs Fertilizer Prices*; additional notes: (a) Average price paid by Australian farmers at 30 June. (b) Bulk price. (c) Bagged price from 1970 to2001; Bulk price from 2002 onwards. (d) Based on a four quarter average for the year ending 30 June.

Table 3.1 *Indicative Australian and New Zealand fertilizer prices*

Product type	Southern WA (AU$)	South Eastern Australia (AU$)	New Zealand (NZ$)
Urea	455	485	620
DAP	570	540	660
MAP	900	960	1,150

*At the time of writing, US$1 = AU$1; US$1 = NZ$1.25.

Source: FRL's market research.

Table 3.2 *Costs and revenues for Adveco's project (in AU$)*

AU$	2012	2013	2016	2019
Revenue				
Fertilizer	5,150,000	17,171,096	47,563,783	84,196,938
Offtake and royalties	6,100,635	11,973,901	43,326,807	60,821,689
Cost of sales				
Fertilizer	3,261,230	8,938,298	24,429,099	45,145,143
Offtake and royalties	2,275,462	7,855,505	14,969,600	21,731,010
Operating expenses	3,481,391	3,698,215	5,239,868	7,428,297
Tons				
Fertilizer	11,000	27,000	80,000	120,000
Offtake and royalties	22,750	44,175	150,625	199,250
Average price				
Fertilizer	468	636	595	702
Offtake and royalties	268	271	288	305
Average cost of sales				
Fertilizer	296	331	305	376
Offtake and royalties	100	178	99	109
Average operating expense	103	52	23	23
Net profit after tax (NPAT)	2,053,000	9,337,000	32,761,000	49,542,000

Source: Company's financial projections.

Table 3.2 was prepared assuming that the plant in New Zealand could produce 10,000 tpy of fertilizers (working one shift) along with a new plant in Australia with a capacity of 30,000 tpy (working one shift).

However, FRL had never achieved the product volumes through its New Zealand pilot plant because of continual problems arising from drying the raw material during processing, indicating that Alan Walker's skepticism about the lack of these skills within the restructured company had some basis. As a result of these problems, FRL failed to supply markets in Australia as promised for two successive growing seasons. FRL's credibility among its potential user community had been severely dented.

Emma estimated that FRL needed about AU$9 million in capital at this point: AU$1.75 million to modify and make fully operative the existing processing facilities in New Zealand, AU$5 million for new developments in Australia (including AU$4 million to build a new processing plant plus AU$1 million in operating costs), and a further AU$2 million to undertake exploration of new Diatomite deposits in New Zealand.

FRL needed cash to continue. It could have obtained this through sales to Mr. Chin, but in the process may have lost control of its intellectual property rights in the Chinese market. It could have gone back to existing shareholders and sought more funding. Or it could have explored other options such as a private buyout or by issuing an initial public offering (IPO).

The End Nears

In December 2013, FRL was placed in administration and receivership by its management and made subject to a deed of arrangement. Neither Australia nor New Zealand has an equivalent to U.S. Chapter 11 Bankruptcy permitting a company to trade out of difficulties; insolvent companies in these jurisdictions are placed in the hands of a controller such as an administrator, receiver, or liquidator (depending on the stage of the process), who then seeks to realize the value for creditors (and owners, if anything remains left) by disposing of the company or its assets.

FRL had chewed through $11 million of equity and debt by this stage. What value remained in the company was access to the mineral asset, knowledge about the manufacturing process, the pilot manufacturing plant, and intellectual property arising from the R&D the company had done. The main asset—access to the mineral—would revert to the State if the company was liquidated, and the creditors and shareholders would be left with, effectively, nothing.

Company management wanted to restructure the company so as to retain the key asset—access to the mineral—and not lose control of the company. A creditors' meeting was arranged in New Zealand for early 2014 to give effect to the proposed restructuring that would see a company with allegedly friendly links to the current management pay out the creditors and assume control of the company for less than $5 million, but leave the former shareholders with nothing. The former shareholders were alarmed at the prospect of this outcome and sought court intervention to prevent the change of control of the company and its assets.

In a case involving over 20 parties (including shareholders, creditors, and directors and managers of FRL)[14], it was said before the court that there was shareholder "apprehension" that to accept the $4.8 million offered by the Plaman Group "would frustrate their efforts to have a loan with FRCN [Pty Ltd] set aside on the basis of breaches of directors' duties by Simon Kember, Timothy Goodacre and the former executive Emma Weston."

In his decision in this case—one of many involving FRL at this stage—Justice Black of the Supreme Court of New South Wales in Australia said: "The proposed proceedings relate to the circumstances in which Featherston Resources issued convertible notes to another entity, FRCN Pty Ltd which is apparently associated with two of Featherston Resources directors [Simon Kember, Timothy Goodacre] and with a former executive, Ms Weston."

The shareholders, who looked likely to lose their investment and any future ownership of FRL, were concerned that the second management team had used an artifice to draw a line under the accumulated losses while allowing them to retain their control and at least some ownership of the enterprise.

By March 2015, the receivership ended, with key assets having been sold by the receiver to Plaman Group for a total of $4.8 million. The

---

[14] In the matter of Featherston Resources Limited (Receiver and Manager Appointed) (Administrators Appointed) [2014] NSWSC 12.

former 217 shareholders were left with nothing for their $11 million investment.

How Did This Happen?

The *first set of managers* were opportunistic entrepreneurs who could see the potential of the asset they had acquired, but they *lacked sufficient business focus and experience in marketing and sales*. Raising and spending some $8 million over 10 years made the enterprise look more like a hobby than a serious business.

There *appeared to be little accountability enforced on the directors and managers* because these people controlled most of the company's equity during much of this time, although their equity stake proportion declined over time as more shareholders entered the register. Nonetheless, most shareholders appeared happy to leave the decisions and direction of the company to the directors, which meant, in effect, that Alan was the guiding force.

However, even the patient shareholders became restless after 10 years and subscribing $8 million for little in terms of tangible results. FRL was a small company that operated using informal processes. A new managing director was selected on the basis of his experience and being known to one of the directors. *Nothing like an executive search was conducted.* And the new managing director soon asserted his authority by sacking the old board and appointing a trusted colleague to run the enterprise.

The new management team became committed to the success of the venture by making a financial investment in the company when most other shareholders had given up being willing to subscribe more capital.

The *second set of managers were strong in finance and marketing, but underestimated the challenges of manufacturing a new product using a new process*, and did not have the support infrastructure of a large business to which they were accustomed. Their optimism overcame their caution as they invested in a pilot plant to prove and refine the manufacturing process, but in doing so they faced problems they were not sufficiently equipped to resolve. And when FRL failed to meet its commitments to customers two growing seasons in a row, it became difficult to hold the threads together as cash ran out.

The options available to the managers became limited. To become insolvent meant that they risked losing the key asset on which the whole enterprise revolved: access to the diatomite mine. The managers chose the route of a voluntary administration in which they could control the disposal of the assets of the company and, if they could keep some semblance of control as a result, then all the better.

Without an injection of capital, which was unlikely to emerge, the company was not going to survive. Probably, the only way to allow the enterprise to move forward, with its key asset retained, in a controlled way, while preserving the other assets of value—the know-how and intellectual property—required a line to be drawn under the old structure. This was done, after numerous court contests, with the shareholders of FRL jettisoned as a result, and losing their entire stake.

There was no one reason why FRL failed; there were a number of reasons arising largely from weaknesses and blind spots—a little different in each case—within each group of managers.

At the start, Alan and David were dazzled by the immense wealth that the diatomite deposit promised to deliver, but failed to give the business clear direction. They did not bother with anything as sophisticated as a business plan to guide their actions. Meanwhile, as David convinced people to invest their money, there was little or no accountability imposed on the managers. They spent money using poorly formed ideas such as an investment in a new business in Australia that lacked goals, targets, and disciplined budgeting. When the management's inability to deliver shareholder value became apparent after 10 long years of trying, a new set of managers was appointed without the benefit of an executive search—changing horses because a change was needed, and hoping that the experience of the new managers would take the business in the right direction.

The new direction involved concentrating on using the mineral as a fertilizer—at least the company now had a clear goal. But the realization of this goal required a specific set of skills for mining the mineral, processing the mineral into a saleable product, marketing and sales, and distribution. A team would be needed, but a team with all the skills was not assembled.

Tim and Emma were good at marketing, but naive in manufacturing; they spent almost all their energy organizing distribution channels and market strategies without paying sufficient attention to the technical aspects of the business. Like the first management team, they did not develop a business plan. Without a clear plan, the business continued to be undercapitalized, while the managers continued making promises to the market they were unable to fulfill because of the manufacturing problems. Finally, and sadly, the moral integrity of the second team of managers was called into question by some of the long-term investors when the company was placed into voluntary administration and eventually liquidated.

What Could FRL Have Done to Avoid Its Demise?

Small companies are difficult to run. They often lack one or more vital skills.

FRL had even greater challenges because it had ambitions to become a vertically integrated company offering a new product, to conservative customers, against entrenched competitors. Stated in these terms, you would need a strong appetite for risk to invest in such a company.

Small companies can become the alter ego of a strong personality and then flourish or flounder based on the founder's abilities. Probably, neither the first management team nor the second had the spectrum of skills needed to make FRL a successful vertically integrated company. Combined, the two sets of management may have possessed all the skills needed for success, but combining them was never going to be an option given the history of the company and the predispositions of the actors involved.

There are many reasons FRL failed: lack of focus, lack of sufficient skills essential for the enterprise, undercapitalization. But start-up companies need something more, something intangible: they need an entrepreneur. They need someone who has, as their first goal, to make the company earn money and pay its way, with dedication and focus to solvency and growth. Once the enterprise has focus and goals and has achieved solvency, there may be a need to ease out the entrepreneur and insert professional management in order to regularize decision making, insert transparency, and remove idiosyncrasy. But FRL never had an entrepreneur with the skills needed to translate the potential the diatomite resource offered into a successful business.

Maybe, the second set of managers have learned enough from their experience with FRL to make the new enterprise work. They are smart enough to do it—it will probably work. But for the original FRL shareholders it is all too late.

Even the brightest and most amazing idea needs a business plan. FRL should have developed a clear plan to articulate what it wanted to achieve with its amazing diatomite resource. The managers should have concentrated on identifying the product with the most profit potential, and then employed the right set of skills to mine, manufacture, distribute, and market that product, and they should have capitalized the business appropriately. With a sound direction given, goals could have been set for the company to achieve its aim.

It may be argued that entrepreneurs, who thrive in situations of uncertainty, make investments without clear goals, and succeed just by assessing the potential within a business and proceeding with the intention of *creating* the opportunity. However, successful entrepreneurs will also make a decision about what is an affordable loss and limit their risk to that figure. The first team of managers at FRL played the entrepreneur game without playing it well. They did not have a clear goal, so they did not know what potential to aim for, and they did not know when to stop.

The second team of managers brought with them professional management credentials, but failed to create the team needed to realize their ambition. Figure 3.2 lists the fundamental elements essential for success. If a new business is deficient in enough of these elements, then it risks compromising its chance of success. Forensic investigation is one approach capable of verifying the presence of sufficient elements: the right people, sound processes, and finances capable of delivering on a business plan or, at the very least, the ability to identify an affordable loss.

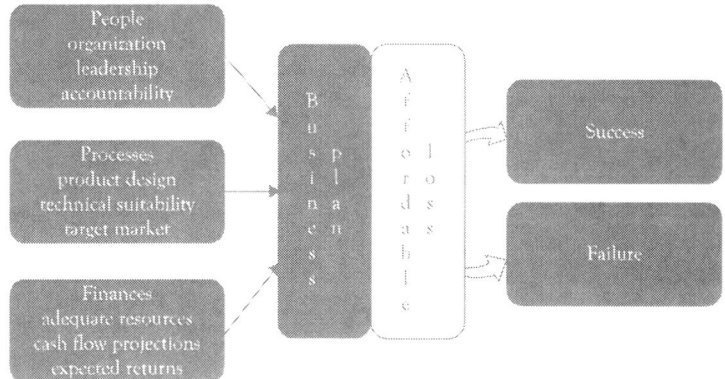

Figure 3.2 Fundamental elements for avoiding business failures

Source: Authors.

CHAPTER 4

Trust and Legal Considerations in Managerial Forensics

Zheng Liu and Alice de Jonge

The Role of Trust in Managerial Forensics

Trust is a broad yet complicated concept, which requires a deep understanding of human experience, psychological factors, and social forces. With globalization and increasing reliance on communication technologies such as the Internet, more and more companies are realizing the importance of trust in reputation building, the maintenance of collaborator relationships, value chain coordination, and risk management. In fact, trust is like a hidden asset, allowing companies to develop internal capabilities and external alliances to achieve competitive advantage.

Trust lies at the very heart of managerial forensics. Managerial forensics is the practice of gathering evidence to uncover corporate malaise or demise. Trust issues affect this process in several fundamental ways. First, and most obviously, healthy corporate relationships are based on trust. Declining levels of trust indicate malaise in the affected relationship. Yet, declining levels of trust also make it less likely that individuals involved in the operations of a corporation—those at the coalface of corporate operations—will respond to forensic inquiries truthfully. In other words, the absence of trust is an indicator of malaise in corporate relations, and also makes it harder to uncover the truth about that malaise. In addition, declining levels of trust make it increasingly unlikely that when evidence of malaise in intra- or intercorporate relations is uncovered, it will be believed, trusted, or acted upon. This is a second aspect of the relationship between trust and the process of managerial forensics. Trust is needed as a basis for allowing the process of evidence collection to go forth, and the results of the managerial forensics process need to be both trustworthy and trusted.

The issue of trust has attracted increasing levels of attention. Management research focuses on the role of trust in intra- and interfirm relationships, while studies from other areas identify the conceptual framework of trust from an economic, sociological, psychological, or cultural perspective (Bachmann 2011). Currently, the literature on trust can be divided into micro and macroviews (Bachmann 2011) as interpreted in Figure 4.1. Intra- and interorganizational trust, regarded as microlevel, can be developed through relationship management, whereas on a macrolevel, external environmental factors can change the value of trust. On a microlevel, trust is regarded as interactions between individual persons (McAllister 1995), or as a continuous learning and upgrading of goodwill between individuals over time (Platts 2005). Intra- and inter-organizational trust are also microlevel trust, as they are extensions of interpersonal trust.

Several researchers have identified the various elements of trustworthiness, such as "Openness, Promise keeping" (Anderson and Narus 1990); "Promise keeping, Relationship equity" (Ring and Van de Ven 1994); "Affect-based, Cognition-based" (McAllister 1995); "Ability, Benevolence, Integrity" (Mayer, Davis, and Schoorman 1995); "Reliability, Predictability, Fairness" (Zaheer, McEvily, and Perrone 1998);

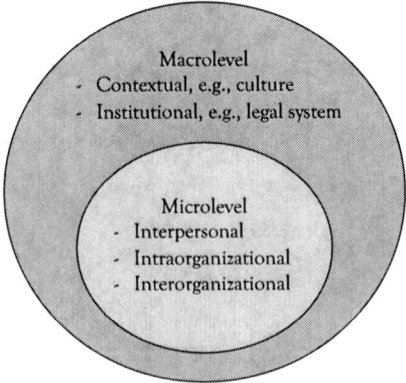

Figure 4.1 Micro- and macrounderstandings of trust

"Calculus-based, Knowledge-based, Identification-based" (Lewicki and Bunker 1996); "Predictability, Competence, Goodwill" (Sako 1998); "Reliability, Fairness, Non-exploitation" (Dyer and Chu 2003); and "Competence, Contract, Goodwill" (Platts 2005). Although different terms are used, these trustworthiness factors can be generally categorized as competence (belief in professionalism or specific skills or both.); reliability (confidence due to contract, fairness, and promise keeping); or goodwill (belief that trust can be developed into long-term benevolence and caring) aspects of trust.

In more practice-oriented studies, trust has been studied in the context of alliance and relationship (Mouzas, Henneberg, and Naude 2007; Welch 2006); supply chain partner selection and management (Laeequd din et al. 2012; Tejpal, Garg, and Sachdeva 2013); supply chain governance (Ghosh and Fedorowicz 2008); and e-commerce (Papadopoulou et al. 2001). On a macrolevel, trust is influenced by factors external to the intra- or interfirm relationship, including contextual factors such as culture and institutional factors such as the legal system. The interaction between trust and culture has been identified such as in Li's (2013) framework of the differences between intra- and intercultural trust. Institutional-based trust can be seen through legal system, law, formal contract, and agreement (Child and Mollering 2003). The role of institutions is suggested as an alternative mechanism for managing uncertainty in organizational relationships (Bachmann 2011); yet current studies mainly focus on how contract influences trust (Arrighetti, Bachmann, and Deakin 1997). There is a requirement for in-depth research on institutional-based trust, especially from policies and legal system perspectives.

Previous studies have examined the elements of trustworthiness in interpersonal and intrafirm relationships. However, there is still a need to understand trust at the microlevel in terms of how it is built and maintained, and how macrolevel legal-system factors can facilitate these processes of trust building and trust maintenance. This chapter therefore focuses on the processes through which trust is built, valued, and maintained, as well as how legal institutions foster such processes. In order to explore in detail the processes through which interfirm trust is formed, an initial investigation was conducted during the period 2007 to 2011. It aimed to analyze interfirm trust from two aspects: the activities or processes of trust building and the attitudes or values needed to build and maintain trust. The study adopted the approach of theory building with qualitative methods as a way to understand significant issues not previously explored in detail (Yin 1994). Nine in-depth case studies were conducted on Chinese and UK animation game industry, with a focus on identifying the detailed stages and activities to build trust in an interfirm collaborated project. Qualitative data were collected through face-to-face interviews, secondary case materials, and documents provided by the company. Details of activities to build interfirm trust were mapped and analyzed according to three stages: trust formation, trust development, and trust continuation (Figure 4.2).

In particular, trust formation refers to the interfirm relationship issues from the moment of selecting partners until the collaborative relationship is formally established. Trust development is correlated with the entire collaborative project, as competence is demonstrated and assessed, and goodwill develops. Trust continuation is the further building of trust after the project, with improvement and deepening of the relationship and further understanding of each partner's areas of competence. To further explore the values related to trust, a survey of personnel from 30 Chinese, U.S., and UK creative-industry companies with experience in building collaborative partnerships was also conducted through face-to-face interviews, telephone interviews, and e-mails. Interviewees were asked to list their highlighted trustworthy factors when selecting and collaborating

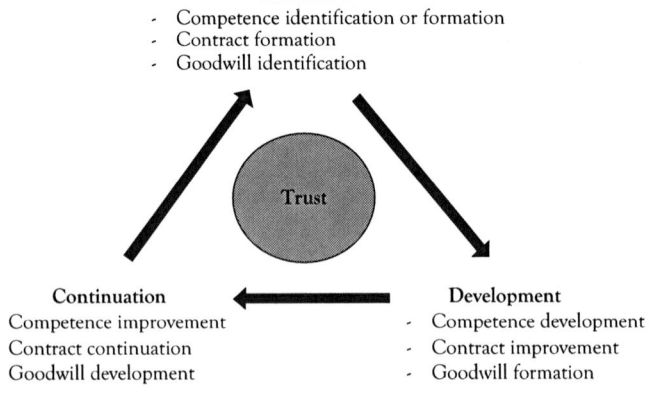

Figure 4.2 A general development process of trust

with business partners, and explain them in detail. From their responses, content analysis was made to categorize the values into dimensions of competence trust, reliability trust, and goodwill trust. The research approach used by Leong (2005) was implemented to generate an evaluation system of interpersonal trust by analyzing comments from individuals who participated in a collaborative team. Questions such as "what do you think is the most important feature of trust"; "what kind of competence do you think is important in order to trust your partners"; "what type of contract can secure the relationship"; and "what is most important to form a good relationship" were asked. To address the interfirm relationship trust, several adaptations were made to these questions during the study, as follows: (1) Instead of looking at individual behavior in a collaborative team, the study looked more into company factors or group-shared factors, such as the technology and managerial capability of a company as a whole, and the communication between companies. (2) Following interviewees' opinions, they were asked "do you think it is your own opinion, or a more general opinion of the company" and "do you think this happens on a particular person from your partner company, or can it represent a general attitude of the company" (3). At the data analysis stage, values and opinions were separated into personal factors, and nonpersonal factors reflecting organizational behavior and culture rather than individual beliefs. Although these two factors are related to each other, they were differentiated as much as possible. (4) Open questions were also used along with the structured questionnaire as a further way to obtain data on companies' values in relation to trust and interfirm relationships.

Interfirm Trust Development Process

It is seen that at the trust formation stage, reputation management, partner selection, assessment of competence, goodwill, and third-party introduction are key decision-making areas. During collaboration, learning, conflict management, and the development of technology, management, and the formation of contractual relationships are key activities. After a collaborative project, companies will focus on technology improvement, resource integration, and further relationship maintenance. A detailed summary of the activities related to three types of trust, namely competence, reliability, and goodwill trust, along with the three stages of trust formation, trust development, and trust continuation were developed (Table 4.1).

Table 4.1 Activities in trust-building-up process

	Trust formation	Trust development	Trust continuation
Competence	• Assess partners on their business plan and ideas, hardware and software technology • Assess partners on quality control, production management skills, IP protection systems • Standard way of selecting partners • Invite external experts to judge partner performance	• Improve technology and management skills including quality control during the project • Allocate tasks according to each partner's competencies • When problems occur, invite external experts • Third-party dispute resolution	• Use the existing technology and management experience for further collaboration • Develop common goal of coevolution over the longer term • Judgment of partner's competence, which will be used for further collaboration • Upgrade hardware and software, continuous improvement
Reliability	• Propose agreement based on role and responsibility, technology pieces, cost, and so on. • Standard contracts	• Contract can be renegotiated based on mutual benefits, and technology challenge • Renegotiate, and cut a big project into small pieces to avoid risk	• New contract building • Renegotiate toward more value-added project
Goodwill	• Assess partners on their friendship, passion, and persistent and innovative spirit • Select longer-term collaborators as partners	• Encourage partners with passion • System of personal communication both top management and project team, formal and informal	• Feelings of gratitude and confidence afterwards • Analyze similarities and differences, and improved understanding of partners.

• Find partners from very close social community • Judge partners on their relation connect with large companies, local government, and so on. • Judge partners based on owner-ship • Find local gov-ernment and big companies for help	• Encourage sharing information, mutual learning • Build teamwork spirit in certain mini-projects • Share resources and send people, facilities to partners • Keep in a common society or cluster with partners • Bring in long-term collaborator to join the project	• Mergers and acqui-sitions sometimes used as a further way of resource integration • Bring in long-term collaborator to join the project

Trust Values and Dimensions

The literature defines competence trust as trust in the specialized skills, management skills, and business judgment of another. In our survey, it was found that *business judgment* refers to the standard criteria and processes used to select partners by business practitioners. The specific competencies and skills sought were viewed as *technological skills*, such as being able to use advanced software. Capabilities including the potential for *idea innovation* were also valued highly as a factor of competence worthy of trust. This aspect of trust also embraced trust in a potential for competence upgrading through learning. In addition, *management styles*, including quality control systems and marketing capabilities, were also found to be significant in the formation of trust. Literature defines reliability as a type of trust based on contract and agreement, with an emphasis on fairness. In the survey, companies showed special preferences for *responsibility, fairness* in the furtherance of mutual interests, *standard-ization* of contract formation, and *negotiable possibility*. Studies suggest that UK contracts are more transparent and clear, whereas in China there are some hidden rules such as to develop personal relationship, social networking, and showing respect to people with strong background. This can make collaboration successful if the real competence and goodwill are there, but sometimes it prevents knowing the truth and thus bad management and a false competence can destroy the collaboration.

The previous literature has pointed out the *friendship* aspects of goodwill, such as integrity and benevolence, caring, are a long-term contribution to the relationship. In our study, goodwill was mentioned by many companies as "passion factors," "openness," and "willingness to share information with others." *Long-term relationships* were expected eventually after the collaborative project. However, the values related to this relationship varied. According to some companies, *social community* is a key element in sustaining the relationship. Also, some companies preferred to develop deeper trust and longer relationships with partners who are strongly connected with large companies or government. This factor can be defined as *relation connection*. Chinese companies use an approach favoring *who you know*, which can be compared to the Western approach of *what you know* in evaluating partners. Resources integration and *coevolution* were also considered key to the development of a deeper relationship.

Based on theory and our survey, a set of 16 dimensions of interfirm trust are developed as shown in Table 4.2. These dimensions can be further used as practical tools to evaluate the actual trusting relationship, as scores can be given against each of the 16 dimensions. For example, there can be three scores: *L* means that the companies are very low or

Table 4.2 Types of interorganizational trust and dimensions

Trust	Details of dimensions
Competence	1. Technological capability—whether the partner has certain artistic skills, technology, or products 2. Management styles—whether the partner has an efficient way of running business, well-structured organizational culture 3. Idea innovation—whether the partner has innovative idea, design, or overall concept 4. Business judgment—whether the focal firm has a systemic way of selecting partners; partner's management skills
Reliability	5. Standard terms—whether the contract follows a standard format, companies have experience of creating contract 6. Negotiable contract—whether the contract is formal, informal, strict, or flexible 7. Role specification—whether the contract specifies each role, technology, and responsibility 8. Agreement fairness—whether the agreement shows equal fair status of participating companies

Goodwill	9. Friendly attitude—whether the partners are tolerant, friendly, or caring about others
	10. Passion factors—whether partners have strong intention of success, passion, persistence, confidence, or feeling
	11. Open-mindedness—whether the company is open to new things, and listen to others
	12. Willingness to share—whether partners work for a common goal, feeling part of a group, teamwork, share information
	13. Long-term vision—whether the company develops partnership for the long term, shares common vision
	14. Coevolution—whether the companies upgrade capability by learning mutually or from external resources, or by mergers and acquisitions, joint venture
	15. Social community—whether the companies stay in a community after the collaboration
	16. Relational background—whether the partner is connected to specific important organizations, for example, government, community

not strong on this dimension, or activities can be observed against it; *H* means that companies show extremely high preferences for these values or several activities can be observed to support this; and *M* means medium, or not mentioned.

Trust and Contract

In addition to the microlevel view of trust, trust interacts largely with macroinstitutional and cultural factors, which include legal issues. In addition to relying on trust factors that are integral to the parties, the quality and maintenance of business relationship also depend on institutional factors such as the quality of formal contractual arrangements and the system that supports and enforces those arrangements. The quality of the contract itself and that of the institutions that support and enforce that contract are both important in determining the nature and quality of the business relationship. Contract is defined here as one or more written agreements between two or more parties that the parties perceive as (or intend to be) binding (Lyons and Metha 1997). The institutions that support and enforce contractual arrangements include the laws governing the contractual relationship, the legal profession, accountants and auditors, dispute-resolution bodies such as arbitrati⟨

tribunals, and, usually as a last resort, the courts of law. This chapter is primarily concerned with the formation, development, and continuation of interfirm trust. This section discusses the various ways in which the formation, presence, revision, improvement, and continuation of the contract both affect and interact with interfirm trust. First, when initiating business relations with a new partner, firms must determine the level of detail (complexity) of the contract that forms part of the new relationship. The choice of a simple or complex contract, in turn, can be critical in fostering, or damaging, the development of trust between the parties (Praxmarer-Carus 2014). The nature of the contractual relationship proposed—short or long term, umbrella agreement, or specific contract—also influences, and interacts with, the process of forming trust. Second, during the trust-development stage of the contract, both trust and formal contract are useful tools in fostering exchange and enhancement of intellectual property, as well as in managing knowledge leakage. In addition, both trust and contract can operate to foster innovation performance. In different cultures, trust and contract may each be more or less important in determining the quality of the relationship that develops. Third, both trust and contract are important in the continuation phase of the business relationship. A business contract oriented to the long term rather than designed simply to maximize short-term gain provides a framework for the stable continuation of the relationship. But even the best of written agreements cannot overcome the need for a solid foundation of trust as the basis of an ongoing relationship.

Contract and Trust Formation

Ferrin, Bligh, and Kohles (2008, 175) emphasize that the development of a trusting relationship is "fundamentally affected by partners' initial moves." In some cultures, trust building alone is viewed as the most appropriate initial stage, even before the beginnings of a formal business relationship are established. Nair and Papadopoulou, for example, wrote in 1998 that "Trusting relationships must precede any Chinese business transaction" (Nair and Stafford 1998, 140). Much has changed in China in the nearly two decades since that time, and much more use is now made of contracts in Chinese business partnerships. It remains true, however, that more attention needs to be paid to the personal aspects of trust building when forming business partnerships in a Chinese cultural context than is typically the case in Western business relationships (Wang, Siu, and Barnes 2008; Wang, Shi, and Barnes 2015). The proposal of a complex contract, compared to a simple contract, is an important initial-move choice affecting the development of a trusting relationship. By proposing a more complex or less complex contract, a firm may influence the initial trust of its partner and thus influence how the relationship develops (Praxmarer-Carus 2014). Some authors theorize that complex contracts are more likely to foster a partner's trust than are simple contracts (e.g., Bennett and Robson 2004; Sitkin and Roth 1993), whereas other researchers argue that complex contracts may damage trust between partners (e.g., Lewicki, McAllister, and Bies 1998; Mahoney, Huff, and Huff 1994; Poppo and Zenger 2002). The empirical literature demonstrates both positive and negative effects arising from the use of a complex contract. Praxmarer-Carus (2014) uses interviews and empirical experiment to identify a number of mediating variables able to explain these positive and negative effects. She finds that:

> Individuals may perceive a complex contract as a safeguard, a signal of a partner's commitment, and a signal of a partner's competence. Each of these perceptions affects individuals' trust positively. Alternatively, individuals may perceive a complex contract as a signal of a partner's distrust and as a restriction. Both perceptions affect individuals' trust negatively. (Praxmarer-Carus 2014, 1427)

Contract and Trust Development

While firms are more likely to use more complex contracts for larger transactions and more complex products, contract complexity still varies when size and product complexities are comparable, and when similar levels of risk are involved (Anderson and Dekker 2005; Mooi and Ghosh 2010; Klein Woolthuis, Hillebrand, and Nooteboom 2005). More closely related to the choice of complex or simple contract are two important factors: culture and legal infrastructure (Hofstede, Hofstede, and Minkov

2010; Wang and Chung 2013; Wang et al. 2005; Yen, Yu, and Barnes 2007). In China, for example, business relationships are formed in a high-context culture in which personal relationships and oral promises are often considered to be more important than legal documentation (Calantone and Zhao 2000; Wang and Chung 2013; Wang, Shi, and Barnes 2015). Cultural norms foster reciprocal obligation (renqing), while a generally underdeveloped commercial legal system means that less reliance is generally placed on written contract. Trust and contract interact with and impact on each other and on the business relationship throughout the term of that relationship. Wang, Yeung, and Zhang (2011) examined the individual and interactive effects of contracts and trust on firms' innovative performance in the context of firm partnering relationships. Based on a survey of Chinese manufacturing firms, they found that there is a positive relationship between trust and firms' innovation performance. They also found an inverted U-shaped relationship between the use of contracts and firms' innovation performance, and that contracts and trust are substitutes. Moreover, they also found that environmental uncertainty enhances the effects of trust, but does not influence the impact of contracts on innovation performance during the development phase of a business relationship. Jiang et al. (2013) also looked at the effect of trust and formal contract during the business relationship, with a focus on managing knowledge leakage. When firms exchange knowledge in interfirm alliances, that knowledge, especially high-value rent-generating knowledge, is susceptible to expropriation hazards. Interfirm knowledge-exchange agreements, such as those that often form part of strategic alliances, have become popular as vehicles for firms to gain access to external technologies and intellectual resources. However, such alliances also subject firms to the risk of knowledge leakage. Based on a survey of 205 partnering firms in China, Jiang et al. (2013) found that goodwill trust has a U-shaped relationship with knowledge leakage, whereas competence trust has a negative impact. Moreover, goodwill trust and competence trust interact differently with formal contracts on knowledge leakage. The interaction between goodwill trust and formal contracts has a statistically significant positive effect on knowledge leakage, while the interaction between competence trust and formal contracts has a significant negative effect. Formal contract alone was found to be not statistically significant in its effect on knowledge leakage. The balance between trust and contract needs to be carefully managed. For those companies seeking to avoid over-reliance on the legal enforceability of strict contractual terms, more time and resources need to be devoted to fostering a trusting relationship. Thus, in a study of project alliances in Australia, Rowlinson et al. (2006) found that the use of *no litigation* contracts required significant investment in relationship management infrastructure to succeed. Managing and fostering trust between partners were important not just for the success of immediate project aims, but also to create opportunities and willingness for further alignment and future alliances (Rowlinson et al. 2006, 79).

Contract and Trust Continuation

The literature demonstrates that firms use different forms of business agreements with their partners. Most often, especially in new relationships, firms use contracts that focus on immediate contractual obligations. In established relationships, however, many firms use umbrella agreements that provide a *set of rules* (Mouzas and Ford 2012, 155) designed to help the partners reach mutual consent in future interactions during the continuing relationship, but without specifying immediate contractual obligations (Mouzas and Ford 2006). These longer term cooperation agreements can form the basis for stable continuation of a trusting business relationship. Del Campo, Pardo, and Perlines (2014) find that trust is a key factor in cooperation agreements, and that partner reputation and prior partnering experience are both crucial to that trust. As in the formation and development stages of the business relationship, therefore, it seems that it is the interaction between trust and contract that serves to enhance or harm the continuation of the business relationship. Rather than focusing on trust or contract alone, therefore, it seems that businesses need to understand the interaction between these variables in managing their partnering arrangements.

Conclusion

This chapter has discussed trust issues in management forensics. It provided an overview of the literature of trust from micro- and macrolevels. On a microlevel, trust is mainly explored as interpersonal, intraorganizational, and interorganizational trust, whereas on a macrolevel, cultural and institutional factors can influence trust values and trust formation and maintenance. It was found that while many researchers have explored the types of trust, there is still a need to understand: (1) the building-up process and stages of trust; (2) values and dimensions of trust in business; (3) the institution-based trust such as legal issues. Therefore, the chapter has highlighted these three issues.

From a recently conducted study, it was seen that trust can be explained as a process of three stages, namely trust formation, development, and continuation, each with key decision-making areas. As for the three types of trust—competence, reliability, and goodwill, the detailed meanings in business practice can be found through interviewees' interpretation. The chapter then provides 16 key dimensions of interorganizational trust, based on literature and practical studies. Along with the three stages of building up trust, the important roles of legal issues and contract are also addressed.

In terms of managerial implementation, the table frameworks (Tables 4.1 and 4.2) provide a comprehensive evaluation system for understanding trust in a business relationship. Companies can use these frameworks as guides when assessing potential partners, identifying capability, building up contracts, and assessing goodwill factors. Tables 4.1 and 4.2 also provide guidance for companies with the intention of upgrading trusting relationship in collaboration. They can use it to set detailed goals, and measure the actual performance in each dimension. Business practitioners can also get comprehensive understanding on the linkage between trust relationship and contract development.

This chapter presents original findings of interfirm trust and trust dimensions based on qualitative methods with case studies and interviews. The sample size was small, and further research is needed. Furthermore, the findings show different patterns and prioritized values of trust in Chinese and UK companies. As business nowadays is increasingly more interactive internationally, further exploration of how trust varies in different national culture contexts is needed. In this chapter, we have analyzed the elements of trust that are most essential at the formation, development, and continuation stages of the trust relationship. In so doing, it is hoped that we have provided insight into how the tools of forensic management can be applied to investigating and understanding the health of trust levels in interfirm relations. It is important that managers are able to monitor and maintain healthy levels of trust in all company relations, and it is the task of managerial forensics to provide them with the tools and insights for doing so.

References

Anderson, S.W., and H.C. Dekker. 2005. "Management Control for Marketing Transactions: The Relation Between Transaction Characteristics, Incomplete Contract Design, and Subsequent Performance." *Management Science* 51, no. 12, pp. 1734–52.

Anderson, J.C., and J.A. Narus. 1990. "A Model of Distributor Firm and Manufacturer Firm Working Partnerships." *Journal of Marketing* 54, no. 1, pp. 42–58.

Arrighetti, A., R. Bachmann, and S. Deakin. 1997. "Contract Law, Social Norms and Inter-Firm Cooperation." *Cambridge Journal of Economics* 21, no. 2, pp. 171–95.

Bachmann, R. 2011. "At the Crossroad: Future Directions in Trust Research." *Journal of Trust Research* 1, no. 2, pp. 203–13.

Bennett, R.J., and P.J. Robson. 2004. "The Role of Trust and Contract in the Supply of Business Advice." *Cambridge Journal of Economics* 28, no. 4, pp. 471–88.

Calantone, R.J., and Y.S. Zhao. 2000. "Joint Ventures in China: A Comparative Study of Japanese, Korean and US Partners." *Journal of International Marketing* 9, no. 1, pp. 1–23.

Child, J., and G. Mollering. 2003. "Contextual Confidence and Active Trust Development in the Chinese Business Environment." *Organization Science* 14, no. 1, pp. 69–80.

del Campo, J.D.S., I.P.G. Pardo, and F.H. Perlines. 2014. "Influence Factors of Trust Building in Cooperation Agreements." *Journal of Business Research* 67, no. 5, pp. 710–14.

Dyer, J.H., and W. Chu. 2003. "The Role of Trustworthiness in Reducing Transaction Costs and Improving Performance: Empirical Evidence from the United States, Japan, and Korea." *Organization Science* 14, no. 1, pp. 57–68.

Ferrin, D.L., M.C. Bligh, and J.C. Kohles. 2008. "It Takes Two to Tango: An Interdependence Analysis of the Spiraling of Perceived Trustworthiness and Cooperation in Interpersonal and Intergroup Relationships." *Organizational Behavior and Human Decision Processes* 107, no. 2, pp. 161–78.

Ghosh, A., and J. Fedorowicz. 2008. "The Role of Trust in Supply Chain Governance." *Business Process Management Journal* 14, no. 4, pp. 453–70.

Hofstede, G., G.J. Hofstede, and M. Minkov. 2010. *Cultures and Organizations: Software of the Mind: Intercultural Cooperation and its Importance for Survival.* New York: McGraw Hill.

Jiang, X., M. Li, S. Gao, Y. Bao, and F. Jiang. 2013. "Managing Knowledge Leakage in Strategic Alliances: The Effects of Trust and Formal Contracts." *Industrial Marketing Management* 42, no. 6, pp. 983–91.

Klein Woolthuis, R., B. Hillebrand, and B. Nooteboom. 2005. "Trust, Contract and Relationship Development." *Organization Studies* 26, no. 6, pp. 813–40.

Laeequddin, M., B.S. Sahay, V. Sahay, and K.A. Waheed. 2012. "Trust Building in Supply Chain Partners Relationship: An Integrated Conceptual Model." *Journal of Management Development* 31, no. 6, pp. 550–64.

Leong. 2005. *Biopharmaceutical Development Networks: Architecture, Dynamic Processes and Evolution* [dissertation]. University of Cambridge.

Lewicki, R.J., and B.B. Bunker. 1996. "Developing and Maintaining Trust in Work Relationships." In *Trust in Organizations: Frontiers of Theory and Research*, eds. R. Kramer and T.R. Tyler, 114–39. Thousand Oaks, CA: Russell Sage Foundation.

Lewicki, R.J., D.J. McAllister, and R.J. Bies. 1998. "Trust and Distrust: New Relationships and Realities." *Academy of Management Review* 23, no. 3, 438–512.

Li, P. 2013. "Intercultural Trust and Trust-Building Process: The Contexts and Strategies of Adaptive Learning in Acculturation." In *Handbook of Advances in Trust Research*, eds. R. Bachmann and Z. Zaheer. Cheltenham, UK: Edward Elgar Publishing.

Lyons, B., and J. Metha. 1997. "Contracts, Opportunism and Trust: Self-Interest and Social Orientation." *Cambridge Journal of Economics* 21, no. 2, 239–57.

Mahoney, J.T., A.S. Huff, and J.O. Huff. 1994. "Toward a New Social Contract Theory in Organizational Science." *Journal of Management Inquiry* 3, no. 2, pp. 153–68.

Mayer, R.C., J.H. Davis, and F.F. Schoorman. 1995. "An Integrative Model of Organizational Trust." *Academy of Management Review* 20, no. 3, 709–34.

McAllister, D.J. 1995. "Affect- and Cognition-Based Trust as Foundations for Interpersonal Cooperation in Organizations." *Academy of Management Journal* 38, no. 1, pp. 24–59.

Mooi, E.A., and M. Ghosh. 2010. "Contract Specificity and Its Performance Implications." *Journal of Marketing* 74, no. 2, pp. 105–20.

Mouzas, S., and D. Ford. 2006. "Managing Relationships in Showery Weather: The Role of Umbrella Agreements." *Journal of Business Research* 59, no. 12, pp. 1248–56.

Mouzas, S., and D. Ford. 2012. "Leveraging Knowledge-Based Resources: The Role of Contracts." *Journal of Business Research* 65, no. 2, pp. 153–61.

Mouzas, S., S. Henneberg, and P. Naude. 2007. "Trust and Reliance in Business Relationships." *European Journal of Marketing* 41, no. 9–10, pp. 1016–32.

Nair, A.S., and E.R. Stafford. 1998. "Strategic Alliances in China: Negotiating the Barriers." *Long Range Planning* 31, no. 1, pp. 139–46.

Papadopoulou, P., A. Andreou, P. Kanellis, and D. Martakos. 2001. "Trust and Relationship Building in Electronic Commerce." *Internet Research: Electronic Networking Applications and Policy* 11, no. 4, pp. 322–32.

Platts, M.J. 2005. "The Roots of Civic Trust." Paper Presented at Management Philosophers Conference, Oxford, UK.

Poppo, L., and T. Zenger. 2002. "Do Formal Contracts and Relational Governance Function as Substitutes or Complements?" *Strategic Management Journal* 23, no. 8, pp. 707–22.

Praxmarer-Carus, S. 2014. "Why the Proposal of a Complex Contract May Harm or Foster a Partner's Trust." *Journal of Business Research* 67, no. 7, pp. 1421–29.

Ring, P., and A. Van de Ven. 1994. "Developmental Processes of Cooperative Interorganizational Relationships." *Academy of Management Review* 19, no. 1, pp. 90–118.

Rowlinson S., F.Y.K. Cheung, R. Simons, and A. Rafferty. 2006. "Alliancing in Australia—No Litigation Contracts: A Tautology?" *Journal of Professional Issues in Engineering Education and Practice* 132, no. 1, pp. 77–81.

Sako, M. 1998. "Does Trust Improve Business Performance?" In *Trust Within and Between Organizations: Conceptual Issues and Empirical Applications*, eds. C. Lane and R. Bachmann, 88–117. Oxford, UK: Oxford University Press.

Sitkin, S.B., and N.L. Roth. 1993. "Explaining the Limited Effectiveness of Legalistic Remedies for Trust/ Distrust." *Organization Science* 4, no. 3, pp. 367–92.

Tejpal, G., R.K. Garg, and A. Sachdeva. 2013. "Trust Among Supply Chain Partners: A Review." *Measuring Business Excellence* 17, no. 1, pp. 51–71.

Wang, C.L., and H. Chung. 2013. "The Moderating Role of Managerial Ties in Market Orientation and Innovation: An Asian perspective." *Journal of Business Research* 66, no. 12, pp. 2431–37.

Wang, C.L., X. Lin, A.K.K. Chan, and Y. Shi. 2005. "Conflict Handling Styles in International Joint Ventures: A Cross-Cultural and Cross-National Comparison." *Management International Review* 45, no. 1, pp. 3–21.

Wang, C.L., Y. Shi, and B.R. Barnes. 2015. "The Role of Satisfaction, Trust and Contractual Obligation on Long-Term Orientation." *Journal of Business Research* 68, no. 3, 473–479.

Wang, C.L., N.Y.M. Siu, and B.R. Barnes. 2008. "The Significance of Trust and Renqing in the Long-Term Orientation of Chinese Business-to-Business Relationships." *Industrial Marketing Management* 37, no. 7, pp. 819–24.

Wang, L., J.H.Y. Yeung, and M. Zhang. 2011. "The Impact of Trust and Contract on Innovation Performance: The Moderating Role of Environmental Uncertainty." *International Journal of Production Economics* 134, no. 1, 114–22.

Welch, M. 2006. "Rethinking Relationship Management: Exploring the Dimension of Trust." *Journal of Communication Management* 10, no. 2, pp. 138–55.

Yen, D.A., Q. Yu, and B.R. Barnes. 2007. "Focusing on Relationship Dimensions to Improve the Quality of Chinese-Western Business-to-Business Exchanges." *Total Quality Management & Business Excellence* 18, no. 8, pp. 889–99.

Yin, R.K. 1994. *Case Study Research Design and Methods*. Beverly Hills, CA: Sage Publications.

Zaheer, A., B. McEvily, and V. Perrone. 1998. "Does Trust Matter? Exploring the Effects of Interorganizational and Interpersonal Trust on Performance." *Organization Science* 9, no. 2, pp. 141–59.

PART II

Tools in Managerial Forensics

CHAPTER 5

Internal Governance Structures and Corporate Behavior

Alice de Jonge

Introduction

Different corporate forms have different strengths and weaknesses. It is the job of managerial forensics to identify potential weaknesses in particular, thereby enabling corporations to guard against them and to diagnose them when they lead to corporate malaise. This chapter examines the particular governance challenges and potential governance weaknesses faced by family firms, state-owned enterprises (SOEs) in particular. Also examined are the governance challenges presented by the changing nature of corporate control over *widely held* transnational corporations. Finally, Part 4 of this chapter examines how new hybrid forms of the corporation, in particular the benefit corporation, are allowing for new understandings to be reached between owners and managers regarding the purposes that the corporation exists to serve. It is argued that the experience of these firms may provide many valuable lessons for scholars of managerial forensics interested in the assessment of corporate performance against social and economic, as well as purely economic criteria.

The health of an organization derives largely from the internal relationships that keep it working. When these internal relationships break down or become corrupted, governance of the corporation is at risk. The primary internal governance relationship in any corporation is the relationship between the owners (shareholders) of the company on the one hand, and management on the other. This relationship is typically mediated through the board of directors, where, at least in theory, the interests and points of view of both management and shareholders (and sometimes other stakeholders) are represented (Davies 2000). The relationship between owners and managers typically differs according to the nature of the firm's ownership structure. In family-owned firms, for example, the relationship between owners and managers is so close that they are often identical, which means that corporate decision making looks very different than it does in, say, the SOE, or the widely held corporation. Understanding the different ways in which relationships of power and influence operate in each of these different corporate owner- ship types furthers the project of managerial forensics by identifying the different pressure points of vulnerability to which each ownership type is most prone. This then aids a more accurate diagnosis of corporate malaise when evidence of strain emerges.

Internal governance relationships also differ according to the nature of the understanding reached between owners and managers regarding the purpose of the corporation. Traditionally, the agreed purpose of the corpo- ration is to carry out business in order to maximize profits for shareholders. But in a world where the single-minded pursuit of profits has been discov- ered to contain within it the seeds of its own destruction in the form of environmental and human rights harms, an increasing number of corpo- rations are seeking to redefine their business purpose. Benefit corporations are now a legal possibility in a growing number of jurisdictions, includ- ing, by the end of 2014, those in 28 American states (Clark and Vranka 2013; http://benefitcorp.net/policymakers/state-by-state-status). Benefit corporations differ from *ordinary* business corporations in at least three important respects: (1) They have a corporate purpose to create a material positive impact on society and the environment; (2) the duties of directors are expanded to require consideration of interests in addition to the finan- cial interests of shareholders; and (3) they are required to report each year on overall social and environmental performance using a comprehensive, credible, independent, and transparent third-party standard (Clark and Babson 2012; Reiser 2011). This altered direction of the corporate rai- son d'etre impacts on the nature of internal governance relationships in a number of important ways, discussed in Part 4. The chapter concludes by arguing that the experience of benefit corporations provides lessons that are useful for all firms seeking to incorporate the principles of the Global Compact, the UN Guiding Principles for Business and Human Rights, the OECD Guidelines for Multinational Enterprises, and other corporate codes on social responsibility that now exist as normative instruments for all business entities.

Family Firms

There are many different definitions of what constitutes a family firm, many of which seek to identify at what level of family-member share ownership a firm can be said to be *family controlled*. The best definitions, however, recognize that higher ownership concentration per se does not capture the ways in which the role of the owner-managers of these firms as members of an (extended) family interacts with their role as business decision makers. Chua, Chrisman, and Sharma (1999, 21, 25, Table 1) enumerate a wide range of definitions of family firms in the literature, ranging from simple measures of ownership of shares to identification of people who form an *emotional kinship group*. Formally, they define a family firm as:

> … a business governed and/or managed with the intention to shape and pursue the vision of the business held by a dominant coalition controlled by members of the same family or a small number of families in a manner that is potentially sustainable across generations of the family or families.

By drawing attention to the political nature of ownership and control (dependent on family relationships and the cohesion of the dominant coalition), this definition also draws attention to the potential points of vulnerability within the family firm that may lead to its demise (Brenes, Madrigal, and Requena 2011).

First, research shows that long-term orientation can be a dimension of family business culture that can contribute to distinct advantages in family firms (Kim and Gao 2013; Lumpkin and Brigham 2011; Zahra, Hayton, and Salvato 2004), and that this dimension increases with the passage of time. Lumpkin and Brigham (2011, 1162) argue that long-term orientation is stronger in older, established and multigenerational family firms. Yet, this same feature can also create the risk that the family business will become hampered by insistence on continuing with a low performing line of business. It can also mean that founders become obsessive about control, which in turn can create conflict within family coalitions.

Because of the personal and emotional nature of family relationships, the potential for conflict (between family members) within family firms is higher, and this is perhaps the primary source of vulnerability in all family firms (Kidwell, Kellermanns, and Eddleton 2012). The personal nature of family relations can also be a strength, however, because it creates possibilities for reciprocal altruism that can serve to insure component parts of the family and its ventures from shocks without the need for formal contracting (Bingham et al. 2011; Chrisman, Chua, and Steier 2011; KPMG and FBA 2013). This enables families to collectively take more risk, form organizations, or perform both with low transaction costs (Sanchez-Bueno and Usero 2014). In other words, some family firms may be able to withstand the storms of economic crisis, but collapse under the pressures of dispute among family members.

Finally, there is the fact that potential employees, investors, and creditors of the family firm may think that any investment they make in the firm will be diverted to the primary purpose of serving family interests, rather than longer term business goals and objectives (Chen, Li, and Su 2005; Chua et al. 2012; Singla, Veliyath, and George 2014). The challenge for family firms is to proactively engage with nonfamily stakeholders in order to create an enabling environment for the future of the enterprise as a business, rather than as a family project (Chua et al. 2012; Singla, Veliyath, and George 2014). Meeting this challenge may, in turn, require the ability to strategically relinquish a certain degree of family ownership and control (Burkart, Panunzi, and Shleifer 2003; Chen, Li, and Su 2005; Sanchez-Bueno and Usero 2014).

State-Owned Enterprises

SOEs still represent a significant share of activity in a number of OECD economies and an even larger share in most emerging market economies (OECD 2006, 2013). In particular, in emerging market economies such as China, India, Brazil, and South Africa, it is not uncommon for the state, as a result of partial privatization or direct intervention, to hold significant equity stakes in large publicly listed companies (OECD 2006, 2013). SOEs face specific difficulties in terms of governance. The job of managerial forensics has been to identify these difficulties and to facilitate the design of appropriate mechanisms for addressing them.

The governance difficulties faced by SOEs derive from a number of characteristics that may be more or less acute depending on a country's administrative traditions, its recent history of state sector reforms, and the degree of liberalization of the economies concerned. In general, however, SOEs tend to share in common a number of characteristics that may lead to governance vulnerabilities.

First, when ownership entities (typically state ministries) exercise rights to appoint one or more directors to the SOE board, this can lead to political interference in the nomination process, and detracts from the need for merit-based board appointments. When posts are filled on the basis of political connections regardless of merit, this can lead to failures of independent oversight of management, allowing less scrupulous officers to take advantage of positional power for personal gain (Huang and Snell 2003; Nguyen and van Dijk, 2012; OECD 2006). In some countries, including New Zealand, Sweden, and Denmark, efforts have been made to establish mechanisms for more independent nomination of potential board appointees (OECD 2004, 2006).

Second, in many instances, SOE boards are not granted the full responsibility generally accorded to boards of joint stock companies in terms of strategic guidance, monitoring of management, and disclosure. The roles and responsibilities of SOE boards are often encroached from both ends, as they are bypassed both by senior management and the ownership entity. When senior management can take a decision directly to the relevant government minister for approval, there can seem little need for the board to become involved. Particularly, when senior management has been appointed by, and even from, a sector ministry with both knowledge of the sector and its own sectoral priorities, both parties have a strong interest in developing a strong and direct link that by-passes the board (OECD 2006).

Some board functions, such as reporting and disclosure, may be duplicated by state control organs, so that boards feel deprived even of their role regarding the completeness, exactness, and fairness of reporting.

Moreover, accounting and disclosure standards for SOEs often do not meet private sector standards but rather become oriented toward public expenditure control, which may at the same time be more burdensome and not fulfill the requirements of timeliness and materiality central to private sector disclosure practices.

Governance difficulties may also derive from the fact that in the case of SOEs there may be no clear ownership entity, but competing owners and stakeholders with widely different policies and other objectives. The ownership entities of the SOE may be political entities (typically ministries) with conflicting objectives. Defining the best interests of the corporation in such a context becomes complicated by the interrelationship among political, economic, and social objectives that may all need to be, or should not be but are, taken into account.

There is also the fact that SOEs are often effectively protected from two important disciplines that are essential in policing management behavior in public corporations: the threat of takeover and the threat of bankruptcy. The consequence has often been demands on the budget for investment and expansion programs and an overreliance on state-backed financing to rescue the firm from the results of lazy or inefficient management decision making (de Jonge 2008; OECD 2006; Shi 2012). Moreover, SOEs often operate in sectors where they are protected from competition, and where state policy objectives can detract from a focus on efficiency and profit maximization. This is not necessarily a bad thing. Many worthy state policy objectives, such as the desire to provide energy and other essential services to remote rural areas, necessarily require state subsidization (OECD 2013). However, the risk with partially privatized SOEs especially is that these policy objectives are not publicly recognized and reported, so that nongovernment stakeholders are effectively excluded from decision making about the proper balance of public purpose versus profit-making effort and investment.

Globalization and Changing Ownership Patterns

It is evident that the structure of firm ownership can affect internal governance relationships in a number of different ways. Moreover, where national systems are dominated by a particular type of ownership structure, the national company law will be more concerned to address the particular agency problems that arise from that particular structure (Armour, Hansmann, and Kraakman 2009; Davies 2000). For example, in countries where company ownership is dispersed (such as in the United States between the 1930s and 1980s), agency costs are generated by the fact that while control of the company lies in the hands of management, no single shareholder has a strong enough incentive to spend resources ensuring management acts in the interests of all shareholders. This increases the risk of mismanagement, thereby generating agency costs (Armour and Gordon 2011). Greater dispersion and mobility of shareholders can also tend to fixate management on short-term financial performance, a fixation that can maintain competitiveness by facilitating rapid adjustments to changing market conditions, but which can also detract from the need for longer-term planning (Armour and Gordon 2011).

Regulators in countries with dispersed company ownership patterns thus tend to focus on addressing agency problems arising out of the relationship between management and the shareholders as a class. They do this in a number of ways. First, by imposing transparency disciplines through disclosure rules, regulations can serve to empower shareholders with knowledge about company performance so that they can more effectively utilize their powers to appoint and remove directors (Armour and Gordon 2011; Armour, Hansmann, and Kraakman 2009). Second, most company law systems now require a minimum number of *independent* directors (variously defined) who have no financial relationship with the company or its manager and are entrusted with the job of impartially scrutinizing management behavior. Third, there are market disciplines imposed through merger and acquisition rules. Market discipline rules are aimed at ensuring that companies that fail to operate competitively over the longer term can be absorbed (taken over) by companies with more visionary leadership (Armour and Gordon 2011; Armour, Hansmann, and Kraakman 2009).

In countries dominated by concentrated firm ownership patterns (including family or state-controlled enterprises or both), regulators are more concerned to address the agency problems arising out of the relationship between majority shareholders and minority shareholders. These agency problems arise because, while the controlling shareholder and (if present) other block shareholders have an incentive monitor management, they also have an incentive to extract benefits from the firm at the expense of minority shareholders. Thus, while the risks of mismanagement decrease, the risks that controlling shareholders will engage in one or more forms of tunneling (transferring assets and profits out of firms for the benefit of those who control the firm) increases (Armour and Gordon 2011). There is also the risk that controlling shareholders may overmonitor management, which may discourage managers from showing initiative (Armour, Hansmann, and Kraakman 2009; Davies 2000).

In practice, the behavior of controlling shareholders will largely depend on their type (e.g., family, state, institutional investors). Regulators in countries with a large proportion of companies controlled by one or more dominant shareholders have developed a range of strategies for addressing agency problems arising from the relationship between majority and minority shareholders (Cheffins 2000; Davies 2000). For example, minorities can be protected by the rules relating to voting at shareholder meetings. Voting rules such as proportional representation or cumulative voting in election of directors can enable minority shareholders to successfully elect one or more representatives to the board (Armour and Gordon 2011; Cheffins 2000). It is also possible for voting rules to impose voting caps on dominant shareholders. This involves a trade-off, however, in that it reduces the ability of dominant shareholders to hold management to account. In countries such as China where the state dominates as shareholder, requiring that all company boards have a minimum number of independent directors has been seen as a way of protecting the interests of minority shareholders (de Jonge 2008; Shi 2012).

Disclosure rules can be used to inform minority shareholders about the management and possible misappropriation of company assets. However, transparency rules, to have *teeth*, need to be combined with the ability for minority shareholders to take legal action in the event that assets are siphoned off to the interests of a controlling shareholder. This can be difficult: first, because allowing individual or minority shareholders to challenge company decisions subverts the collective nature of the company (Armour and Gordon 2011; Cheffins 2000;) and, second, because allowing minority shareholder actions involves the courts being asked to determine the balance of advantage between controllers and noncontrollers, something rarely welcomed by either business or judges (Armour and Gordon 2011; Cheffins 2000). In China, however, regulators have successfully adapted the traditional shareholders deriv- ative lawsuit to suit the particular circumstances of China's state-dom- inated share markets (Clarke and Howson 2011). In particular, the statutory scheme established by the 2005 *Company Law* permits *hori- zontal* claims against controlling or oppressive shareholders in addition to *vertical* claims against orthodox insiders and fiduciaries (managers and directors). In fact, as Clarke and Howson (2011) find, use of the deriva- tive action in China has been almost entirely limited to the closely held form of corporation.

Armour and Gordon (2011) identify a significant shift in global corporate control since the early 1990s, which has resulted in a concen- tration of control over global equities in the hands of a small number of key institutional investors. Vitali, Glattfelder, and Battiston (2011) analyze ownership stakes in a network of over 43,000 transnational corporations and find that a core of 147 global firms, mostly invest- ment funds and banks, together control (as at 2011) 40 percent of total assets in the network. A total of 737 firms control 80 percent of assets. This concentration of ownership has brought about a convergence of regulatory concerns, which are now much more likely to be shared in common between previously much more diverse markets (Armour and Gordon 2011; Patoski and Prakash 2004; Pauly and Reich 1997; Spitzeck 2009). These shared concerns are further discussed in Part 4.

Controller or Stakeholder Agency Problems and the Emergence of the Benefit Corporation

Regulators in many jurisdictions have also become increasingly concerned to protect nonshareholder stakeholder interests. Traditionally, the only stakeholder interest protected in Anglo-American systems was that of creditors, whose interests are not only protected by imposing a duty on the board to advance the interests of the company, but also by their rights to become involved in the control of the company in the event of insolvency (Armour and Gordon 2011; Cheffins 2000). Creditors may also, of course, bargain for board representation either on the basis of their interests as creditors alone, or, as increasingly is the case, by becoming involved as block shareholders (Armour and Gordon 2011; Cheffins 2000).

In many civil law systems, a second nonshareholder stakeholder interest that has long been protected is that of employees. Use of company law to regulate the company–employee agency problem typically involves ensuring that employee interests are directly represented at the board level. Appointment rights for employees to one third or less of the board are quite widespread in Europe, and also found in China (Armour, Hansmann, and Kraakman 2009; Davies 2000; de Jonge 2008).

Evidence from Germany, where the matter has been extensively researched, suggests that employee representation at the board level acts to reduce the principal–agent problem as between company and employees, but also serves to reduce the effectiveness of the board in regulating the management–shareholders-as-a-class agency problem (Davies 2000). One way of ensuring that both shareholder and employee interests are most effectively represented at the board level is to establish a two-tier board structure—a management board and a supervisory board. In the European Union, corporate governance codes recommend a dual-board system in 10 countries (Austria, Czech Republic, Denmark, Estonia, Germany, Latvia, Netherlands, Poland, Slovakia, Slovenia) and a unitary-board system in eight countries (Belgium, Cyprus, Greece, Ireland, Malta, Spain, Sweden, and the United Kingdom). In the remaining nine countries (Bulgaria, Croatia, Finland, France, Hungary, Italy, Lithuania, Luxembourg, and Romania) a hybrid system applies and companies can choose between a one- or two-tier approach (European Union 2012). In Asia, both China and Indonesia (which inherited it from their former Dutch colonizer) follow a two-tier board structure (de Jonge 2008; Jaswadi 2013; Tumbuan 2005), while those jurisdictions that were former British colonies (Australia, Hong Kong, Malaysia, New Zealand, and Singapore) have inherited a unitary-board system (http://www.asx.com.au; http://www.hkex.hk; http://www.nzx.com/; http://www.sgx.com/). Both unitary and two-tier board structures have their advantages and disadvantages, but the lesson from China, Indonesia, and elsewhere is that for two-tier board systems to live up to their potential for ensuring that management board decision making is properly scrutinized and boards are held to account, the supervisory board needs to have strong powers, including the power to call a general meeting of shareholders whenever it deems such a meeting to be necessary (Jaswadi 2013; Shi 2012).

In tandem with growing concern at the international level about the too-often detrimental impacts of for-profit corporate activities on the environment and on human rights, there has been a growing movement of for-profit social entrepreneurship, social investing, and sustainable business supporters that has now gained critical mass. Social entrepreneurs believe that the interests of society, as the most basic stakeholder impacted by all corporate actions, can be protected and enhanced as part of profitable business activity (Clark and Vranka 2013; Reiser 2011). Traditionally, however, those seeking to ensure that the interests of society and the environment are recognized and protected in all corporate decision making have faced both practical and legal pressures to favor profit maximization over other goals whenever there is conflict, as inevitably sometimes occurs (Clark and Babson 2012; Reiser 2011). A growing movement of sustainable and social entrepreneurs has therefore demanded access to hybrid forms of organizations able to accommodate social goals and environmental sustainability while also protecting the interests of traditionally recognized stakeholders (shareholders, creditors, and employees).

A growing number of jurisdictions have attempted to meet this demand by enabling new hybrid organizational forms. These include the low-profit limited liability company available in some U.S. states, the community interest company (CIC) available in the United Kingdom and the B-Corp, which gains its status from private certification with the nonprofit B Lab (Clark and Vranka 2013). The Benefit Corporation differs from each of these in several respects, especially in its use of third-party standard-setting organizations to vet the social good bona fides of potential incorporators (Clark and Babson 2012). Other mechanisms built into legislation aimed at protecting the dual purposes of the benefit include reporting and transparency obligations, and the naming of specialized benefit director position and appointment of a benefit officer (Clark and

Babson 2012). The benefit director is typically charged with the task of including in the annual report her own statement assessing whether the company and its directors have acted in compliance with the benefit purposes of the corporation during the relevant period. If a benefit director opines that the benefit corporation has failed to meet the requirements of the law, she must describe these failures (Clark and Babson 2012). The benefit officer is typically an optional position, which benefit companies may create in order to oversee the fulfillment of the public benefit purposes for which the company was created. Identifying at least one and perhaps two roles with clear responsibility for tracking and assessing public benefit provides additional monitoring and enforcement resources over mandated disclosure alone (Clark and Babson 2012; Reiser 2011). As a third mechanism for protecting stakeholder interests, many benefit corporation jurisdictions also offer a special right of action, often called a *benefit enforcement proceeding*, to enforce the special duties of benefit corporation directors and officers and the public benefit purposes of the corporation.

There are two main ways in which the purposes and operations of a benefit corporation can fail to align or fall out of alignment, providing evidence of corporate malaise. The first relates to the articulation of an appropriate, achievable, and sustainable dual mission, and the second relates to the enforcement of that dual mission.

It has always been difficult for the traditional for-profit enterprise to incorporate social purposes in their articles of association. This is because of the potential for conflict with the fiduciary duty owed by directors and officers always to act in the best interests of the company (and by extension its owners). By expressly requiring directors and officers to take considerations other than profit maximization into account, the benefit corporation allows them to openly pursue social and environmental aims while also keeping profits in mind. When one thinks more deeply about how a dual purpose will be articulated in a benefit corporation, however, doubts emerge. The requirement of general public benefit is vague, and statutes currently offer little guidance on the thorny issue of how profit and social good should be balanced (Clark and Babson 2012; Reiser 2011). The benefit corporate form offers the strength of flexibility in that directors have wide discretion to forego profit maximization in favor of social good production or vice versa, according to changing circumstances. Yet, this same flexibility is also a source of weakness in the absence of guidance for directors on how to exercise their high degree of discretion. In most cases, directors are merely subject to an obligation to consider the impact of decisions on a wide range of stakeholder constituencies. Not only does this leave a lot of room for disagreeing on what those impacts actually are, it also leaves a great deal of potential for disagreement over how to define the acceptable limits of impacts, either good or bad on any one or more groups of interests. While diversity of opinions and vigorous debate can be a source of strength, it can be argued that the benefit corporation contains within it much greater potential for discord, dissention, and different visions about what the corporate purpose should be from the traditional company model. When profit is the sole object, at least directors are united behind a single cause. When *social good* is declared as an objective, consensus on what this actually means in practice is going to be far, far harder to achieve.

If current standards for articulation of their dual mission by benefit corporations are weak, current standards for enforcement of that mission are even weaker. Even where a particular social or environmental interest is named as beneficiary in a benefit corporation's founding documents, that social group or interest has no standing, and thus no legal recourse in the event that the corporation fails in its social purpose aims (Clark and Babson 2012; Reiser 2011). The only mechanisms currently available for enforcement of the corporation's dual mission are transparency, third-party certification and verification, and shareholder actions. Shareholders in benefit corporations retain the informational, voting, and litigation rights of ordinary shareholders. Any of these rights can theoretically be used to enforce dual mission. Shareholders can access meetings and other records to determine how a particular mission conflict was resolved or decision made. They may vote out directors who fail to sufficiently pursue their preferred balance of mission and profit, or even sue directors for failure to meet their fiduciary obligations. Yet, shareholders are unlikely to be either assiduous or consistent enforcers (Clark and Babson 2012; Reiser 2011). Given the broad nature of decision-making discretion given to directors and officers under existing benefit corporation legislation, establishing a failure of fiduciary duties in decision making becomes very difficult indeed. Moreover, shareholders receive a pecuniary benefit whenever directors veer toward the profit-making side of their dual-mission mandate and away from the social purpose side. Shareholders will typically have an incentive *not* to monitor or enforce the social purpose achievements of the corporation. In the United Kingdom, an effort has been made to overcome this problem in the case of CICs by establishing a specialized CIC regulator with broad authority to investigate, remove fiduciaries, and even terminate CICs found out of compliance (Clark and Babson 2012; Clark and Vranka 2013). No similar agency yet exists for benefit corporations, and even if one were to be established, it would, in many cases, still face the same problems of determining an appropriate balance between profit and public benefits that those involved in benefit corporations face today. The main solution to these difficulties will be experience and the working out of more sophisticated tools for measuring the value of social and environmental capital.

The experience of benefit corporations may well provide a valuable contribution to expanding the boundaries of accepted understandings about corporate purpose and corporate governance. Just as accountants are expanding accepted understandings of what can, or should, be defined as capital (Gleeson-White 2014) so also are management teachers, through the Principles for Responsible Management Education (http://www.unprme.org), expanding the accepted curriculum boundaries for management students. Changing the legal definition of what constitutes valid corporate purposes both contributes to, and draws upon, both of these movements. In addition, a growing number of traditional businesses are expanding the definition of corporate purpose less directly, by signing up to the Global Compact (http://www.globalcompact.org); the Global Reporting Initiative (https://www.globalreporting.org); the ISO 26000 Guidance on Social Responsibility (http://www.iso.org/iso/home/standards/iso26000.htm); the International Integrated Reporting Initiative of the IIRC (http://integratedreporting.org/); and the UN Guiding Principles on Business and Human Rights (http://www.ohchr.org/Documents/Publications/GuidingPrinciplesBusinessHR_EN.pdf). All of these standards and guidelines are aimed at assisting corporations in measuring and monitoring their performance against social and environmental criteria as well as against traditional economic criteria. Measuring and monitoring corporate performance is the essence of managerial forensics, and so all of these emerging developments in understanding who, how, what, where, and why corporations behave as they do open up new horizons in this important field.

Conclusion

All organizations are political. They are political because the way they operate depends on relationships of control and influence. This is as true of the local football club as it is of the largest corporation. Healthy organizations have systems that connect them to the real world, and processes to help them develop their people (Gettler 2005). The effective, long-term maintenance of such systems and processes depends largely on achieving an optimal degree of understanding and balance of power between owners (shareholders) on the one hand, and management on the other. This chapter has examined how ownership structure affects the internal governance of different types of corporations, with a particular focus on the family firm and the SOE, as well as a look at the changing nature of the *widely held* corporation. The chapter also examines how new hybrid forms of the corporation, in particular the benefit corporation, allow for new understandings to be reached between owners and managers regarding the purposes for which the corporation exists. All of these different types of corporations have their own strength and weaknesses when it comes to sustainable governance, and it is the job of managerial forensics to understand the weaknesses in particular, in order to enable corporations to guard against them, and to diagnose them when they lead to corporate malaise. Managers in any corporation should ensure that assessments are made on a regular basis of the ownership structure of the firm, and of any potential conflict of interest challenges presented by that structure. All firms need to be aware of governance risks and of what mechanisms the firm has in place for managing such risks. In a rapidly changing world, regular reassessments of potential governance risks arising, for example, from any concentrations of power (of any type) in a single or small number of individuals, are needed. The three key principles for any company seeking to manage its governance risks are *transparency*, which should be as full as possible (Schouten 2009); *monitoring*, which should be as independent as possible, and *accountability*, which should be as direct, immediate, and remedial as possible. New risk-assessment and risk-management tools are regularly developed and updated by international, sectoral, and national bodies. Examples include the IFC's Human Rights Impact Assessment Tool (for all companies), and the *Dohas* (Irish Association of Non-Governmental Development Organisations) Corporate Governance Assessment Tool. Managers should regularly assess whether the firm is making full and efficient use of such tools.

References

Armour, J., and J.N. Gordon. 2011. "The Berle-Means Corporation in the 21st Century." Incomplete draft. http://www.law.yale.edu/documents/pdf/Intellectual_Life/Armour_BerleMeansCorp091021.pdf

Armour, J., H. Hansmann, and R. Kraakman. July 2009. "Agency Problems, Legal Strategies and Enforcement." Discussion Paper No. 644. John M. Olin Center for Law, Economics and Business.

Bingham, J.B., W.G. Dyer, I. Smith, and G.L. Adams. 2011. "A Stakeholder Identity Approach to Corporate Social Performance in Family Firms." *Journal of Business Ethics* 99, no. 4, pp. 565–85.

Brenes, E.R., K. Madrigal, and B. Requena. 2011. "Corporate Governance and Family Business Performance." *Journal of Business Research* 64, no. 3, pp. 280–85.

Burkart, M., F. Panunzi, and A. Shleifer. 2003. "Family Firms." *The Journal of Finance* 58, no. 5, pp. 2167–202.

Cheffins, B.R. September 2000. "Does Law Matter?: The Separation of Ownership and Control in the United Kingdom." ESRC Centre for Business Research, University of Cambridge Working Paper No. 172.

Chen, C.J.P., Z.Q. Li, and X.J. Su. 2005. "Rent Seeking Incentives, Political Connections and Organizational Structure: Empirical Evidence from Listed Family Firms in China." The Chinese University of Hong Kong Business School. http://www.bschool.cuhk.edu.hk/research/cig/pdf_download/ChenLiSu.pdf

Chrisman, J.J., J.H. Chua, and L.P. Steier. November 2011. "Resilience of Family Firms: An Introduction." Theories of family enterprise series. *Entrepreneurship Theory and Practice* 35, no. 6, 1107–19.

Chua, J.H., J.J. Chrisman, and P. Sharma. 1999. "Defining the Family Business by Behavior." *Entrepreneurship Theory and Practice* 23, no. 4, pp. 19–39.

Chua, J.H., J.J. Chrisman, J.P. Steier and S.B. Rau. November 2012. Sources of Heterogeneity in Family Firms: An Introduction. Theories of family enterprise series. *Entrepreneurship Theory and Practice* 36, no. 6, pp. 1103–13.

Clark, W.H., Jr., and E.K. Babson. 2012. "How Benefit Corporations Are Redefining the Purpose of Business Corporations." *William Mitchell Law Review* 38, no. 2, pp. 817–51.

Clark, W.H., Jr., and L. Vranka. 2013. "The Need and Rationale for the Benefit Corporation: Why It Is the Legal form That Best Addresses the Needs of Social Entrepreneurs, Investors, and, Ultimately, the Public." White Paper. Benefit Corp Information Centre. http://benefitcorp.net/storage/documents/Benecit_Corporation_White_Paper_1_18_2013.pdf

Clarke, D.C., and N.C. Howson. August 31, 2011. "Pathway to Minority Shareholder Protection: Derivative Actions in the People's Republic of China." Working Paper, University of Michigan Law School, George Washington University Law School. http://ssrn.com/abstract=1968732

Davies, P.L. December 7–8, 2000. "The Board of Directors: Composition, Structure, Duties and Powers." Paper on Company Law Reform in OECD Countries: A Comparative Outlook of Current Trends. Stockholm, Sweden.

de Jonge, A. 2008. *Corporate Governance and China's H-Share Market.* Cheltenham, UK: Edward Elgar.

European Union. 2012. Women in Economic Decision Making in the EU: Progress Report. Luxembourg. http://ec.europa.eu/justice/gender-equality/files/women-on-boards_en.pdf

Gettler, L. 2005. *Organisations Behaving Badly: A Greek Tragedy of Corporate Pathology.* Australia: John Wiley & Sons.

Gleeson-White, J. 2014. *Six Capitals, or Can Accountants Save the Planet?: Rethinking Capitalism for the Twenty-First Century.* Sydney: Allen & Unwin.

Huang, L.J., and R.S. Snell. 2003. "Turnaround, Corruption and Mediocrity: Leadership and Governance in Three State Owned Enterprises in Mainland China." *Journal of Business Ethics* 43, no. 1/2, pp. 111–24.

Jaswadi, J. 2013. *Corporate Governance and Accounting Irregularities: Evidence from the Two-Tier Board Structure in Indonesia* [Degree thesis]. Victoria University. http://vuir.vu.edu.au/22352/

Kidwell, R.E., F.W. Kellermanns, and K.A. Eddleton. 2012. "Harmony, Justice, Confusion, and Conflict in Family Firms: Implications for Ethical Climate and the "Fredo Effect." *Journal of Business Ethics* 106, no. 4, pp. 503–17.

Kim, Y., and F.Y. Gao. 2013. "Does Family Involvement Increase Business Performance? Family-Longevity Goals' Moderating Role in Chinese Family Firms." *Journal of Business Research* 66, no. 2 pp. 265–74.

KPMG and FBA (Family Business Australia). 2013. "Performers: Resilient, Adaptable, Sustainable." Family Business Survey 2013. The University of Adelaide.

Lumpkin, G.T., and K.H. Brigham. 2011. "Long-Term Orientation and Intertemporal Choice in Family Firms." *Entrepreneurship Theory and Practice* 35, no. 6, pp. 1149–69. Baylor University.

Nguyen, T.T., and M.A. van Dijk. 2012. "Corruption, Growth, and Governance: Private vs. State-Owned Firms in Vietnam." *Journal of Banking and Finance* 36, no. 11, pp. 2935–48.

OECD. 2004. *OECD Principles of Corporate Governance.* Paris: OECD Publishing.

OECD. 2006. *Corporate Governance of State-Owned Enterprises: A Survey of OECD Countries.* Paris: OECD Publishing. http://dx.doi.org/10.1787/ 9789264009431-en

OECD. 2013. *State-Owned Enterprises in the Middle East and North Africa: Engines of Development and Competiveness?* Paris: OECD Publishing. http:// dx.doi.org/10.1787/9789264202979-2-en

Patoski, M., and A. Prakash. 2004. "Regulatory Convergence in Nongovernmental Regimes?" *Journal of Politics* 66, no. 3, pp. 885–905.

Pauly, L.W., and S. Reich. 1997. "National Structures and Multinational Corporate Behavior: Enduring Differences in the Age of Globalization." *International Organization* 51, no. 1, pp. 1–30.

Reiser, D.B. 2011. "Benefit Corporations—A Sustainable form of Organization?" *Wake Forest Law Review* 46, pp. 591–625.

Sanchez-Bueno, M.J., and B. Usero. 2014. "How May the Nature of Family Firms Explain the Decisions Concerning International Diversification?" *Journal of Business Research* 67, no. 7, pp. 1311–20.

Schouten, M.C. April 2009. "The Case for Mandatory Ownership Disclosure." MPRA Paper No. 14880. University of Amsterdam, Columbia Law School. http://mpra.ub.uni-muenchen.de/14880

Shi, C.X. 2012. *Political Determinants of Corporate Governance in China.* London: Routledge.

Singla, C., R. Veliyath, and R. George. 2014. "Family Firms and Internationalization-Governance Relationships: Evidence of Secondary Agency Issues." *Strategic Management Journal* 35, no. 4, pp. 606–16.

Spitzeck, H. 2009. "The Development of Governance Structures for Corporate Responsibility." *Corporate Governance: The International Journal of Business in Society* 9, no. 4, pp. 495–505.

Tumbuan, F.B.G. September 7, 2005. "The Two-Tier Board and Corporate Governance (Pointers for Discussion)." Presented at a One-day Seminar on Capital Market and Corporate Governance Issues in Indonesia, Intercontinental Hotel, Jimbaran, Bali. http://www.oecd.org/daf/ca/ corporategovernanceprinciples/35550189.pdf

Vitali, S., J.B. Glattfelder, and S. Battiston. 2011. "The Network of Global Corporate Control." *PLoS ONE* 6, no. 10, p. e25995. doi: 10.1371/journal. pone.0025995. http://journals.plos.org/plosone/article?id=10.1371/journal. pone.0025995

Zahra, S.A., J.C. Hayton, and C. Salvato. 2004. "Entrepreneurship in Family vs. Non-Family Firms: A Resource Based Analysis of the Effect of Organizational Culture." *Entrepreneurship Theory and Practice* 28, no. 4 pp. 363–81.

CHAPTER 6

Assessing Leadership Preparedness

Anthony Liberatore and J. Mark Munoz

Introduction

Leaders are in many ways their own consultants. Most are continuous learners, sponges of information, monitors of changes and trends, networkers, direction setters, problem solvers, coaches, and disciplinarians. Much needs to get done. They cannot help but form strong opinions about people and motivation and how to run a business. But times come when changes are needed: when sales or profits fall, when competition picks up, when technologies change, or in more positive circumstances, when opportunities arise. Each and every new challenge tests leadership preparedness. This chapter offers some ideas that have proven useful in helping leaders understand, plan for, and orchestrate changes to improve company performance.

This discussion centers on a company that customizes vehicles for field work in gas and electric generation and distribution, public service departments, plumbing and electrical services, and other utility vehicles. For ease of reference, it will be called Company X.

The work process begins with drawings and estimates for a proposed vehicle(s) retrofit or in company terms "quotes," through the ordering of chassis and bodies, to fabrication, assembly, testing, and delivery. The ordering and inventorying of chassis, bodies, and parts originate in various departments and are joined together in the production process.

Over the years, the company developed a data management system to coordinate, monitor, and track the production process. This system is organized around job numbers and tracks jobs, due dates, start dates, and percent completed by tracking hours on each job relative to hours bid. The production management team meets weekly to monitor job progress and deal with the typical array of production management issues ranging from prioritizing work, dealing with deficiencies, delays, to job changes, and so on.

The production work is coordinated, monitored, and supervised by a manager who also works on vehicles. Production crews track hours daily by job and task. These are submitted at the end of each day and entered into the data management system, which then updates the hours used relative to bid for use in the production management meetings. One of the management difficulties in this process is coordinating multiple small-run projects. Jobs range from single units to perhaps 30 vehicles at the upper end, across vehicle types from panel trucks to large boom trucks. Time required for jobs ranges from a few hours to hundreds of hours.

Financial performance is reviewed monthly by the senior management team. Over the past couple of years, the production operation has moved from losing money to breaking even, a positive, but unstainable performance. The issue is how to improve performance without damaging the company's hard-work ethic, positive work environment, and low employee turnover.

The culture of the company values teamwork. Many improvements have percolated up from the production teams. There is a strong aversion to command and control approaches in both management and production.

Anyone who has had training and experience in these types of operations would immediately have a number of suggestions and solutions to improve efficiency. However, in a broader context, the leadership, management, and staff need more than solutions; they need a unifying understanding of the situation that helps diagnose and prescribe. Everyone would benefit from having a clear picture of the issues in the process in order to understand and frame the work ahead.

Diagnosing and Prescribing

It is a natural and practical step to set production goals for increasing efficiency and profitability, but understanding and framing the issues for improving performance is a developed leadership and management skill. One simple tool the authors found useful over the years to help frame, diagnose, and prescribe at a business unit or operational level is what is called the *Leadership Circle* (LC). This visual tool helps organize thinking and supports analysis and action. (It is reminiscent of Deming's (1986) Plan, Do, Check, Act, but with more of a focus on managerial behavior).

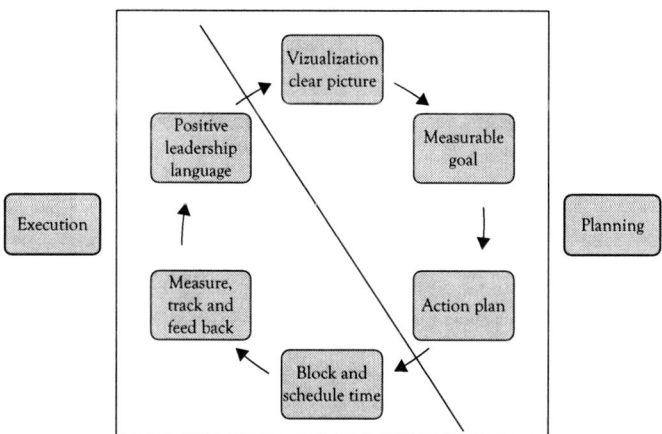

The LC is divided into two sections: Planning on the right and Executing on the left. Planning begins with a clear picture of the situation. Visualizing, thinking, and writing to clarify thought lead to action. It is unlikely that desired results will be achieved without a clear and motivating picture. Take delegation as a sin ple example. Differing pictures lead to differing results. Leaders and managers must clarify thinking by developing common pictures and understandings of the issues and results desired.

Next, set measurable goals and action plans (Myers 1994). Most people do this naturally. However, the vast majority do not commit goals and plans to paper, especially personal ones. Keeping goals and plans in mind only and communicating verbally make it easy to change one's mind and direction. It conditions employees to wait for direction and go with the flow. Leaders and managers should commit goals to paper in order to clarify thinking, make commitments, and improve the leadership environment for employees.

Getting results requires good execution. Teams, committees, groups, taskforces, and so on come together and often get the planning straight but fail to execute. It is a common frustration in most organizations that meetings proliferate while actions linger. Planning does not get the job done. Moving from planning to action requires commitment. If goals are important and believed in, block and schedule time to work on them. It is not unusual for people to get engulfed by crises and trapped by daily tasks, and to relegate important activities to when time is available. Psychologists tell us that procrastination is common (Ariely 2010). To get better results schedule activities not just deadlines. In a production environment, production meetings should be scheduled at regular times and not missed. Otherwise, nature takes over. Whether it is a large formal project plan or a simple several-step process, write it down and schedule the time to make it happen. Make commitments.

One of the commonly missed links in the leadership and management chain of events is Measuring, Tracking, and Feeding Back what is being managed. Make a distinction here between policing and managing. In today's world, with extensive and customized data management systems, large amounts of data are collected, stored, and analyzed. It is relatively easy to query the systems and run reports. Ask a company, leader, or manager if they are measuring, tracking, and feeding back information, and inevitably the answer is yes. Querying a system to analyze, say, the number or mortgages processed, or the pounds of laundry washed, or the experience with a client or a number of clients provides useful but not adequate information. Querying a set of data is not tracking, it is policing. Searching for information provides a snapshot of a process that is generating data. Managing requires a continuous flow of data that reveal performance and patterns, a time picture rather than a snapshot. A report of the number of mortgages processed per day or week, say, broken down by type—received, withdrawn, denied, approved, and funded—and the associated hours and cost of the work relative to goal would provide to managers, leaders, and staff measurable results of performance and help the team analyze performance, identify and diagnose issues, and provide a basis for prescribing action. And, importantly, it would provide measurable data of good performance and a basis for celebrating work well done—an exercise of Positive Leadership Language.

In organizations where people work hard but do not measure and track performance, there sometimes tends to be an inverse pattern of communication—Quiet, Quiet, Negative (QQN). Normality is expected. When things are going right, it is expected behavior and not much is said. When things are not going right, they draw negative comments—a QQN pattern. As is sometimes the case in parenting, the normally good behavior of children goes unproclaimed, but the bad behavior draws fire. Measuring and tracking create a basis for identifying and reinforcing good performance, promoting more of a quiet, quiet, positive communication pattern.

What leaders say has a tremendous impact on personal attitudes and the work environment. To drive this point home, think of something negative your parents said to you when you were young. These memories and feelings hang on forever. This is so in the workplace. What the boss says matters and negative feedback can undermine the workplace. A leader or manager who always points out mistakes may be asking for trouble. Consider a processing or manufacturing plant where people are working hard and want to feel appreciated. When a boss shows up and sees something out of place he or she immediately asks why. The expectation of getting good comments for working hard is shattered with the negative comment. This is not to say that accountability and command are not important. They are. However, the point is to create a system that measures and tracks.

Positive Leadership Language in the LC refers to external talk, talking to others, and to internal talk, talking to yourself. Thinking drives behavior. Thinking that managing is identifying and solving problems leads to certain behavior patterns. How one thinks about developing people, improving systems, and creating accountability leads to certain types of behaviors. McGregor's work on attitudes and management success, Theory Y and Theory X (McGregor 1960), clearly demonstrates the notion that managers and leaders get what they believe. Thinking that people are basically incompatible with work, and that they need a strong hand and direct supervision at all times, leads to a particular work environment and attitudes that go with it. Measuring, tracking, and feeding back results in a purposeful environment that focuses on results lead to a much different work environment.

Observations

The LC is a guide for asking questions in an orderly way and for sizing up business situations. Moving around the circle raises a number of questions in an ordered fashion. Is there a clear picture of the issue, problem, desired result, and so on? Is there a common vision of success? Or are the pictures fuzzy, or do the people involved have divergent pictures? Are measurable goals associated with these pictures? Or are they mostly hopes and wishes? Is there a written action plan? It may be a set of simple steps, but it is written? Or, are goals and action steps only verbalized? A majority of people tend to keep goals and plans in their heads, but the danger here is the temptation and ease of changing thinking and direction. When thoughts, goals, and plans are clarified on paper, a clearer common path is formed for leading and following.

The second half of the circle asks about execution. Teams come together to plan and then disperse to do the work but often fail to execute well (Bossidy, Charan, and Burck 2002). If it is important to achieve, block the time to make it happen. Has time been blocked and scheduled? Or does the team expect to work on it when they have the time? Inevitably, daily crises intervene to push the needed work to the brink of a crisis. Will the team be dominated by reacting to daily pressures, or will it organize to act before crises emerge? The next stage is typically a weak link in many organizations. If the work is important, is measuring progress, tracking over time (managing not policing), and feeding it back to everyone to keep work goals focused happening? Or are people just working hard while managers expect results without baseline productivity measures?

Finally, is the language appropriate? Has the *why* been stated? Do participants have the right attitude about the work? Are they thinking and saying the right things to themselves about the work? Or have managers just assumed that this is the case? Are managers and leaders establishing metrics so that they have opportunities for giving positive feedback for achieving results, for moving toward and hitting goals?

Doctors are able to diagnose and prescribe for a patient based on understanding human systems and processes. Leaders and managers similarly diagnose and prescribe based on an understanding of business systems and processes. The LC helps organize thinking about the chains of events and it raises questions to ask about business practices: It is a useful tool for analyzing business operations and prescribing steps for improving performance.

Nearly everyone in an organization has an opinion about what ails the company and what solutions are needed. These opinions are framed by individual experiences and positions in the organization. The more responsibility one has in an organization, generally speaking, the broader and more integrated are the perspectives of operations and performance. What may appear as the problem or solution at one level may not match up with the diagnoses and prescriptions at a higher managerial level in an organization. It is advisable to avoid jumping to solutions too quickly for fear of locking in opinions and causing unintended conflicts. The LC model sets a stage of inquiry by focusing on a chain of necessary events for strong performance. A weak or missing link spells trouble. Use it to guide inquiry, to identify strengths and weaknesses, to prioritize solutions, and to educate those involved in the process.

Analysis

The following need to be considered in analyzing leadership in an organization.

Productivity

How might Company X increase productivity, reduce cost, and increase contribution to profit? And what area(s) of the operation might be tackled first?

Clarity of Success Vision

Follow the LC line of questioning. Is the vision of success clear? Are the pictures of achievement clear, shared, and understood? Company X wants to increase the profitability of truck equipment jobs. It is understood throughout the company. The seasoned employees uniformly want to improve results. There is a sense of pride in the work and in the company.

However, the goals are not as well understood. Zero profit in the division is not acceptable but what is desired is not clear. Just improving the performance is not a goal, it is more of a priority or wish. A specific profit improvement goal would help focus and prioritize action plans.

Specific Plans

The charge of the production team is to improve performance, but without measurable goals, specific action plans linked to performance will be difficult to articulate. Much of the observed efforts of the production team revolve around reacting to the needs of particular jobs, not to system improvements. Management would be improved with written action plans set on measurable goals to guide the production team and provide a basis for judging effectiveness.

Measuring and Monitoring

The team is charged with improving performance. The next link in the LC enters into executing work—Measuring, Tracking, and Feedback. Are there systems for measuring the productivity and the quality of the work? Are these tracked over time so that performance patterns and trends are visually evident? Or are they intermittent policing reports? Is the information fed back to everyone to assess performance?

Information Quality

Information fed back into the work process should affect thinking, meaning it should be formatted to show trends easily and continuously. The data tracking and management systems of Company X have grown and improved over time. Daily time sheets are booked against jobs. Weekly production management meetings track hours used relative to the labor hours estimated for each job. Using half of the bid hours does not imply that the job is half completed, nor does it suggest where or how any particular phase of production is behind or ahead of schedule. Managers focus on the list to coordinate priorities and solve problems. The meetings tend to focus on immediate needs and reactions to problems. So how does this system meet the needs for measuring, tracking, and feeding back? What is being measured? Does it provide a time perspective and can those doing the work monitor and track performance relative to goal?

Prompt Feedback

There are several deficiencies in the management link. Time used on a job is tracked but it is not fed back to the work teams on a consistent basis. Teams have a general idea about performance on a job as does the production manager, but that is as far as it goes. Nevertheless, mechanics work hard, have a good culture, and work to the hours bid on the job. (Work time is not measured by the stage or progress of production. While it might seem advisable to do so, production is not a continuous flow process. Jobs are mostly in small chunks of hours, say 20 to 100 hours. Measuring overall hours may be all that could be hoped for in such a job shop.) The performance of production is monitored by the quoting department where again the feedback loop is missing. There is no systematic tracking over time of hours by job relative to quoting meaning that production is not getting the feedback needed for monitoring performance over time nor is there a formal link between production and quoting to assess the accuracy of quoting. Even if production does make changes, it would be difficult to measure effectiveness without some base lines of productivity. Other feedback loops are missing. The stages of the overall process—quoting, ordering and taking inventory, production, delivery, and service—are essentially separated in activity, authority, and management. Ordering and inventory get misaligned with production. The response is reactive—fix the problem. There is no systematic assessment of the deficiencies nor systematic feedback between the departments to change processes.

Work and Information Flow

Service is after the sale and at the other end of the chain. When problems occur, the service department reacts to correct the problem. Service tickets are written and handed off to accounting where warranty claims are processed. But again, the information does not flow back into production, nor is the information tracked on a continuous basis to identify useful changes in operating procedures.

Effectiveness of Communication

The final link in the LC is Positive Leadership Language. This refers to a complex set of ideas about communication, responsibility, and accountability. For this case, we ask if the underlying metrics support communicating about performance. Is there solid data to judge performance? Do workers have enough information to judge their efforts or do they just work hard? Is the work goal focused?

Breadth of Perspective

Individual teams know the hours quoted on a job and perhaps there is enough. Stepping back from the individual work team to the production meetings exposes some opportunities for improvement. The production meetings focus on individual jobs but rarely look at the bigger picture of performance. Although reports could be created for tracking jobs and grouping them by contract or type, they are not. The production team does not get to see the patterns in performance over time nor is it fed to other departments. Communication then cannot form around the work. Conversations drop back to pleasantries and departments mind their own business. Grumblings, frustrations, and even suggestions for improvement are diverted away from the departments toward the owner. Department heads focus on the owner rather than on the management team. Production teams focus on the production manager rather than on the work teams. The production manager ends up being the only person to have a big picture of production. Staff continuously ask what to do next. It is a hectic managerial style.

Management Structure

Having an owner or manager at the center of communications creates a hub-and-spoke management structure, which in itself affects team performance and reinforces existing communication patterns. Metrics, communication patterns, and management structure are inextricably interwoven.

Prescriptions

In the mentioned case, Company X is in a slowly evolving crisis. It is making money but only barely on its primary truck equipment business. It is potentially losing its edge in the market. Losing a major customer would create significant immediate financial problems. Action is needed now to create a better trajectory toward sustained profitability. But where is the starting point? It is difficult to see from inside the company when everything is connected. Possible solutions: change leadership, change management, change systems, change something.

There are no easy answers or silver bullets. It is a complex social organization with its own culture, political forces, and structure, and any changes will have to take these into account to be successful. What the LC provides is an organizing view of its operating system, how it gets work done. Leadership can learn, frame issues, and ask questions through the LC lens. Is the vision clear? There is a stated and widely understood vision of success—improve the profitability of the company to ensure its survival. Maintain major customers and add new customers and lines of equipment to diversify the customer base.

Beyond that it is a bit fuzzy. Are the goals clear? Add measurable goals to vision to focus effort and prioritize actions. Create a few achievable measurable goals to support the vision of success, such as moving from break-even to 5 percent net income in truck equipment.

Strategize Through the Leadership Circle

Getting to action plans is problematic. The balkanized, hub-and-spoke management structure is an impediment. Bring the department managers together and charge them with achieving the profit goal. Visualizing and understanding the LC will give the team a common map and understanding of the task at hand.

Engage Stakeholders

Leadership is a social process (Day 2001). Leadership is never a solo act. Responsible leadership needs to include multiple stakeholders in the decision-making process (Waldman and Siegel 2008). Leadership is about strong internal and external relationships and effective collaboration (Mittal and Dorfman 2012). Collaboration with diverse stakeholders is key. Shared leadership or distributing influence across several team members is essential (Carson, Tesluk, and Marrone 2007). The direct involvement of leadership to set goals and change the established lines of communication is absolutely necessary. If not, the process will slowly fail as the inertia of established practices takes over. This is a major shift in culture and structure that can only happen with leadership at the helm. Leadership must be willing and committed to the process.

Develop Appropriate Attitude and Plans

Leadership approaches influence followers' conduct relating to work, attitudes, and performance (Liden et al. 2014). The right attitude gets results. Attributes such as courage, learning, empowering mindset, and management of complexities are essential for leaders to transform organizations (Tichy and Devanna 1986). Once the team is in place and charged with hitting goals, begin working on action plans. From specific goals will come specific plans, actions, and priorities. Smaller companies generally do not have the experience or expertise to manage such an improvement process. It is a good idea to get some help.

Establish Clear Metrics and Prioritize

Organizations and business environments are constantly changing. Transformational leaders encourage members to prepare for change and adapt to the environment (Jung, Chow, and Wu 2003). Contemporary leaders need to be strategically agile (Doz and Kosonen 2008). If it is important, schedule the time for the teamwork and do not put it off. There are always interruptions, crises, or excuses to fall back to old habits. At the same time, the team can begin filling in gaps in measurements and metrics. What has been and is the performance of truck production, not on a one-time basis, but trending over time? A shift from policing to managing is required. Asking how to improve performance will lead to identification of bottlenecks in production, improvements in parts and inventory, adjustments in quoting, and feedback from servicing after the sale to corrective and preventive actions in production processes.

Tie-in Plans to Your Vision

Always check back to the vision. Communication ability is a key attribute of an effective leader (Nemanich and Keller 2007). Articulating the vision and goals clearly sets the stage for concerted action and minimizes misunderstandings and conflicts. If those involved in the process do not present a clear common picture of what is being pursued and why, if they have dissonant and competing pictures, the process will limp along, fail to deliver measurable results, disappoint, and eventually disband. The team will fail to make the progress it expected or hoped for. More drastic measures will be needed next time.

The LC provides an organizing view, a common map for leadership and management to organize and mobilize action on improving performance. It gives leaders and managers a tool for analyzing and prescribing. It provides a learning and reference point for those involved in the process.

Summary

To summarize, if an organization was likened to a human body, assessing leadership would require examining two parts: Head (Ability to Plan) and the Hands (Ability to Execute).

The "Head"

In examining the *Head*, it is important to assess the leader's ability to know a path forward. Three characteristics need to be considered: (1) clarity of vision, (2) measurability of goals, and (3) actionability of plans.

The key questions to ask are shown in the following table:

Clarity of vision	*Does the leader have a clear understanding of the organization's current position? Does the leader have a notion of where the organization wants to go and when? And is the vision shareable? Is the vision clear enough to generate a common picture with the management team?*
Measurability of goals	*Does the leader have a clear understanding of the goals? Are the goals written? Are the goals measurable?*
Actionability of plans	*Does the leader understand the resources and competencies needed to transform goals to reality? Are the action plans clear and understood by everyone?*

The second half of the circle represents the execution, the hands on the activity.

The "Hands"

In examining the *Hands*, it is important to assess the leader's ability to execute and make the plans happen. Three characteristics need to be considered: (1) efficiency, (2) traceability, and (3) positivity.

The key questions to ask are shown in the following table:

Efficiency of execution	*Blocking and scheduling. Are schedules and timelines well timed? Is time blocked? Is the scheduled and blocked time adhered to? Are contingencies in place? Are the plans executed with the right resources by the right people? Do the people selected have the discipline to execute?*
Traceability of actions	*Is proper monitoring in place? Is the measuring and tracking done over time showing patterns and trends? Is it policing or managing? Are feedback channels provided and are measurements shared with all in the process?*
Positivity	*Does the leader cultivate the proper organizational attitude and motivation for the achievement of goals? Are all team members inspired to do their best? Is performance being monitored? Does it allow for positive contemporaneous feedback on improved performance?*

In any business enterprise, leadership plays a pivotal role that leads to failure or success. The sooner one sees the true color the better. In the view of Cummings (2007), leadership is much like beauty, it is appreciated after you have seen it.

References

Ariely, D. 2010. *Predictably Irrational.* New York: Harper Perennial.

Bossidy, L., R. Charan, and C. Burck. 2002. *Execution: The Discipline of Getting Things Done.* New York: Random House Books.

Carson, J., P. Tesluk, and J. Marrone. 2007. "Shared Leadership in Teams: An Investigation of Antecedent Conditions and Performance." *Academy of Management Journal* 50, no. 5, pp. 1217–34.

Cummings, R.L. 2007. "Can Modern Media Inform Leadership Education and Development?" *Advances in Developing Human Resources* 9, no. 2, pp. 143–45.

Day, D.V. 2001. "Leadership Development: A Review in Context." *Leadership Quarterly* 11, no. 4, pp. 581–613.

Deming, W.E. 1986. *Out of the Crisis.* Cambridge: Cambridge University Press.

Doz, Y., and M. Kosonen. Spring 2008. "The Dynamics of Strategic Agility: Nokia's Rollercoaster Experience." *California Management Review* 50, no. 3, pp. 95–118.

Jung, D., C. Chow, and A. Wu. 2003. "The Role of Transformational Leadership in Enhancing Organizational Innovation: Hypotheses and Some Preliminary Findings." *Leadership Quarterly* 14, no. 4, pp. 525–44.

Liden, R.C., A. Panaccio, J.D. Meuser, J. Hu, and S.J. Wayne. 2014. "Servant Leadership: Antecedents, Processes, and Outcomes." In *The Oxford Handbook of Leadership and Organizations*, ed. D.V. Day, 357–79. Oxford, England: Oxford University Press.

McGregor, D. 1960. *The Human Side of Enterprise.* New York: McGraw Hill.

Mittal, R., and P.W. Dorfman. 2012. "Servant Leadership Across Cultures." *Journal of World Business* 47, no. 4, pp. 555–70.

Myers, P.J. 1994. *Effective Personal Productivity.* Waco, TX: Leadership Management Inc.

Nemanich, L.A., and R.T. Keller. 2007. "Transformational Leadership in an Acquisition: A Field Study of Employees." *The Leadership Quarterly* 18, no. 1, pp. 49–68.

Tichy, N.M., and M.A. Devanna. 1986. "The Transformational Leader." *Training & Development Journal* 40, no. 7, pp. 27–32.

Waldman, D.A., and D. Siegel. 2008. "Defining the Socially Responsible Leader." *The Leadership Quarterly* 19, no. 1, pp. 117–31.

CHAPTER 7

Developing Evidence-Based Data on Ethics and Culture

Duane Windsor

Introduction: Ethics and Culture

Three key instances of corporate ethics failures since 2000 have been Enron (together with its external auditor Arthur Andersen), Satyam, and Siemens. These failures are well documented in the literature. Ethics failure reflects failures as well in corporate governance and internal controls, and sometimes external public and private regulation. At Satyam, an Indian information technology (IT) firm, the founder orchestrated a falsification of the firm's true economic situation. That founder confessed to the fraud, and with some others, was in 2015 sentenced to prison. Another Indian IT firm took over and absorbed Satyam (which means *truth*). German and U.S. authorities alleged a global bribery scheme by the German firm Siemens, focused particularly on telecommunications equipment. Siemens paid at least $2 billion in fines and internal transformation costs. At Enron, senior officers allegedly orchestrated a series of maneuvers for expanding the firm and concealing underlying problems; various senior officers went to prison, the firm disappeared in bankruptcy, and its external auditor Arthur Andersen also disappeared as a result of federal prosecution. (The U.S. Supreme Court later overturned the criminal conviction of Arthur Andersen.)

The failure or governmental rescue of a number of major firms in the investment banking industry—such as American Insurance Group, Bear Stearns, and Lehman Brothers—both involved poor risk management with respect to residential mortgages, associated with defects in culture and ethics. Recent scandals in the global banking industry—including foreign currency exchange, money laundering, and manipulation of interbank lending rate reports—do reflect similar defects in culture and ethics.

Scope

A key recommendation for improving the practice of business ethics is that executives and boards of directors should foster an ethical culture and climate for the guidance of employees. However, no definite measurement system for assessing culture and climate presently exists. There is a lot of fragmented practice, and academic and consulting literature. Organizations do internal and external surveys, develop codes of conduct, maintain hotlines, collect statistics, and provide training in ethics and values combined with often annual completion tests required for employees. There are various questionnaire instruments designed to obtain data about climate, culture, ethics, and values. Emphasis should be placed on three dimensions: organizational culture, ethical culture, and ethical climate.

Organizational Culture

Because the concept can denote a wide range of social phenomena, organizational culture has multiple definitions in the academic literature; and there is little concurrence on either precise definition or methodology for observation and measurement (Scott et al. 2003). Multiple dimensions of a culture may influence the personal values and ethical behavior of organizational members.

> Organizational culture defines an organization's uniqueness and identity. It is made up of basic assumptions (values, beliefs, attitudes, norms, and characteristic patterns of behavior) that are shared and adopted by individuals in an organization to cope with internal and external pressure. Understanding an organization's culture helps us to understand why organizations do what they do and achieve what they achieve. (Kets De Vries Institute [KDVI] 2015)

Organizational culture is one of the supporting pillars for many activities, such as total quality management (TQM). Zeitz, Johannesson, and Ritchie (1997) designed a 113-item survey instrument to measure 13 a priori dimensions of TQM and 10 a priori dimensions of organizational culture as experienced by members of the organization. Analysis of a sample of 886 respondents found that seven of the TQM dimensions and five of the culture dimensions explained most of the variance and were significantly related to stages of a formal TQM program.

Ethical Culture

The basic definition for ethical culture is the set of attitudes, beliefs, norms, and values that prevail within the organization (Stoner 1989). These attributes may shape the expectations and behaviors of employees. A significant problem for both research and management forensics is the possible existence of subcultures (organizational and ethical). While one can say that an organization has a culture at some level of abstraction, parts of the organization may have distinct subcultures (Hofstede 1998). Hofstede measured the organizational culture of a Danish insurance company (with 3,400 employees) using employee answers to 18 questions concerning work practices. Analyzing 131 work groups within the firm, he concluded that there were three distinct subcultures: professional, administrative, and customer interface. Additionally, these cultural differences affected practices with tangible effects.

Ethical Climate

Key (1999) raised a fundamental question about ethical culture, as defined earlier: Survey methods can ascertain individual perceptions, but not shared beliefs about ethical culture (unless all individuals have precisely the same beliefs, a condition that seems unlikely). This criticism does not rule out discovering ethical climate, which is about individual experiences. The basic definition for ethical climate is how employees individually experience ethical culture within an organization. Windsor (forthcoming) is a conceptual study of what we know about how employee perceptions of organizational politics, values, and management practices influence employee behaviors (see Windsor [2015], on conceptualizing the ethics of power and politics in organizations). One article proposes seven mechanisms by which organizational leaders can establish a value-based culture and climate; leaders at different levels likely will rely on different mechanisms to communicate value and expectations. Emphasizing the importance of ethical values can help shape shared understandings and perceptions (Grojean et al. 2004; Wittmer and Coursey 1996).

Analysis

For the practice of managerial forensics, the complexity of culture and climate means that there are significant methodological difficulties in establishing key metrics. Competing or complementary definitions and multiple dimensions require *triangulation* through multiple qualitative and quantitative measures to establish a reasonable picture of an organization's culture (Duncan 1989). A good guide is available in the 20 case studies published by Hofstede et al. (1990).

In an organization, values influence practices, which in turn interact with external expectations and generate external impacts. The model for culture and ethics forensics outlined in Figure 7.1 suggests the following:

1. *Internal assessment*—Managerial forensics is partly concerned with the attributes and effects on employee behaviors and organizational practices of the organizational culture, the ethics culture, and the ethics climate. The two major dimensions—organizational culture and ethics culture and climate—interact in possibly complicated

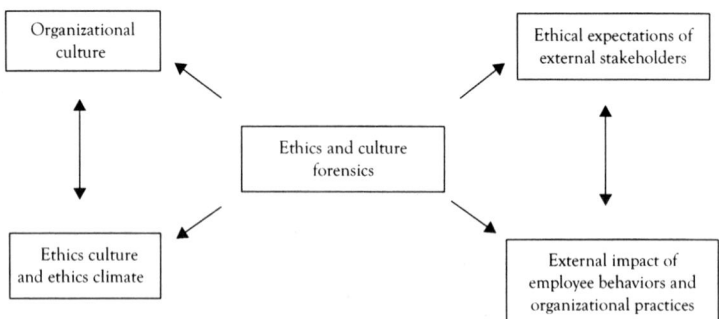

Figure 7.1 Model for ethics and culture forensics

ways so as to change over time, with consequent effects on behaviors and practices.

2. *External assessment*—Managerial forensics is also concerned with the ethical expectations of external stakeholders and the impacts on those stakeholders of employee behaviors and organizational practices. As with the internal assessment, the two major dimensions—expectations and impacts—interact in possibly complicated ways to affect organizational performance outcomes. There are approaches proposed for assessing corporate social performance and corporate social responsibility (CSR) or corporate citizenship (Abbott and Monsen 1979; Maignan and Ferrell 2000; Turker 2009; Wood 2010). Turker collected data from 269 business professionals working in Turkey. Factor analysis yielded a four-dimensional structure of CSR: to social and nonsocial stakeholders, employees, customers, and government. Maignan and Ferrell surveyed 210 U.S. and 120 French managers concerning corporate citizenship, defined in terms of the four dimensions of economic, legal, ethical, and discretionary conduct typically used also in the CSR literature. These approaches can be associated with stakeholder relationships. Sirgy (2002) models organizational performance (i.e., long-term survival and growth) in terms of quality of relationships among internal stakeholders (internal service quality), between internal and external stakeholders (external service quality), and between internal and more distal stakeholders (company goodwill). Corporate performance measurement identifies strengths and weaknesses in these relationships.

3. *Feedback effects on organizational performance outcomes*—Figure 7.1 does not explicitly depict feedback effects on internal behaviors and practices from external impacts. There are two general possibilities for study. Positive impacts may enhance organizational reputation and performance outcomes. Blunders and failures generate negative impacts adversely affecting reputation and performance outcomes. Enron, Arthur Andersen, and Satyam disappeared as companies. Siemens underwent a major internal transformation in culture and controls.

Expert Opinion

Academic literature and practitioner views support the importance of aligning organizational culture, ethical culture and ethical climate, and external expectations for sustainable organizational performance outcomes (Barney 1986; Duncan, Gintei, and Swayne 1998). There is support for the argument that values underlie culture in organizations (*Business Week* 1980; Wiener 1988). A number of useful reviews of both literature and questionnaires are available, and such reviews provide access to a wide range of questionnaires for psychometric data collection (see Delobbe, Haccoun, and Vandenberghe 2002; Jung et al. 2009; Scott et al. 2003).

The Ethics Resource Center (2013) reports that large companies (defined as 90,000 or more employees) can dramatically improve integrity through effective ethics and compliance programs and strong ethics cultures to reduce employee misconduct, improve workplace behaviors, and increase employee reporting of observed misconduct. At such companies, effective programs cut violations in half and significantly reduce retaliation expectations and pressure to compromise standards.

The National Business Ethics Survey, conducted biannually since 2007 by the Ethics Resource Center (2013), is a survey of a large representative sample of U.S. workers. The observed misconduct is down to a historic low. A comparison of the 2013 and 2011 surveys shows some progress across key dimensions: percentage of workers reporting observed misconduct on the job declined to 41 percent from 45 percent (the record high was 55 percent six years previously); this decline occurred systematically between 2011 and 2013 across 26 specific categories of misconduct; perceived pressure to compromise personal standards fell from 13 percent to 9 percent. The proportion of workers reporting misconduct they observed was about the same (65 percent in 2011 versus 63 percent in 2013, and 63 percent in 2009). About one-third did not report observed misconduct. Reported expectation of some form of retribution for reporting misconduct was at a high value of 22 percent in 2011, barely down to 21 percent in 2013. About 87 percent who observe violations in large companies with effective programs report those violations, compared to 32 percent when programs are lacking. However,

60 percent of misconduct reportedly involved someone at the supervisory level or above. About 24 percent of observed misdeeds involved senior managers. About 26 percent of misconduct was reported as ongoing within the firm, and 12 percent was occurring company-wide.

Recommendations and Tools

Based on the instances of misconduct noted and the research literature, some recommendations can be offered to practitioners assessing leadership preparedness for developing evidence-based data on ethics and culture. Leaders can conduct internal and external culture and climate audits. Organizational members conduct an internal audit. Academics or consultants conduct an external audit.

There is a well-established tradition of culture auditing (Wilkins 1983). "A culture audit can be used to measure how far an organization's behavior matches its expressed values" in order to align values and practices with strategy (KDVI 2015). The KDVI Organizational Culture Audit consists of a questionnaire covering 12 proposed dimensions of organizational culture: competitiveness, social responsibility, client or stakeholder orientation, change orientation, teamwork, fun, responsibility and accountability, trust, learning environment, results orientation, respect for the individual, and entrepreneurship. A different Cultural Audit Instrument, a 17-page questionnaire, measures each respondent's perceptions of his or her own and others' situations in relationship to their ideal situation—summarizing the findings as four bipolar descriptive types: homogeneous versus heterogeneous, enriched versus managed, developing versus stationary, and balanced versus dissonnant (Fletcher and Jones 1992). Monitoring of culture and climate through identifiable indicators is vital to measure change and improve management practices (Franzoni 2014). A review of 20 organizational culture questionnaires by Delobbe, Haccoun, and Vandenberghe (2002) finds that despite conceptual overlaps on a few core dimensions of organizational culture psychometric support for most instruments is weak. The authors thus proposed a different measure to capture commonality among cultural dimensions, and tested the instrument on a large sample of respondents across firms and industrial sectors.

It is arguably more difficult to identify ethical culture, and there is a considerable literature concerning measurement of ethics within organizations (Bernard 2013; Davis, Anderson, and Curtis 2001; Kaptein et al. 2005; Lather and Singh 2009). One example of an ethics audit instrument is *The Social Work Ethics Audit* (Reamer 2001). The goal is to help assess the adequacy of ethics-related policies, practices, and procedures with respect to customers, employees, documentation, and decision making. Kaptein et al. (2005) propose periodic surveys to discover information about employee misconduct and internal control defects. Managers can use detailed cultural and ethical information (Kaptein et al. 2005). However, there are reportedly more than 250 global and local ethical indices schemes including the FTSE4 Good and the Dow Jones Sustainability Index (Johnson 2013).

There is also, in any ethical assessment, a fundamental problem of where to assign responsibility. Kish-Gephart, Harrison, and Treviño (2010) distinguish among three antecedents of unethical choices, based on a meta-analysis of literature. These three antecedents concern the individual employee ("bad apple"), a difficult moral issue ("bad case"), and a difficult organizational environment ("bad barrel").

Both culture audits and ethics audits can be focused on specific considerations such as workplace safety (Molenaar, Park, and Washington 2009) or ethical trade (Sedex 2012); or specific functions, units, or personnel such as sales and marketing employees (Valentine and Barnett 2007). And such audits can be adapted to nonprofit organizations (National Council of Nonprofits 2011) and government entities (U.S. General Accounting Office [GAO] 1992; Svensson, Wood, and Callaghan 2010). Some studies demonstrate the feasibility of cross-national studies (Akram and Azad 2011; Vigoda-Gadot and Kapun 2005). Schwartz and Weber (2006) validated a survey instrument across nations for various kinds of organizations: academia, business, social or ethical investment, business ethics organizations, government activity, social activist groups, and media coverage.

Relatively little is known about the effectiveness of ethics training on ethical culture and climate. Valentine and Fleischman (2004) examined the relationship between ethics training and employee perception in a national sample of 313 business professionals working in the United

States. The findings suggest that formalized ethics training programs improve employees' perceptions of company ethical context and improve job satisfaction (Al-Zu'bi 2010).

Corporate citizenship and CSR remain controversial with respect to definition and specification (Turker 2009). Windsor (2013b) emphasizes that preventing corporate social irresponsibility is arguably much more important than promoting voluntary corporate citizenship. The analysis separates the general rubric of CSR into the three dimensions of avoiding harm to others (business ethics), legal compliance (obeying criminal and civil laws), and voluntary corporate citizenship. For purposes of ethical auditing, business ethics and legal compliance overlap considerably; the vital difference is between moral integrity (Windsor 2013a) and respect for law.

Table 7.1 provides suggested evaluation questions for ethics and culture forensics. The table is structured for organizational culture, ethical culture, ethical climate, and organizational preparedness.

Table 7.1 Evaluation questions for ethics and culture forensics

Organizational culture	What are the dimensions and attributes of the organization's culture in relationship to strategy and performance? What are the important subcultures within the organization? How do these dimensions, attributes, and subcultures affect the ethical culture of the organization and in turn organizational performance outcomes? What are the operational indicators to track on a continuing basis? How should such tracking and internal or external audits be conducted, and by whom?
Ethical culture	What are the dimensions and attributes of the organization's ethical culture? Are there subcultures with respect to ethics? What are the operational indicators to track on a continuing basis? How should such tracking and internal or external audits be conducted, and by whom? Is the organization's culture consistent with the desired and expected ethical culture? Have there been blunders and failures during the reporting period? Who are moral exemplars in the organization?
Ethical climate	What do periodic surveys of employees and external stakeholders reveal about their experiences of the organization's ethical culture? Is the reported ethical climate consistent with the desired and expected ethical culture? What happens to whistleblowers in the organization?
Organizational preparedness	Does the organization have the requisite systems, controls, policies, instruments, and other resources to foster and support an ethical culture? What are the leadership's values and motives? Is leadership committed to fostering a strongly ethical culture and a positive ethical climate for employees and external stakeholders?

Inattention to culture and climate can result in the kinds of problems noted in the Introduction for companies such as Enron and Arthur Andersen, Satyam, and Siemens. A reported instance of inattention is a German–Hong Kong joint venture Accuform (Woo, Lau, and Wong 2006). The joint venture produced chemical coatings for garments sold in China through a local partner. An unauthorized Chinese firm—in reality owned by an Accuform employee—stole a coating and sold treated clothing camouflaged to appear as if an Accuform product. Allergic reactions in some children in China revealed the misconduct through wide media coverage. Investigation revealed a tangled web of money laundering, theft of assets, illegitimate rebates, and bribery. The case reveals key differences in practices and values among German, Hong Kong, and Chinese firms (Wong and Ellis 2002).

Ethical challenges exist in organizations worldwide. The effective practice of managerial forensics helps to minimize potential risks to the organization and potential harms to external stakeholders, and thus contributes to the organization's future success.

References

Abbott, W.F., and R.J. Monsen. 1979. "On the Measurement of Corporate Social Responsibility: Self-reported Disclosures as a Method of Measuring Corporate Social Involvement." *Academy of Management Journal* 22, no. 3, pp. 501–15.

Akram, M., and M.K.K. Azad. 2011. "Attitude Towards Business Ethics: Comparison of Public and Private Organizations in Pakistan." *International Journal of Economics and Management Sciences* 1, no. 3, pp. 73–77.

Al-Zu'bi, H.A. 2010. "A Study of Relationship Between Organizational Justice and Job Satisfaction." *International Journal of Business and Management* 5, no. 12, pp. 102–09.

Barney, J.B. 1986. "Organizational Culture: Can It Be a Source of Sustained Competitive Advantage." *Academy of Management Review* 11, no. 3, pp. 656–65.

Bernard, C. April 9, 2013. "Ethical Assurances to Avoid Reputational Damage: The Ethical Audit." http://www.kpmgfamilybusiness.com/ethical-assurances-avoid-reputational-damage-ethical-audit/

Business Week. 1980. "Corporate Culture: The Hard-to-Change Values That Spell Success or Failure." no. 2660, pp. 148–54, 158–60.

Davis, M.A., M.G. Anderson, and M.B. Curtis. 2001. "Measuring Ethical Ideology in Business Ethics: A Critical Analysis of the Ethics Position Questionnaire." *Journal of Business Ethics* 32, no. 1, pp. 35–53.

Delobbe, N., R.R. Haccoun, and C. Vandenberghe. 2002. "Measuring Core Dimensions of Organizational Culture: A Review of Research and Development of a New Instrument." Working Paper 53-2002. https://www.uclouvain.be/cps/ucl/doc/iag/documents/WP_53_Delobbe.pdf

Duncan, W.J. 1989. "Organizational Culture: 'Getting a Fix' on an Elusive Concept." *Academy of Management Executive* 3, no. 3, pp. 229–36.

Duncan, W.J., P.M. Gintei, and L.E. Swayne. 1998. "Competitive Advantage and Internal Organizational Assessment." *Academy of Management Executive* 12, no. 3, pp. 6–16.

Ethics Resource Center. 2013. The State of Ethics in Large Companies: *A Research Report from the National Business Ethics Survey® (NBES®) and* About the National Business Ethics Survey of the U.S Workforce 2013. http://www.ethics.org/nbes/

Fletcher, B., and F. Jones. 1992. "Measuring Organizational Culture: The Cultural Audit." *Managerial Auditing Journal* 7, no. 6, pp. 30–36.

Franzoni, S. 2014. "Measuring Corporate Culture." *Corporate Ownership & Control* 10, pp. 308–16.

Grojean, M.W., C.J. Resick, M.W. Dickson, and D.B. Smith. 2004. "Leaders, Values, and Organizational Climate: Examining Leadership Strategies for Establishing an Organizational Climate Regarding Ethics." *Journal of Business Ethics* 55, no. 3, pp. 223–41.

Hofstede, G. 1998. "Identifying Organizational Subcultures: An Empirical Approach." *Journal of Management Studies* 35, no. 1, pp. 1–12.

Hofstede, G., B. Neuijen, D.D. Ohayv, and G. Sanders. 1990. "Johnson, D. October 30, 2013. "While Ethical Indices Can Sometimes Cause Confusion, They Provide a Useful Benchmark for Companies and Stakeholders." http://www.ethicalcorp.com/business-strategy/ethical-indices-how-do-you-measure-ethics

Jung, T., T. Scott, H.T.O. Davies, P. Bower, D. Whalley, R. Mcnally, and R. Mannion. 2009. "Instruments for Exploring Organizational Culture: A Review of the Literature." *Public Administration Review* 69, no. 6, pp. 1087–96.

Kaptein, M., L. Huberts, S. Avelino, and K. Lasthuizen. 2005. "Demonstrating Ethical Leadership by Measuring Ethics: A Survey of U.S. Public Servants." *Public Integrity* 7, no. 4, pp. 299–311.

Kets De Vries Institute (KDVI). 2015. "Organizational Cultural Audit." http://www.kdvi.com/Page/Organizational_Cultural_Audit

Key, S. 1999. "Organizational Ethical Culture: Real or Imagined?" *Journal of Business Ethics* 20, no. 3, pp. 217–25.

Kish-Gephart, J.J., D.A. Harrison, and L.K. Treviño. 2010. "Bad Apples, Bad Cases, and Bad Barrels: Meta-Analytic Evidence About Sources of Unethical Decisions at Work." *Journal of Applied Psychology* 95, no. 1, pp. 1–31.

Lather, A.S., and G.G. Singh. 2009. "Measuring the Ethical Quotient of Corporations: The Case of Small And Medium Enterprises in India." Forum on Public Policy. http://forumonpublicpolicy.com/spring09papers/archivespr09/lather.pdf

Maignan, I., and O.C. Ferrell. 2000. "Measuring Corporate Citizenship in Two Countries: The Case of the United States and France." *Journal of Business Ethics* 23, no. 3, pp. 283–97.

Molenaar, K., J. Park, and S. Washington. 2009. "Framework for Measuring Corporate Safety Culture and Its Impact on Construction Safety Performance." *Journal of Construction Engineering and Management* 135, no. 6, pp. 488–96.

National Council of Nonprofits. 2011. "Conducting an Ethics Audit at Your Nonprofit." https://www.councilofnonprofits.org/sites/default/files/documents/Conducting%20an%20Ethics%20Audit%20at%20Your%20Nonprofit.pdf

Reamer, F.G. 2001. *The Social Work Ethics Audit: A Risk Management Tool.* Washington, DC: NASW Press.

Schwartz, M.S., and J. Weber. 2006. "A Business Ethics National Index (BENI): Measuring Business Ethics Activity Around the World." *Business & Society* 45, no. 3, pp. 382–405.

Scott, T., R. Mannion, H. Davies, and M. Marshall. 2003. "The Quantitative Measurement of Organizational Culture in Health Care: A Review of the Available Instruments." *HSR: Health Services Research* 38, no. 3, pp. 923–45.

Sedex. 2012. "Sedex Members Ethical Trade Audit (SMETA) Best Practice Guidance." Version 4.0 May 2012 (Replaces V. 2.2. Sept 2010). http://www.sedex.org.uk/sedex/_website/pdf/smeta_best_practice_guidance.pdf

Sirgy, M.J. 2002. "Measuring Corporate Performance by Building on the Stakeholders Model of Business Ethics." *Journal of Business Ethics* 35, no. 3, pp. 143–62.

Stoner, C.R. 1989. "The Foundations of Business Ethics: Exploring the Relationship Between Organization Culture, Moral Values, and Actions." *SAM Advanced Management Journal* 54, no. 3, pp. 38–43.

Svensson, G., G. Wood, and M. Callaghan. 2010. "A Comparison of Business Ethics Commitment in Private and Public Sector Organizations in Sweden." *Business Ethics: A European Review* 19, no. 2, pp. 213–32.

Turker, D. 2009. "Measuring Corporate Social Responsibility: A Scale Development Study." *Journal of Business Ethics* 85, no. 4, pp. 411–27.

U.S. General Accounting Office (GAO). February 1992. Organizational Culture: Techniques Companies Use to Perpetuate or Change Beliefs and Values. Report to the Chairman, Committee on Governmental Affairs, U.S. Senate. Washington, DC. http://www.gao.gov/assets/220/215677.pdf

Valentine, S., and T. Barnett. 2007. "Perceived Organizational Ethics and the Ethical Decisions of Sales and Marketing Personnel." *Journal of Personal Selling & Sales Management* 27, no. 4, pp. 373–88.

Valentine, S., and G. Fleischman. 2004. "Ethics Training and Businesspersons' Perceptions of Organizational Ethics." *Journal of Business Ethics* 52, no. 4, pp. 391–400.

Vigoda-Gadot, E., and D. Kapun. 2005. "Perceptions of Politics and Perceived Performance in Public and Private Organizations: A Test of One Model Across Two Sectors." *Politics & Policy* 33, no. 2, pp. 251–76.

Wiener, Y. 1988. "Forms of Value Systems: A Focus on Organizational Effectiveness and Cultural Change and Maintenance." *Academy of Management Review* 13, no. 4, pp. 534–45.

Wilkins, A. 1983. "The Culture Audit: A Tool for Understanding Organizations." *Organizational Dynamics* 12, no. 2, pp. 24–38.

Windsor, D. 2013a. "A Typology of Moral Exemplars in Business." In *Moral Saints and Moral Exemplars, Vol. 10 of Research in Ethical Issues in Organizations*, eds. M. Schwartz and H. Harris, 63–95. Bingley, UK: Emerald Group Publishing.

Windsor, D. 2013b. "Corporate Social Responsibility and Irresponsibility: A Positive Theory Approach." *Journal of Business Research* 66, no. 10, pp. 1937–44.

Windsor, D. 2015. "Ethics in Managing Corporate Power and Politics." In *Teaching Ethics across the Management Curriculum: A Handbook for Faculty*, ed. K. Ogunyemi, 277–302. New York: Business Expert Press.

Windsor, D. (forthcoming 2016). "The Ethical Sphere: Organizational Politics, Fairness and Justice." In *Handbook of Organizational Politics: Looking Back and to the Future*, eds. E. Vigoda-Gadot and A. Drory, 2nd ed. Cheltenham, UK and Northampton, MA: Edward Elgar Publishing (accepted June 11, 2015).

Wittmer, D., and D. Coursey. 1996. "Ethical Work Climates: Comparing Top Managers in Public and Private Organizations." *Journal of Public Administration Research and Theory* 6, no. 4, pp. 559–72.

Wong, P.L., and P. Ellis. 2002. "Social Ties and Partner Identification in Sino-Hong Kong International Joint Ventures." *Journal of International Business Studies* 33, no. 2, pp. 267–89.

Woo, C.H.L., A. Lau, and R. Wong. 2006. Accuform: Ethical Leadership and Its Challenges in the Era of Globalization. Hong Kong University Case HKU622.

Wood, D.J. 2010. "Measuring Corporate Social Performance: A Review." *International Journal of Management Reviews* 12, no. 1, pp. 50–84.

Zeitz, G., R. Johannesson, and J.E. Ritchie. 1997. "An Employee Survey Measuring Total Quality Management Practices and Culture Development and Validation." *Group & Organization Management* 22, no. 4, pp. 414–44.

CHAPTER 8

How to Perform an Autopsy on Marketing Strategy

Donald E. Sexton

Introduction: How Companies Fail

Business failures ultimately are due to lack of cash flow. Decisions in many areas such as production, finance, research and development, human resources, and marketing affect cash flow. However, if cash flow is the blood of a company, then marketing—managing products or services successfully—is the heart.

Marketing decisions impacts cash flow through value as perceived by the customer, which directly affects quantity sold and prices paid. Value as perceived by the customer is determined by many marketing decisions such as choice of target market, positioning of product or service, pricing, advertising spend, advertising message, promotion, distribution, and customer service.

This chapter establishes a framework for a marketing autopsy by explaining how marketing drives financial outcomes such as revenue, contribution, profit, and cash flow through perceived value. The key marketing metrics for understanding failures of products or services are perceived value and customer value added (CVA®). Perceived value predicts revenue and CVA® predicts contribution. Decreases in perceived value and CVA® can lead to the death of a product or service. A marketing autopsy examines not only the behavior of perceived value and CVA® but what marketing actions may have caused decreases in perceived value and CVA®.

A systematic plan for performing a marketing autopsy on a failed product or service is described. The autopsy consists of analysis of decreases in the perceived value and CVA® associated with a product or service, what marketing actions may have led to those decreases, and how those decreases adversely affected financial performance. The autopsy provides answers to the question: What marketing mistakes led to the financial failure of a product or service?

Financial Flows Within an Organization

The heart of a business strategy—arguably—is the marketing function (Sexton 2009b). Successful marketing pumps cash flow through an organization, as a heart pumps blood (Exhibit 8.1).

Consider a start-up company planning to produce a new product or a new service. Initially, capital must be amassed from both loans and the sale of shares in the company. That capital is utilized as needed to purchase equipment and materials and to hire people. Then, the operations begin. If the company produces a product, then there are inventories to manage, including raw materials and semifinished products as well as finished products. However, no money comes in until a sale is made.

What is the one thing one needs to have a company? A customer. Factories, warehouses, and employees do not make a living company. One needs customers who purchase the product or service.

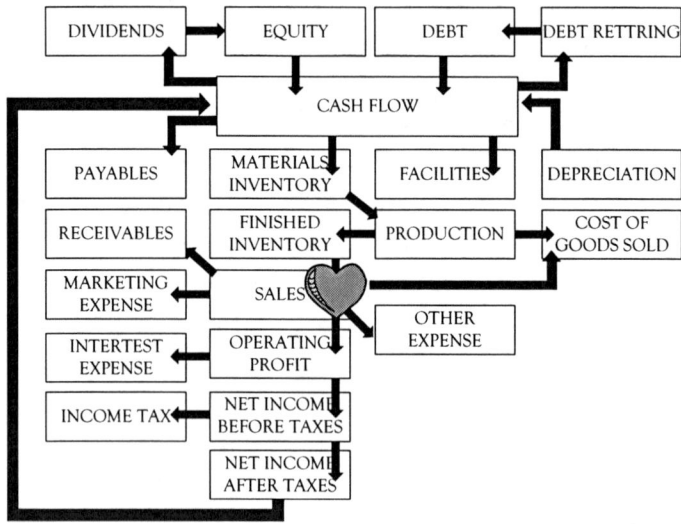

Exhibit 8.1 Cash flows within a company

Source: Don Sexton, Marketing ROI, presentation, CEIBS, Shanghai, 2011.

Net Income After Taxes

If all the marketing activities are successful, then sales are made and revenue is generated for the product or service. That revenue is then applied to pay the costs of goods sold (or services provided) and selling expenses. Deducting variable costs from revenue leads to variable margin or contribution. Deduction of the fixed costs of goods sold and the fixed selling expenses yields operating profit or operating income. Deducting interest expense and income taxes gives net income after taxes for the product or service.

Cash Flow

Profit is not the only financial outcome of interest. Many analysts pay considerable attention to cash flow. As the business operates, there are entries on the balance sheet that affect cash flow. In addition, depreciation is a noncash cost and hence needs to be added to operating income to show the true cash-flow position of the company. Profit is an *opinion* as it depends on how revenue, depreciation, and other accounting metrics are treated; cash flow is a *fact* and determines whether an organization lives or dies.

That entire cash flow diminishes if the product or service does not generate revenue as expected or if costs are higher than expected. If cash flow drops below a certain level, then the company no longer has cash to grow or even to maintain its current position. If the diminished cash flow continues for sufficient time, the company goes out of business unless it receives a transfusion of capital from some other source.

Perceived Value

One can understand how marketing drives financial performance with a single metric: perceived value (Gaber et al. 2009; Sexton 2009a, 2010a, 2010b).

Perceived value is the most a customer will pay for one unit of a product or service. Perceived value, also known as willingness to pay, represents the ceiling on price for a given customer. If the quoted price is above the perceived value of that customer, he or she will not buy. If the quoted price is below the perceived value of that customer, the probability that he or she will buy is associated with the distance between the perceived value and the price—the "incentive" for the customer to purchase (Exhibit 8.2). The larger the distance, the larger the incentive to buy.

The likelihood of the purchase depends not just on the size of the incentive to purchase a company's product or service. It depends also on the incentives to purchase competing products or services that the target customer is considering. Other things equal, the customer will purchase the product or service with the highest incentive. For example, if the incentive—perceived value less price—for one smartphone is $120 and for another smartphone is $90 and if the customer is considering only these two smart phones, then the customer is more likely to purchase the first smartphone.

Perceived value has been found by many researchers working with different data sets to be strongly associated with financial metrics such as revenue, contribution, profit, cash flow, and shareholder value (Aaker and Jacobson 1994, 2001; Agres 1996; Aksoy et al. 2008; Buzzell and Gale 1987, 2003, 2005; Mizik and Jacobson 2008; Sexton 2010b;). Understanding perceived value and how it is determined is therefore a prerequisite to understanding the reasons for the success or failure of a product or service. Unfortunately, many companies ignore perceived value in their diagnoses of business success or failure (Association of National Advertisers [ANA] 2004, 2005, 2006, 2007a, 2007b, 2007c,

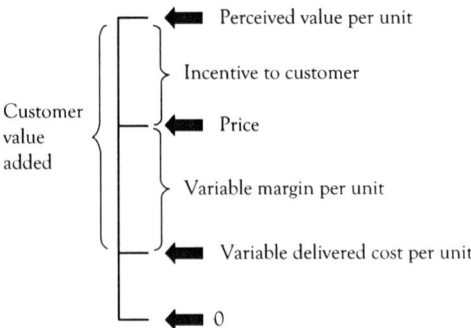

Exhibit 8.2 Perceived value and CVA®

Source: Sexton (2009c).

2008, 2009; Beaman, Guy, and Sexton 2008; Sexton and Rogers 2012; See 2008; Sexton 2015; Sexton and ANA 2013).

Determinants of Perceived Value

Perceived value depends on the benefits associated with the product or service by the customer. However, these are not necessarily the actual benefits but, rather, are the benefits *as perceived* by the customer—the perceived benefits. That means that the management of perceived value involves both the actual benefits given to the customer as determined by design activities—as well as the customer's perceptions of those benefits as determined by communications activities.

Perceived value is not additive—one cannot simply add evaluations of the benefits weighted by the relative importance of the benefits to derive perceived value. Perceived value varies by customer—including different members of a decision-making unit (DMU) within a company or within a family. Perceived value can also be expected to vary across market segments.

There are several ways to estimate perceived value, including:

Direct customer response
Indirect customer response
Price experiments
Value-in-use
Judgment

See Sexton (2009c) for explanations of these techniques.

How Marketing Manages Perceived Value

There are many ways to manage perceived value. Some involve changing the product or service; most of them involve communications (Sexton 2006).

Improve Performance of Product or Service and Inform Customers

At the core of a successful marketing strategy is a product or service that meets or surpasses customer expectations. That is the design of the product or service (Sexton 2009b). Design depends on marketing input in the form of customer insights but design also depends on research and development, on engineering, and on operations. Sometimes, products or services appear that do not fit well with customer expectations. If there is time, those products or services can be redesigned or, at least, their performance improved. This makes the most sense when the additional cost of the product or service improvement is less than the resulting increase in perceived value.

Note that if a product or service is improved, then customers must also be informed of those improvements—otherwise the increased actual value will have no impact on the purchase decision since perceived value will not have changed.

Add Benefits to Product or Service and Inform Customers

Similar to improving product or service performance, one can add benefits to a product or service—hopefully benefits of high priority to the target customer so that they will potentially add more to perceived value than they will add to costs. In turn, those benefits must be communicated to the target customers if those additions are to have an impact on the perceived value of the product or service.

Close Gaps Between Actual Performance and Perceived Performance

A product or service may in fact meet or surpass customer needs but they may not realize it, resulting in a communication gap between actual and perceived values on certain benefit dimensions. This gap can be filled with adroit use of communications. The key decisions are the choice of communication media, the choice of message, and the choice of creative—the way the message is conveyed.

Change Benefit Priorities of the Target Customer

If a target customer does not consider important those benefits where the product or service delivers superior performance, then one can try to change those customer priorities to favor the benefits where the product or service is superior. This approach faces the challenge of trying to reorder the target customer's priorities, which may be somewhat more difficult than informing him or her of improved performance on existing benefits or the addition of benefits.

Improve the Brand Position with the Target Customer

The position of a brand should consist of one, two, or three of the key benefits that the product or service provides (Sexton 2008). More than three benefits become difficult for the customer to remember (Sexton 2006). Strong brands are built with consistency of message—in all forms of communication over time. The brand image plays a role in perceived value. For example, for luxury products such as fragrances, the brand reputation may represent 98 percent of perceived value while for products that are more like commodities such as semiconductors, the brand may represent only 5 percent of perceived value.

Brands may be weak because the associations between the brand identifiers such as the name or logo and the key benefits of the brand position may not be clear to the target customer. Again, this situation can be solved with consistent, targeted communications. One concern: The benefits communicated as associated with the brand must actually be delivered by the product or service. Brands are promises and, if the promise of a brand is not fulfilled, then the brand will not be viewed positively and the perceived value will likely decrease.

Change the Target Competitor

Perceived value depends on what competing products or services are being considered by the target customer. Sometimes, it is possible for a company to influence the competing products or services involved in the target customer's decision process. For example, a company can develop communications that invite comparisons with certain competitors (without mentioning them by name). This can be done by making comparisons to other processes or materials or by referring to general classes of competitors. A large auto-repair chain might invite comparisons with a *typical* local automobile shop.

Change the Target Member of the DMU

If the product or service does not appeal to the target member of the decision-making unit, then one might consider changing that DMU member and find a DMU member who wants the benefits that the product or service provides. For example, if a specific piece of machinery produces high output but is somewhat expensive compared to alternatives, then the target DMU member should be someone who values output and is not so concerned with purchase cost.

Change the Target Segment

If the product or service does not appeal to the target segment, then one might consider selecting another group of customers as the market segment. A hotel that provides a full-service business center should likely not focus on the family vacation market but should likely concentrate their efforts on the business traveler market. The specific services they offer—such as administrative help—might even suggest a target market segment narrower than all business travelers—perhaps just senior executive travelers.

Perceived Value and Revenue

Perceived value is directly associated with revenue. The higher the perceived value, the greater the demand for the product or service, other things such as distribution equal.

Specifically, a higher perceived value leads to either a higher price or a higher number of units sold or both. Perceived value has been shown to predict revenue with R-squared values in excess of 0.90 (Sexton 2010a).

The relationship between changes in perceived value and changes in revenue is surprisingly simple. It is given by Sexton's Revenue Law (Sexton 2009c): The relative change in revenue is equal to the square of the relative change in perceived value.

The impact of changes in perceived value—up or down—is exponential. That is why product or services sometimes find themselves in a vicious downward spiral where falling revenues lead to cost-cutting, which causes falls in perceived value that lead to even larger decreases in revenue, starting the chain of events again.

Customer Value Added and Contribution

Contribution or variable margin is revenue less variable costs, a major component of operating profit. While perceived value predicts revenue, CVA® predicts contribution, also with high accuracy.

CVA® is the difference between perceived value per unit of the product or service and the variable cost per unit (Exhibit 8.2). The larger the CVA®, the more the contribution. CVA® is *not* margin since perceived value is not price. CVA® is the net amount of value per unit—as perceived by customers—which the organization is providing society. Whenever perceived value per unit is below variable cost per unit, then CVA® is negative and the company is spending more to produce a unit of product or service than anyone will be willing to pay for it. If unsupported by other sources of funds, such a product or service will fail. Conversely, the more value—as perceived by customers—that a company provides above the cost of a unit, the higher the value of CVA® and the more successful is that product or service.

CVA® predicts contribution in a way similar to how perceived value predicts revenue. CVA® increases when perceived value per unit increases or when costs per unit decrease. If perceived value increases, then demand in units increases. If costs per unit decrease, then margin per unit increases. Both cases would lead to higher contribution.

However, other changes are also possible. For example, suppose that performance on benefits of a product or service is improved. That improvement might increase costs per unit. The net effect on CVA® might be positive or negative. If CVA® increases due to the improvements, then contribution increases other things equal. If CVA® decreases due to the improvements, then contribution decreases other things equal.

The relationship between changes in CVA® and to changes in contribution is similar to the relationship between changes in perceived value and changes in revenue. The relationship is described by Sexton's Contribution Law (Sexton 2009c): The relative change in contribution is equal to the square of the relative change in CVA®.

Since contribution is a key component of cash flow, Sexton's Contribution Law shows clearly that managers must manage not just costs and not just perceived value but their *difference*—CVA®. If CVA®

is not managed properly, then a company may fall into a financial death spiral.

Performing a Marketing Autopsy on a Failed Product or Service

When a heart stops beating, one looks for the reason. The heart of a marketing strategy is the perceived value to the target customer. When the perceived value per unit falls below costs per unit, then the heart of the marketing strategy has stopped beating. In other words, when CVA® becomes negative, the product or service no longer is a viable, stand-alone business.

A marketing autopsy then must begin with evaluations of perceived value per unit, costs per unit, and CVA® (Sexton 2014b). These metrics can be estimated over time and they can be estimated post hoc with sales data and cost data. After inspection of the behavior of perceived value, costs, and CVA®, the autopsy proceeds to the reasons for any changes in those metrics—changes due to decisions regarding marketing strategies and marketing programs. The autopsy then should be able to provide an explanation as to why the marketing heart stopped beating.

The marketing autopsy consists of four stages (Exhibit 8.3):

1. Evaluation of perceived value, costs per unit, and CVA®.
2. Evaluation of the marketing strategy choices.
3. Evaluation of the marketing program choices.
4. Evaluation of the details of the marketing program choices.

Autopsy Stage I: Evaluation of Customer Value Added

Three questions:

1. How did perceived value per unit trend over time? Did it decrease and, if so, why?
2. How did costs per unit trend over time? Did it increase and, if so, why?
3. How did CVA® per unit trend over time? Did it decrease and, if so, why?

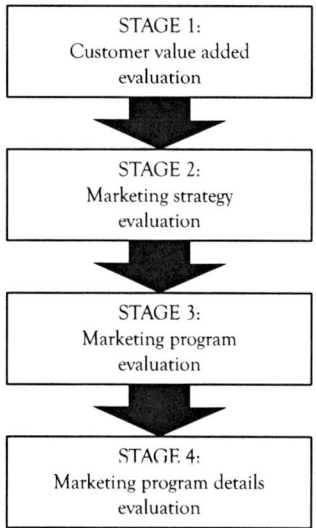

Exhibit 8.3 Stages in a marketing autopsy

Perceived value changes may be due to marketing strategy or marketing program actions considered in stages II and III of the autopsy (Sexton 2010). However, changes in perceived value (and CVA®) may also be due to competitors. During the autopsy, perceived value (and, if possible, CVA®) for competitors should be examined with the same perceived value estimation techniques used for the product or service under investigation.

The impact of competitor-perceived value changes can be summarized with the strategic theme chart (Exhibit 8.4). The vertical axis of the chart is perceived value per unit. The horizontal axis is variable cost per unit. Products or services that win are in the upper right hand corner of the chart—high perceived value or low cost, high perceived value or acceptable cost, or acceptable perceived value or low cost. Each of those locations on the chart would correspond to a product or service with a high CVA®. Locations elsewhere on the chart would describe mediocre or losing strategies. Perceived value for a product or service may have increased but may not have increased sufficiently against a powerful competitor. The same situation may occur with CVA®—even though the CVA® of a product or service has increased, the CVA® of one or more competitors may still be greater.

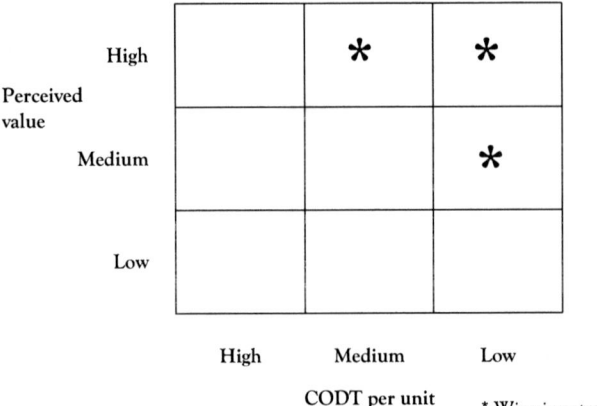

Exhibit 8.4 Strategic themes
Source: Sexton (2009c).

Incremental cost per unit should also be examined during stage I of the autopsy. This is not always easy since many companies do not provide managers with incremental or variable cost information. Often costs include overhead or other indirect costs and are presented as average cost per unit, which mixes variable and fixed costs together. In turn, average cost per unit is typically used to determine gross margin per unit, which is not at all the same as variable margin per unit. Gross margin per unit is always lower than variable margin per unit since it includes average fixed costs. Nonetheless, if gross margin or average cost per unit is all that is available, that is what must be used. However, if the business failure is due to a cost problem, it will be difficult to identify the specific reasons for that cost problem if only average cost per unit information is available.

Autopsy Stage II: Evaluations of Marketing Strategy Choices

Marketing strategy actions that affect perceived value include the target market decision and the positioning decision. To determine whether those are the reasons for decreases in perceived value, one can consider questions such as the following:

1. What were the most important benefits desired by the target segment?
2. What was the ideal product or service for that target segment?

3. What was the actual position of our product or service with respect to those benefits

4. Was there a design gap?

An efficient way to answer the aforementioned questions and examine the target market and positioning decisions is by using a perceptual map (Exhibit 8.5). A perceptual map shows the key benefits of a product or service for target customers. In Exhibit 8.5, assume that the two benefits pictured are the two most important benefits to customers in a specific target segment. The ideal point in the upper right represents the expectations of these target customers regarding those two benefits. The solid circle marked "A" is the position of the product or service being diagnosed. The other solid circles represent the positions of competitors Q, R, S, and T. As can be seen, there is a design gap between the actual value of product or service A and what the target customers want. That may account for the failure of product or service—it was not what the target customer wanted. In turn, that could have been a failure of the design or a failure of the target market choice.

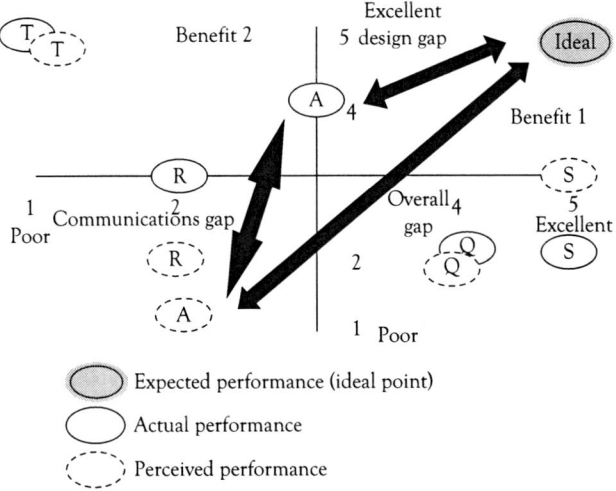

Exhibit 8.5 Perceptual map

Source: Sexton (2009a).

Autopsy Stage III: Evaluations of Marketing Program Choices

Many marketing tactics affect perceived value and CVA®. Such actions include:

- Traditional advertising
- Nontraditional advertising
- Social media
- Public relations
- Personal selling
- Customer service
- Brand expression
- Packaging
- Distribution
- Pricing.

Many of these tactics concern communications and suggest the following questions:

1. What was the actual position of our product or service?
2. What was the position of our product or service as perceived by the target customer?
3. Was there a communications gap?

The perceptual map (Exhibit 8.5) can also be utilized to identify communication gaps. The dotted circles represent the perceived positions of the product or service. Circle A is the perceived position of the product or service being autopsied while the other circles Q, R, S, and T are the perceived positions of the competing products or services. The perceived position of product or service A is in the southwest quadrant of the map shown in Exhibit 8.5—far away from its actual position. That distance represents a communication gap and may have caused the death of the product or service.

The two remaining tactics (not directly related to communications) are distribution and pricing. Distribution can be evaluated by looking at metrics concerned with distribution coverage. Even if CVA® is high, a product or service can fail because customers cannot find it to purchase.

Pricing can be evaluated by comparing price to perceived value per unit (Jain et al. 2015). The closer the price is to perceived value, the lesser the incentive for the customer to purchase. The further the price is from the perceived value, the more the incentive for the customer to purchase. In addition, the incentive for the product or service under scrutiny must be compared to the incentives for the competing products or services. Competitors may be offering more net value (incentive) to the target customer.

Autopsy Stage IV: Evaluations of the Details of the Marketing Program Choices

Each of the marketing programs listed in the previous section consists of several specific decisions such as amount spent on different types of advertising, magnitude of promotion, creative expression of message, and types of resellers (Mela, Gupta, and Lehmann 1997; Sexton 1970, 2012). The fatal mistake for the product or servi e could have occurred at this level. However, these programs are too diverse to articulate specific questions for each program in this chapter. In general, the questions at this stage of the autopsy would include the following:

1. Was the program on-strategy—did the program implement the target market and positioning decisions?
2. Was the level of support of the program sufficient—was enough money spent for the desired impact?
3. Was the program executed properly? For example, did the advertisements appear at the right times? Were the customer service representatives properly trained? Were the sales people allocated optimally?

One way to evaluate the marketing decisions at this level of detail is to use some form of the purchase funnel (Colley 1961; Lavidge and Steiner 1961). However, one should keep in mind that the order of the stages of the purchase funnel—awareness, knowledge, liking, preference, conviction, and purchase—has been found to vary according to the category of the product or service (Sexton 2005; Vakratsas and Ambler 1999). Some purchase funnel behavior—behavior for many consumer packaged goods, for example—begins with conviction to purchase rather than knowledge.

Numerous specific metrics for examining tactical details and evaluations of the usefulness of each metric by managers can be found in Beaman, Guy, and Sexton (2008).

Recommendations

The suggested steps for a marketing autopsy represent a *top-down* process. The marketing autopsy begins with the large question: Why did this product or service die? That question can be answered by inspecting the behavior of CVA® over time. All product or service failures due to marketing can be traced to a fall in CVA®—due to declines in perceived value or increases in costs or both. Next, the specific reasons for the observed changes in perceived value or costs are determined. In turn, that diagnosis leads to the examination of specific marketing decisions at either the strategic level, such as target markets and positioning, or the program level, communications, distribution, and pricing. Finally, the details of programs can be assessed to identify specific errors that may have led to the death of the product or service.

This approach to a marketing autopsy, moving from the major strategic choices to tactical choices to the details of tactical choices, constitutes a systematic way to determine the reasons for the death of a product or service (Sexton 2014b). Hopefully, such analyses of the remains of failed products or services can yield insights so that the same mistakes are not repeated and that other products or services are given an improved chance to live.

References

Aaker, D.A., and R. Jacobson. May 1994. "The Financial Information Content of Perceived Quality." *Journal of Marketing Research* 31, no. 2, pp. 191–201.

Aaker, D.A., and R. Jacobson. November 2001. "The Value Relevance of Brand Attitudes in High-Technology Markets." *Journal of Marketing Research* 38, no. 4, pp. 485–93.

Agres, S. 1996. *Brand Asset Valuator® Presentation.* New York: Columbia Business School.

Aksoy, L., B. Cooil, C. Groening, T.L. Keiningham, and A.Yalçin. July 2008. "The Long-Term Stock Market Valuation of Customer Satisfaction." *Journal of Marketing* 72, no. 4, pp. 105–22.

ANA (Association of National Advertisers). June 2004. *The State of Marketing Accountability Study.* New York: ANA.

ANA. 2005. *The Path to Marketing Accountability Study.* New York: ANA.

ANA. 2006. *The Path to Marketing Accountability Study.* New York: ANA.

ANA. April 2007a. *Brand Deterioration: How to Identify, Measure, and Respond.* New York: ANA.

ANA. July 2007b. *Marketing Accountability Study.* New York: ANA.

ANA. 2007c. *Survey of Finance Managers.* New York: ANA.

ANA. 2008. *Marketing Accountability Study.* New York: ANA.

ANA. 2009. *Marketing Accountability Study.* New York: ANA.

Beaman, K., G.R. Guy, and D.E. Sexton. 2008. *Managing and Measuring Return on Marketing Investment, Research Report 1435-08-RR.* New York: The Conference Board.

Buzzell, R.D., and B.T. Gale. 1987. *The PIMS* Principles.* New York: Free Press.

Colley, R. 1961. *Defined Advertising Goals for Measured Advertising Results.* New York: Association of National Advertisers

Gaber, S., D.E. Sexton, K. Sen, and A. Lin. 2009. "Customer Input: Incorporate Perceived Value into Marketing Strategy." *Business Digest,* October, pp. 3–9.

Gregory, J.R. 2003. *The Best of Branding.* New York: McGraw-Hill.

Gregory, J.R. 2005. *Driving Brand Equity.* New York: ANA.

Jain, R., V. Gorti, K. Sen, and D.E. Sexton. March 2015. "Brand Equity & Optimal Pricing." Paper presented at ReThink Conference, ARF, New York.

Lavidge, R., and G. Steiner. October 1961. "A Model for Predictive Measurements of Advertising Effectiveness." *Journal of Marketing,* 25, no. 5, pp. 137–69.

Mela, C.F., S. Gupta, and D.R. Lehmann. May 1997. "The Long-Term Impact of Promotion and Advertising on Consumer Brand Choice." *Journal of Marketing Research* 34, no.2, pp. 248–64.

Mizik, N., and R. Jacobson. February 2008. "The Financial Value Impact Perceptual Brand Attributes." *Journal of Marketing Research* 45, no. 1 pp. 15–32.

See, ed. 2008. "What Marketing Thinks . . . ." *Association of National Advertisers.* New York: ANA.

Sexton, D.E., August 1970. "Estimation of Marketing Policy Effects on Sales." *Journal of Marketing Research* 7, no. 4, pp. 338–47.

Sexton, D.E. February 2005. "Building the Brand Scorecard." *The Advertiser,* pp. 54–58.

Sexton, D.E. April 2006. "Pricing, Perceived Value, and Communications." *The Advertiser,* pp. 56–58.

Sexton, D.E. 2008. *Branding 101.* Hoboken, NJ: John Wiley & Sons, Inc.

Sexton, .E. August 2009a. "Achieving Marketing Nirvana." *The Advertiser,* p. 18.

Sexton, D.E. 2009b. *Marketing 101.* 2nd ed. Hoboken, NJ: John Wiley & Sons, Inc.

Sexton, D.E. April 2009c. *Value Above Cost: How CVA® Drives Financial Performance, the Most Important Metric You've Never Used.* Upper Saddle River, NJ: FT Press (Wharton School Publishing).

Sexton, D.E. February 2010a. "Competing with Customer Value Added." *Effective Executive,* pp. 44–47.

Sexton, D.E. 2010b. "How to Determine Marketing ROI." *ANA Insight Brief.* New York: ANA.

Sexton, D.E. 2014a. "Implementing Marketing Metrics in Organizations: Opportunities and Challenges," In *The Handbook of Customer Equity,* eds. V. Kumar and D. Shah, 431–47. Northampton, MA: Elgar Publishing.

Sexton, D.E. 2014b. "Using Metrics to Achieve 'Steering Control' of Your Marketing Actions." *Effective Executive* 17, no. 1, pp. 7–19.

Sexton, D.E. January 2015. "International Marketing Strategy—Does Country Nationality Matter?" *Journal of Marketing Trends,*(January), 7, no. 1, pp. 47–54.

Sexton, D.E., and Association of National Advertisers. 2013. *Global Brand Equity Survey Results.* New York: ANA.

Sexton, D.E., K. Sen, and V. Gorti. 2010. "Determining Marketing Accountability." *Journal of Marketing Trends* 2, no. 1,pp. 39–45.

Sexton, D.E., and D. Rogers. March 2012. *Marketing ROI in the Era of Big Data.* New York: BRITE and NY AMA.

Vakratsas, D., and T. Ambler. January 1999. "Advertising Effects." *Journal of Marketing* 36, no. 1, pp. 26–43.

CHAPTER 9

Forensic Accounting: Show Me the Money

Scott P. McHone and Tricia-Ann Smith DaSilva

Introduction

Business owners, managers, and accountants are faced with many financial questions and concerns. Why are our revenue and profits not meeting expectations? Why are our expenses so much? Why is productivity down and what happened to our cash flow? They also seek to know what their business is worth and how to control risk. These are just a few of the questions forensic accountants can help answer. In addition, forensic accountants can aid the legal team with expert opinions and help uncover and document fraud. No matter what type of business you have or what type of accountant you are, having forensic accounting in your tool kit will be a great addition.

This chapter is organized in sections to help facilitate the understanding of the material and to assist as a practical toolkit. These sections are: the history of forensic accounting, what is forensic accounting, who needs forensic accounting and why, the implementation of forensic accounting recommendations, and the future of forensic accounting. Also provided are two case studies with a conclusion.

History of Forensic Accounting

People have been using forensic accounting for thousands of years. Surprised? How far back in time do you think there have been disagreements over money or trading? How long ago did someone take something that was not hid or hers or over charge for an item or service?

Let us roll back the clock a few thousand years and then move forward to demonstrate how long forensic accounting has been part of our society and culture.

Let us start with the Ten Commandments. Number eight of the top ten is, "thou shalt not steal" and number 10 is "thou shalt not covet." Most likely, stealing and envious behavior go back much further than Moses, but this is a good place to start. What do you think they were stealing and envious of back in the days of Moses? Most likely a lot of things like gold, jewels, money, fancy houses and chariots, sheep, cattle, art work, people, popularity, and so on. Kind of like the same type of stuff as today with a different sense of relevance. Even back then, people were rationalizing why they should steal, over charge for stuff, and there must have been opportunity to do so.

Now let us jump forward a little bit to Roman times. We have seen many instances of fraud during these times in the sense of overtaxing the citizens, theft from the temples, fake artwork, unlawful charges, and the like, but with the advancement of government and the introduction of law and corporate structure, fraud and legal issues became a new area to be concerned with.

Moving even more forward through time we see horse stealing in the Wild West, counterfeiting, overcharging for goods, fraudulent records, and poor accounting in the government.

With the creation of new business structures and reporting and the advent of selling and buying stock and business, we see poor accounting; lack of standards, which leads to inflated earnings; and fictitious reporting.

It is safe to say that forensic accounting and fraud analysis have been part of combating these issues and reporting them to the proper individual from present day all the way back to the days of Moses and before. Once we discovered that people will steal stuff, overvalue their work and worth, and provide inaccurate and false reports, then requests for controls to help prevent this were needed and put into place. A good rule of thumb in forensic accounting is most needed where we see monetary growth with a lack of checks and balances. Show me where the money is and there is where forensic accountants need to be. Forensic accountants are a good tool to have to not only discover and report financial problems, but to also help prevent them.

What Is Forensic Accounting?

The *Oxford Dictionary* defines forensic accounting as, "The use of accounting skills to investigate fraud or embezzlement and to analyze financial information for use in legal proceedings" (*Oxford* 2015, 1).

Most of us know that forensic accountants work in areas where fraudulent activity may have taken place and also support the legal team, but forensic accounting in itself is much more.

Forensic accounting is a tool that can be used by individuals to help them not only uncover and report improper acts, but also to locate and help prevent errors. Accountants, especially auditors, like to delineate between errors and irregularities. Irregularities occur when rules or laws are breached. An accounting irregularity is a term for behavior that does not conform to the rules of the accounting profession and the individual's intent is to deceive or defraud. Accounting irregularities can consist of intentionally misrepresenting accounting and other information in financial reports, whereas errors are unintentional mistakes made by an individual with no intent to deceive or defraud.

Individuals and business can utilize the skills of a forensic accountant to help increase the chances of success by controlling known problem areas, and possibly reducing the chance of having financial errors in their reporting. Forensic accounting teaches accountants how to look for numbers that do not fit in for some apparent or not so apparent reason.

Forensic accountants can value a business and let you know how much it could sell for. Knowing what a business is worth is not only important for a sale, but equally important is to know if your business is going up or down in value and why. It is a good idea for business owners and investors to have a business valued each year.

Who Needs Forensic Accounting and Why?

A forensic accountant tends to communicate financial information concisely in a courtroom with the main areas of focus including the analysis of financial evidence through the application of computerized systems. In the aftermath of accounting scandals, the field of forensic accounting continues to define itself. The need for forensic accountants by various professionals such as accountants, attorneys, and business owners continues to grow. Forensic accountants are tasked with solving the financial riddle after financial crimes. They reconstruct data, expose fraud of a financial nature, and discover concealed assets.

Forensic accountants assist individuals and business enterprises by analyzing data for entities with multiple transactions and complex systems. A forensic accountant can be of assistance in various ways, including forensic investigations related to criminal investigations on behalf of the government and the security forces, execution of investigations to assist with tracing funds, identifying and recovering assets, and performing due diligence reviews. In addition, a forensic accountant may assist with matrimonial disputes and employee fraud investigations and the performance of reviews surrounding professional negligence including technical assessments or loss quantifications.

However, if you were to ask someone if he or she needed a forensic accountant, the most common response would be, only if there is fraud happening. This basic misconception that forensic accountants are only used for fraud analysis is reasonable. The word forensic as defined earlier does relate to fraud and legal proceedings, but forensic accounting is so much more.

Let us look at the definition for *forensic* and then the definition of *accounting* to see how those definitions compare to the definition of forensic accounting. This is not an exercise on play of words, but more of a word analysis. As one of my attorney friends told his kids many times at the dinner table, every word has a meaning. The full definition of forensics by *Merriam-Webster* is "belonging to, used in or suitable to courts of judicature or to public discussion and debate" (*Merriam* 2015, 1). The part to focus on is public discussion and debate. Accounting is defined as "the system of recording and summarizing business and financial transactions and analyzing, verifying, and reporting the results" (*Merriam* 2015, 1).

With the two words merged outside of current public perception and vernacular, forensic accounting is the discussion, debate, or both of financial information. This level of debate is most common in the legal setting, but is used outside of this venue. This definition is more in line with a detailed financial analysis report that requires communication and debate when putting some of the pieces of the puzzle together. In this sense, it sounds more like auditing.

Business owners want to know how their company is performing and how much it is worth. In that sense, forensic accounting is an added benefit to the accounting department. The mindset of a forensic accountant is like an amplified version of an internal auditor. Forensic accountants are looking at the details with a sense of skepticism. With this mind-set and the natural conservative behavior of accountants, more realistic reporting is achieved. In addition, errors are more readily identified. One of the worst things a business owner can have is accounting information that provides a false sense of comfort. If business owners feel that they are doing better than they really are and their company is worth more than it really is, then the end result could be business failure. Accounting information needs to be useful and relevant to make business decisions on.

According to The U.S. Chamber of Commerce, occupational frauds have cost U.S. companies billions of dollars annually (Mohlenhoff and Uhl 2013, 1). Numerous occupational frauds have been detected internationally. In addition, indicators demonstrate that no business is protected from fraud that often leads to considerable financial losses. In recent times, the media has been overflowing with news of significant corporate frauds. For instance, the Internal Revenue Service reported that 108 investigations have been initiated related to financial institution frauds during 2014 with a rate of 64.8 percent incarcerations (IRS 2015, 1).

As such, business professionals may choose to appoint a forensic accountant to interview the employees with responsibilities for key processes, review the design and operational effectiveness of internal controls, and review the monthly financial statements for unusual patterns. An organization that is proactive in assessing the internal controls would avoid the pitfalls of fraudulent conversion of its assets and financial statements.

These issues also are at play with executives and managers who work for the owners or stockholders. They too need accurate information to make decisions. Fortunately or unfortunately, executives and managers, a number of times, have their compensation based on performance. This is where the potential conflicts come into play between looking good for a bonus versus operating in reality so that the correct decisions can be made to help grow a company.

Out of all of the people whom forensic accounting can help, it is the accountants whom it will help the most. Accountants who have forensic accounting in their toolkit develop an enhanced mindset on how to view and record transactions.

Implementation of Forensic Accounting Recommendations

Each forensic accounting assignment is distinctive with the actual approach adopted being specific to the scope of the service. The client defines a precise goal and the forensic accountant may use certain examination techniques to achieve the objective. For instance, the forensic accountant may examine paperwork and other supporting evidence to corroborate the calculations needed to estimate economic losses resulting from a motor vehicle accident. Subsequently, the forensic accountant's report may be used to facilitate a settlement, claim, or jury award by removing the financial component as an area of continuing debate.

Upon engaging a forensic accountant, the client will anticipate a report containing the findings of the investigation, including a summary of evidence and a conclusion as to the amount of loss suffered as a result of the fraud. The report should also discuss the observations made surrounding how the fraudster established the fraud scheme, and which internal controls, if any, were circumvented. In addition, an area of focus included in the report may include recommendations to improve controls within the organization to prevent any similar frauds occurring in the future.

As you may imagine, a forensic accounting assignment may include numerous phases. However, key segments of an assignment include the execution of a preliminary investigation and the completion of a report outlining the findings and recommendations. It is often useful to conduct a preliminary investigation prior to the development of a detailed plan of action. This plan will take into account the information gained through client meetings and initial investigations.

This will aid in the establishment of objectives and the utilization of appropriate methodology to accomplish the relevant goals. One should, however, note that the actual analysis performed will be dependent on the nature of the assignment and may involve, for example, summarizing a bulky quantity of transactions, carrying out present value calculations utilizing suitable discount rates, and utilizing charts and graphics to clarify the analysis.

As an output of the assignment, often a report will be arranged, which may include segments on the nature of the project, range of the examination, approach utilized, limitations of scope, and findings. The report may include schedules and graphics necessary to properly support and explain the outcomes. Within the report, the forensic accountant may present deficiencies noted in this report, along with comments and recommendations regarding these matters. The findings that are contained in the report may be broadly classified based on risk rating. The risk rating given to each deficiency is a reflection of the potential impact the weakness has upon the engagement. For instance, deficiencies may be identified and flagged as priority items, which could result in significant adverse impact on an entity's cash flows, customers, and reputation.

A well-written report can be a vital tool in litigation as it helps to collect the expert's thoughts, and express them in a clear manner. The jury may be predominantly impressed with a convincing report. However, if the report is not thoroughly discussed before it is written, the persuasiveness may be fragile. An expert should, however, never draft a written report unless he or she has been expressly engaged to do so. Furthermore, the sufficiency of the engagement is solely the responsibility of the engaging party.

The Future of Forensic Accounting

As mentioned earlier in this chapter, forensic accounting has been around for a long time. With that said, it has had a unique place in the accounting industry. There have been a lot of rules and standards developed over the years for general accounting along with governmental accounting and auditing, but the forensic accounting discipline has not had as much attention. During the past 25 years, there has been movement in this area. In the late 1980s and early 1990s, there were organizations that were created to help in the area of fraud and forensic accounting. The Association of Fraud Examiners and the American Board of Forensic Accounting were two of the first. These are two different and separate organizations, one of which focuses on fraud whereas the other is forensic accounting oriented. Then, about 15 years later, the American Institute of Certified Public Accountants developed the Certification in Financial Forensics. There are now numerous programs and organizations that teach and train individuals to become more knowledgeable in fraud and forensic accounting. But what will the future be like for this long-standing but unique part of the accounting profession?

We have seen a resurgence of interest in forensic accounting during the past 10 years. With the advancement of technology and the increase in fraud and the need to improve internal controls for businesses, the need to have more individuals able to help reduce risk and implement preventative measures is much needed.

The traditional role of a forensic accountant will stay the same. There will always be a need for business valuations, legal support, and fraud analysis, but the future forensic accountant will be used in more situations as a preventative measure and to help reduce errors and waste. Even during times when businesses are not doing as well, the need for forensic accounting will continue to be in demand in addition to traditional accounting services such as preparing financial statements, tax returns, and so on.

A lot of companies are using tools such as Six Sigma and Lean Management to help identify and reduce known errors and reduce the chances of unknown errors from occurring. Having a forensic accountant or two on the team will increase the chances of having a successful engagement. It is the mindset of a forensic accountant to find and prevent errors and irregularities.

With the increases in technology we have seen more cyber-based crimes and the need for cyber security. A forensic accountant with an information technology background will be a useful tool to help reduce risk in these areas. Forensic accountants so equipped can utilize software tools to analyze large amounts of data in a fraction of the time it would take manually.

Another area is in the medical field. Health care fraud is increasing and so is waste. Having a forensic accounting background is valuable for health care auditors and individuals who are entrusted with the custodianship of these resources.

As you can tell, there is a pattern here. Forensic accounting is great as a stand-alone tool, but can also be very useful when added to another discipline. Forensic accounting plus almost anything in the areas of risk management, business operations, auditing, law, information technology, and so on can enhance and enrich the areas it is coupled with.

Currently, forensic accounting and the area of forensics as a whole are one of the fastest growing in the business sector. With so much growth, even the media and entertainment industry has turned to forensics for subject matter material. Along with CSI and all the other programs and shows out there, there is currently some interest in forensic accountants playing a lead role among the characters for both television and movies. Do not be surprised if you see a TV series or a motion picture with a forensic accountant playing the lead role or even the star. Solving mysteries is fun and when it deals with money, it touches many emotions.

Accounting is the language of money, and forensic accounting speaks to the need for accounting to become more prospective in helping find the money. Show me where the money is and you will see the need for a forensic accountant to be there.

Case Study One

Kathy has been the bookkeeper for her company for almost 10 years. She enjoys her job and is a great person to work with. She is energetic and works hard to get her jobs done on time and help the company succeed. As a bookkeeper for a smaller construction company, she is asked to help in other areas outside of accounting. She prepares bids, answers the phone, makes appointments, and helps find new work. Her husband is also an employee of the company and their family is well vested to make sure this business does well. She is friends with the two owners Jake and Tom. The group even socializes together and goes to sporting events and has fun being with each other. This is a great job for Kathy and she is admired and trusted by Jake and Tom.

Jake and Tom are great business people who know the construction industry well. They are not good at bookkeeping, but they know and trust Kathy. Without Kathy, the business would not be where it is. Jake and Tom share different roles. Jake stays in the office more and works with Kathy, whereas Tom is more out on-site helping the crews. Being in the office allows Jake to spend a little time away. His brother is a professional football player and Jake helps him in his career and goes to as many games as he can. Kathy likes football too and once in awhile she and her husband get to go to a game and watch Jake's brother play. He is a quarterback. This is a fun lifestyle for Kathy and again she likes her job.

Jake decides that he wants to spend more time with his family and help his brother in his football career. He also wants to do a few new business ventures that he thinks can make him some money. He has always been interested in setting up an oilfield company and he has some money set aside plus his brother may want to help get this started. Jake approaches Tom and asks if Tom would be interested in buying him out. Tom says yes and they work out the terms. This gives Jake some extra money to live off of and help set up his oilfield business. He asked Tom if it is okay to have Kathy help him part time with bookkeeping and administrative duties while she is working full time for Tom's construction company. Tom agrees. Both Jake and Tom have been friends and business partners for many years. They even went to the same high school. Tom knows that Jake and Kathy are close and wants to help lighten the load and hires another person at the office to help Kathy. Furthermore, Tom's business is growing and they need the help anyway.

Jake's oilfield business starts to grow right out of the chute. Jake is a great businessperson and it seems everything he touches turns into gold. Kathy likes working for Jake more than Tom, though she likes Tom also. Kathy is close to Jake's family and is like a sister to Jake. About a year or so after the sale of Jake's interest in the construction company to Tom, things start to change. Jake seems a little irritated at Tom at times and does not like the way he spends his money. When Tom bought out Jake, Jake allowed him to make payments. When Jake sees what he thinks is Tom wasting money, he thinks this is more money Tom could have paid me. Kathy agrees with Jake and tells him he should do something about it. Over the next year, things get a little tense between Jake and Tom and Kathy seems to be in the middle of it.

Kathy is still the full-time bookkeeper for Tom, but she is not in the office as much and spends more time in Jake's company. Kathy's heart is not in her work as much with Tom even though her husband still works there. Virginia, the new bookkeeper who was hired to help Kathy with Tom's business, is now working full time and taking on a larger role. With Kathy absent more as time goes on, she starts to do some of the work that only Kathy did. Kathy did almost everything and recorded all the information in the system, reconciled the information, and even would sign checks and make transfers. This is a lot of work to cover, and Kathy was very helpful and well trusted by all.

One afternoon, Virginia decided to do some of the bank reconciliations. Kathy had not been able to do them for the past month or so and Tom needed to know how much the balances are and needed some updated reports. It was not that long into the process of doing this work that Virginia noticed that there were some items clearing the system oddly and it just was not making sense. After further investigation, she noticed that payments posted into the system were not matching up and there were a lot of electronic transfers that seemed to be questionable. After some research and analysis, she was afraid that something wrong was going on and asked Tom if he could look at it. After some review the two of them came to the conclusion that their friend and bookkeeper Kathy had made some mistakes in the books that needed explanation so that the books could get back in balance. There was also the chance that maybe there was something else going on. Not sure if these were errors or something else, they decided to reach out to a forensic accountant to see if the pieces of the puzzle could be put together and tell if this was a bunch of errors or something worse.

After a lot of work and analysis, the forensic accountant was able to tell that Kathy had been stealing from the construction company. Not only since the sale took place, but several years prior to that. Kathy is very close to Jake and was concerned how this would all end up and work out. Once Kathy confessed what she had done, everyone was hurt. Not Kathy, she is our friend. Tom was disappointed as was Jake. After Kathy's husband found out, the disappointment ran deep. After some time and negotiation, Tom and Jake were able to work out something that would restore what was lost from Kathy.

So how does something like this happen? What can be done to prevent it? Could this happen again? Let us take a look at these questions and see how to respond. There is an old saying: Internal controls keep honest people honest. Short of Billy Graham or the pope, most people give in to their sinful nature in these types of situations. I am not saying these two guys are perfect, but they are good.

The first problem from this case study is that Kathy did everything. She handled the books, reconciled the bank, opened the mail, and had authority to sign and make transfers along with other things. It is easy for smaller business to have and need a person like this. They trust them and know them well and there is no way they would ever steal. They are like family and are paid a good wage.

Second, we see that Jake and Tom are doing well. Their lifestyles are fun and especially Jake's with his connection to his brother and the Professional Football League. Kathy loves to socialize with them and they seem to always pick up the tab. However, Kathy feels that she works just as hard and even though she gets to have some fun, her family is not making enough to keep up with what she wants for them and she wants the best, just like Jake and Tom have. Even though Kathy's husband works for Tom, it still is not enough. Kathy likes to buy things and spend money and shop. She is not a gambler or a drinker and does not have a drug habit. She just likes to spend money and have some of the nicer things in life. She feels she deserves it. Plus the company is making money and she is one of the reasons why.

Well, as you can see Kathy has rationalized that this is okay to do, she has the opportunity, and the desire. She thinks nobody will catch this, and since she has been doing this for a couple of years, it all makes sense to her that she is in the clear.

The big lesson here is do not let someone have this type of control. No matter how small the company is, have someone else come in and look at the details. Companies of this size should never have the bookkeeper as a signer and should have some internal controls in place. If this company had taken preventive measures, this would not have happened. And if something were to happen, it would have been noticed.

Business owners should review your bank statements and reconciliations or have someone independent do it in detail. Check the details once a month. Have a forensic accountant come out and do a risk analysis and let you know where you may be vulnerable. Spending a few thousand dollars a year in these areas will save a lot of money down the road. Furthermore, with a forensic accountant on board, they can also identify other areas of potential waste or errors and help save you money that is not theft related.

Case Study Two

ABC Limited—The Sea of Fraud
Inappropriate segregation of duties has rippling effects.

Background

The management of ABC Limited was concerned that revenues and margins were falling and that this trend was unusual based on the nature of the business and management's expectation of the profitability of businesses in that industry. Our company, Forensic Accounting Inc, was required to perform a review of the risk and controls over the key business processes in order to identify any underlying internal control weaknesses that may have contributed to the trends noted regarding revenues and margins.

Goals

The proposed scope of the review was to perform a forensic review for various areas affecting revenue. This forensic review would determine the fraud and business risks inherent in the processes and whether the design of the existing controls was likely to mitigate these risks.

Our Role

The offender was a trusted manager and was given a wide range of functions. This resulted in poor segregation of duties. From Forensic Accounting Inc's forensic review, the following anomalies were noted:

- Rates were entered when customer accounts were created. Based on our observation of the rate-maintenance process, the updates of the rates in the system were done by the warehouse manager; however, there was no independent review of these rate changes. In addition, the warehouse manager maintained the passwords to the Inventory System in a notebook that was always kept on her desk.

- Based on our discussions with management and a walkthrough of the inventory process, we noted instances of inappropriate segregation of duties for the warehouse manager. The warehouse manager performed stock counts, updated the system for the results of stock counts, and maintained the system for addition or modification of customer profiles and product rates.

- At the end of each day, the warehouse manager reconciled the received notes to the security tally sheets. If there were any discrepancies, immediately on detection, the warehouse manager followed up with the relevant personnel. We however noted that there was no evidence or record of these checks.

- Exception reports for unusual transactions were not generated in the application. For example, the identification of changes to standing data (such as rates and customer details), identification of orders processed for delinquent customers, and so on.

- We also observed that the password and account lockout controls were not enabled for the General Ledger application used by ABC Limited. In addition, auditing was not configured on the application.

- The warehouse manager also had access to cut cashier's checks including access to the checkbook and the ability to post to the General ledger.

What Else Went Wrong?

1. The perpetrator opened fictitious payable accounts and made false payments to herself.

2. She created fictitious Accounts Receivable accounts to manipulate revenue

3. Using a restricted feature in the system, she modified the system's date to allow for the transfer of amounts between accounts and various periods of time.

4. She also manipulated suspense accounts. These accounts were not reviewed periodically.

Results

When ABC Limited asked for our assistance, it had uncovered that there was a shortage of revenue of U.S.$25,000. By the time Forensic Accounting Inc concluded our investigation, we had identified further revenue loss of U.S.$150,000 and misappropriation of inventory of U.S.$20,000. In order to help ABC Limited understand how the aforementioned losses occurred, we provided them with guidance as to methods to improve their internal controls to prevent future losses. ABC Limited subsequently submitted our report to the authorities to assist in prosecuting the perpetrator.

Conclusion

Both of these case studies provide valuable insight on what went wrong, and why it went wrong. As mentioned earlier, most organizations are smaller and one person does all of the accounting work. In general, organizations usually have less than five individuals in the accounting and finance department. For these organizations, preventative measures in the form of stronger outside controls, such as having a forensic accountant perform some checking, testing, and analysis of transactions, are very beneficial. In addition, owners should look at their bank statements and other reports periodically with their accountant and their forensic accountant.

For larger organizations, the separation of duties is still an issue. Even though one person may not do everything, we have seen larger organizations have an individual perform two functions that are incompatible. All it takes is someone to receive a payment from a customer and have the ability to write off the balance for problems to start, or it can be the person who prepares the check to pay a vendor and has access to the accounting software and mails the payment. Larger organizations need to think more globally and allocate resources to internal control audits, password and software protection and security, and strengthening the internal auditing department within their own company. One way to strengthen the internal audit department is to have their internal auditors add a tool to their toolkit by becoming certified forensic accountants.

References

IRS. 2015. "Statistical Data—Financial Institution Fraud." http://www.irs.gov/uac/Statistical-Data-Financial-Institution-Fraud (accessed June 4).

Merriam. 2015. http://www.merriam-webster.com/dictionary/forensic (accessed June 4, 2015).

Mohlenhoff, B., and G. Uhl. Fall/Winter 2013. "Occupational Fraud & Abuse a Clear and Present Danger to Your Business." USLAW. www.uslaw.org (accessed June 10, 2015).

Oxford. 2015. "Forensic Accounting." http://www.oxforddictionaries.com/us/definition/american_english/forensic-accounting (accessed June 4).

CHAPTER 10

Forensic Accounting for Governmental Entities

D. Larry Crumbley[1]

Forensic accounting is not a new discipline, but it is one that is rapidly developing and gaining status in the accounting communities. In today's climate, all accountants—internal, external, corporate, governmental, and forensic-accounting specialists—must develop forensic competencies. After reading and studying this chapter, I hope people cannot say this about governmental accountants:

> "The bad guys are a lot slicker today," I said.
> "They ain't no slicker son. The good guys are just dumber."[2]

Forensic accounting is fairly well established in the private sector, but there still have been Enron, WorldCom, HealthSouth, Bernard Madoff, Adelphia, FIFA, Home Depot, Siemens, Target, BCCI, and other massive frauds. Yet, little has been written about this subject for government entities, and forensic techniques should be used by accountants in all governmental entities. This chapter examines how forensic accounting techniques can be used in governmental entities.

---

[1] Professor Crumbley is Emeritus and Adjunct Professor at Louisiana State University, and the editor of the *Journal of Forensic and Investigative Accounting*. He is the major coauthor of Commerce Clearing House's *Forensic and Investigative Accounting*, 7th edition, and 13 teaching novels, many with a forensic accountant as the main character.

[2] Burke, J.L. 2001. *Bitterroot, 171. New York: Pocket Books.*

Introduction—Some
Failures of Governmental Auditors

Bell, California, had approximately 38,000 citizens, with an average per capita income of $24,800. Yet, the city had higher property tax rates than in Beverly Hills, and the city officials were drawing extremely high salaries.

Seven former elected city officials were convicted of multiple counts of misappropriating public funds by paying themselves huge salaries while raising taxes. Ex-city administrator Robert Rizzo had a salary of $800,000, and his annual pension benefits would have been $971,771 (reaching $1 million after two years).

Where were the auditors? The audit firm Mayer Hoffman of Irvine, California, "committed repeated acts of negligence on more than one occasion in the 2009 audit of the city of Bell and The Bell Community Redevelopment Agency (Bell CRA) that departed from professional standards," and the Certified Public Accounting (CPA) firm "insufficiently documented its audit of Bell and the Bell CRA for the year ending June 30, 2009."[3] The California Board of Accountancy suspended the CPA's firm license for six months stayed (with two years' probation). The board imposed an administrative penalty of $300,000 along with other enforcement actions.

Same song, second verse. The city of Dixon, Illinois, with a population of 15,733 had an annual budget of around $9 million. Rita Crundwell, former controller of Dixon, embezzled $53 million over 22 years. With an annual salary around $83,000, she had one of the most prestigious quarter horse operations in the world.[4] She used the embezzled money to finance 400 quarter horses and a lavish lifestyle, such as two farms, Lexus, Hummer, 1967 Corvette, trucks, 20-foot boat, $259,000 horse trailer, $2.1 million motor home, Florida home, and much more.

---

[3] *Update.* Fall 2012. "California Board of Accountancy." Issue #70, p. 23.

[4] Metcalfe, J. April 18, 2012. "Rita Crundwell vs. Harriette Walters: Who Embezzled Better?" The Atlantic Cities. http://www.citylab.com/politics/2012/04/rita-crundwell-vs-harriette-walters-who-embezzled-best/1797/

How could this modest municipal worker with a high-school education steal so much money? At night "she was a diamond-bedazzled high roller, the doyenne of a world that was million miles in glamor and several million dollars in wealth from the cornfields and cattle farms of Illinois."[5]

The city had six authentic bank accounts and one fraudulent bank account. The fraudulent checking account titled City of Dixon and RSCDA at Fifth Third bank was opened in 1990. Ms. Crundwell would transfer money from five of the city bank accounts into a sixth legitimate bank account (#7503). Next, she created fictitious capital projects, and fabricated only 179 invoices for the nonexistent capital projects. Transfers from the legitimate bank account (7503) to the fraudulent account (9530) were documented as payments of these fictitious invoices. The checks from the legitimate account (7503) were made payable to "Treasurer" that she would sign and deposit into the fraudulent 9530 bank account. She could then write personal checks from account 9530 at any time.[6]

There was a lack of segregation of duties, which is needed in all entities, including governmental entities. Crundwell picked up the daily mail, made all of the deposits, updated the journals and ledgers, prepared and signed checks, moved investment monies, and reconciled the bank accounts.

Clifton Larson Allen audited the city of Dixon from 1993 to 2005, issuing unqualified opinions each year. From 2006 to 2011, Clifton entered into an agreement with Samuel S. Card, CPA, that they would audit the city of Dixon, and Card would issue unqualified opinions each year. Clifton performed the audit work for $30,000 claiming to be doing compilation services, and Card received $6,000 to $7,000 annually to sign off on the audit.[7]

---

[5] Smith, B. 2012. "Rita Crundwell and the Dixon Embezzlement." *Chicago: Politics and City Life,* September 24. http://www.chicagomag.com/Chicago-Magazine/December-2012/Rita-Crundwell-and-the-Dixon-Embezzlement/

[6] Apostolou, B., N. Apostolou, and G. Thibadoux. January–June 2015. "Horseplay in Dixon: Lessons Learned from the Rita Crundwell Fraud." *Journal of Forensic and Investigative Accounting 7,* no. 1, pp. 253–74.

[7] Apostolou, B., N. Apostolou, and G. Thibadoux. January–June 2015. "Horseplay in Dixon: Lessons Learned from the Rita Crundwell Fraud." *Journal of Forensic and Investigative Accounting 7,* no. 1, pp. 253–74.

Clifton also prepared Crundwell's personal tax returns that showed expenses not justified by her $83,000 salary. Mr. Ronald Blaine, a Clifton partner, admitted at deposition that he sought dates with Rita, and that she played softball on the Clifton team. The auditors noted the absence of segregation of duties and the risk of management override, but they made no attempts to address these problems. The firm's audit report did not report these material weaknesses, and even noted that the Internal Controls over Financial Reporting (ICFR) were effective.[8] The audit of ICFR is an integral part of an audit.[9]

Forensic accounting techniques are not needed only in smaller governmental entities. As an example, the Governmental Accounting Office (GAO) found that U.S. federal agencies made $105.8 billion of improper payments in 2013. The GAO believes that the improper payments are even higher than the $105.8 billion reported. In its own words, "the federal government's inability to determine the full extent to which improper payments occur" represents a serious limitation on their estimates. Some programs, such as the Temporary Assistance for Needy Families welfare program do not bother reporting estimates of improper payments at all. The most errors came from Medicare, which represented almost half of all overpayments. Just two entitlement programs—Medicare and Medicaid—accounted for a whopping $64 billion in total improper payments. Currently, government-run health care programs gobble up one out of every five dollars the government spends, which obviously constricts the government's ability to meet other responsibilities.[10]

Another example is the quasigovernmental agency, the U.S. Federal Reserve. On May 12, 2009, Inspector General Elizabeth Coleman could not explain the $1 trillion plus expansion of the Federal Reserve's

---

[8] Apostolou, B., N. Apostolou, and G. Thibadoux. January–June 2015. "Horseplay in Dixon: Lessons Learned from the Rita Crundwell Fraud." *Journal of Forensic and Investigative Accounting* 7, no. 1, pp. 253–74.

[9] Auditing Standards No. 5. "*An Audit of Internal Control Over Financial Reporting That Is Integrated with an Audit of Financial Statements.*" PCAOB Release No. 2007-005A.

[10] Chougule, A. 2014. "Unfathomable Billions in Government Waste." *The Washington Times*, July 18. http://www.washingtontimes.com/news/2014/jul/18/chougule-unfathomable-billions-in-government-waste/

balance sheet since September 2008. While testifying before Congress, Coleman said that the Inspector General does not have jurisdiction to audit the Federal Reserve. The U.S. Federal Reserve could not account for $9 trillion in off-balance sheet transactions. Furthermore, no one at the Federal Reserve had any idea what were the losses on its $2 trillion portfolio. If a U.S. business lost $9 trillion or created $9 trillion on its balance sheet, it would suffer severe penalties.[11]

Then there is the plight of the Internal Revenue Service. Linda Avila admits to obtaining more than $7.2 million from 1,700 fraudulent tax returns. She used stolen identifications used by illegal immigrants. Avila operated a landscaping and cleaning business in the tiny town of Parksley, Virginia.[12]

The District of Columbia tax assessment manager received a 17-and-a-half year's prison sentence for fraudulently approving around $8.1 million in improper tax refunds over two decades. She used only about 92 payments to dummy corporations in the embezzlement. The fraud was not noticed by city officials and internal and external auditors. Apparently, auditors never examined why the city property tax refunds were steadily increasing. During sentencing, she told the judge that her office was still not safe from embezzlement. "If you put me back in there today, I could get each of you a check."[13] There are only three things certain in life: death, taxes, and fraud.

Internal Control in the Federal Government

Fraudulent financial statements may not be as great a problem in governmental units as in the private and not-for-profit sectors. But

---

[11] Crawshaw, J. May 12, 2009. "Federal Reserve Cannot Account for $9 Trillion." Newsmax.com.   http://www.newsmax.com/Finance/FinanceNews/feds-lost-nine-trillion/2009/05/12/id/330048/

[12] Ward, K. November 25, 2014. "Virginia Woman Admits $7.2 Million Child-Credit Scam." foxnews.com   http://www.foxnews.com/politics/2014/11/25/virginia-woman-admits-72-million-child-credit-tax-scam/

[13] Wiber, D.Q. 2009. "Tax Scam Leader Gets More Than 17 Years." *Washington Post*, July 1. http://www.washingtonpost.com/wp-dyn/content/article/2009/06/30/AR2009063001652.html

government units also may face incentives and pressures to achieve certain levels of financial performance to satisfy their stakeholders. Furthermore, "corruption and incidents of fraud are much greater in the public sectors than in private industry because people are not dealing with their own money."[14]

Most surveys and experts indicate that effective internal controls are essential to deter and detect fraud and abuse. PricewaterhouseCoopers 2009 Global Economic Crime Survey provides the frauds reported by various types of organizations with percentage of reported frauds:

- Government- and state-owned enterprises: 37 percent
- Listed on the stock exchange: 31 percent
- Private sector: 28 percent
- Others: 21 percent

In their *2014 Global Economic Crime Survey*, PWC found that 41 percent of government- and state-owned enterprises said they had been crime victims. Only three other sectors had more crime. KPMG's 2013 Integrity Survey, involving the prevalence of fraud, found that the government or public sector ranked second at 79 percent, only behind the consumer market at 82 percent.

Donald L. Cressey indicated that if these three factors are present in organization, fraud will probably occur: pressure, opportunity, and rationalization.[15] The AICPA in SAS 99 (now AU 316) developed his three factors into incentives and pressures, opportunity, and rationalization and attitude. The absence or lapse of internal controls in a governmental organization is a tempting open door or opportunity for fraud. When linked with the lack of integrity or with the ability to rationalize criminal behavior, this absence or lapse completes the so-called fraud triangle and

---

[14] Kahan, S. December 5, 2002. "Minding the Store." *Practical Accountant*. www.electronicaccountant.com/practicalaccountant/index

[15] Cressey, D.L. 1973. *Other People's Money*, 30. Montclair, NJ: Patterson Smith; Crumbley, D.L. Fall 2009. "So What is Forensic Accounting." The ABO Report. http://www.bus.lsu.edu/accounting/faculty/lcrumbley/abo_fa2009.html

allows an individual to engage in fraudulent behavior, without admitting to be a criminal.[16]

Standards for Internal Control in the Federal Government

The U.S. Government Accountability Office (GAO) in September 2014 issued new *Standards for Internal Control in the Federal Government* (*Green Book* hereafter), which includes this definition of fraud:[17]

> Fraud involves obtaining something of value through willful misrepresentation. Whether an act is in fact fraud is a determination to be made through the judicial or other adjudicative system and is beyond management's professional responsibility for assessing risk.

The revised *Green Book* adopts SAS 99's (now AU 316) types of fraud:[18]

- *Fraudulent financial reporting*—Intentional misstatements or omissions of amounts or disclosures in financial statements to deceive financial statement users. This type could include intentional alteration of accounting records, misrepresentations of transactions, or intentional misapplication of accounting principles.

---

[16] Crumbley, D.L., L.E. Heitger, and G.S. Smith. 2015. *Forensic and Investigative Accounting*, 3–25. Chicago, IL: Commerce Clearing House; Crumbley, D.L., E.D. Fenton, Jr., and D.E. Ziegenfuss. 2014. *The Big R: A Forensic Accounting Action Adventure*, 3rd ed. Durham, NC: Carolina Academic Press.

[17] Comptroller General of the U.S. September 2014. *Standards for Internal Control in the Federal Government*, 40. GAO (called the *Green Book*).

[18] Ibid., p. 40. PWC *2014 Global Economic Crime Survey* indicates that the types of fraud with percentages are asset misappropriation (69 percent), procurement fraud (29 percent), bribery and corruption (27 percent), cybercrime (24 percent), and accounting fraud (22 percent).

- *Misappropriation of assets*—Theft of an entity's assets. This type could include theft of property, embezzlement of receipts, or fraudulent payments.
- *Corruption*—Bribery and other illegal acts.

Since Committee on Sponsoring Organizations (COSO) updated the COSO Cube, the comptroller general issued a new *Standards for Internal Control in the Federal Government* in September 2014. Essentially, the *Green Book* adopts the COSO principles for a government environment— five components and 17 principles.[19]

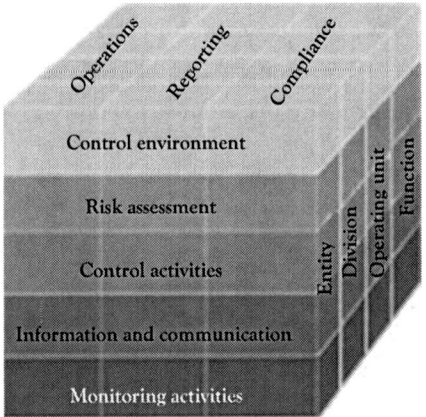

The five components represent the highest level of the hierarchy of standards for internal control in the federal government. The five components of internal control must be effectively designed, implemented, and operating, and operating together in an integrated manner, for an internal control system to be effective. The five components of internal control are as follows:

- Control environment—The foundation for an internal control system. It provides the discipline and structure to help an entity achieve its objectives.

---

[19] GAO. September 2014. *Standards for Internal Control in the Federal Government*, 1. http://www.gao.gov/products/GAO-14-704G

- Risk assessment—Assesses the risks facing the entity as it seeks to achieve its objectives. This assessment provides the basis for developing appropriate risk responses.
- Control activities—The actions management establishes through policies and procedures to achieve objectives and respond to risks in the internal control system, which includes the entity's information system.
- Information and communication—The quality information management and personnel communicate and use to support the internal control system.
- Monitoring—Activities management establishes and operates to assess the quality of performance over time and promptly resolve the findings of audits and other reviews.[20]

The Five Components and 17 Principles of Internal Control

Control Environment

1. Demonstrates commitment to integrity and ethical values
2. Exercises oversight responsibility
3. Establishes structure, authority, and responsibility
4. Demonstrates commitment to competence
5. Enforces accountability

Risk Assessment

6. Specifies suitable objectives
7. Identifies and analyzes risk
8. Assesses fraud risk
9. Identifies and analyzes significant change

Control Activities

10. Selects and develops control activities

---

[20] GAO.gov/greenbook

11. Selects and develops general controls over technology

12. Deploys through policies and procedures

Information and Communication

13. Uses relevant information

14. Communicates internally

15. Communicates externally

Monitoring Activities

16. Conducts ongoing, separate, or both evaluations

17. Evaluates and communicates deficiencies.[21]

Principle 8 requires management to consider the potential for fraud when identifying, analyzing, and responding to risks. Fraud is defined as "obtaining something of value through willful misrepresentation. The determination of fraud is made through the judicial or other adjudicative system and is beyond management's professional responsibility for assessing risk."[22] Waste and abuse do not necessarily involve fraud or illegal acts, but management is to consider these other forms of misconduct.

Principle 10 involves the selection and development of controls. There are generally three major types of controls: preventive, detective, and corrective controls.[23] Preventive controls are first in line to prevent errors, omissions, or misappropriation of assets from occurring. This type of controls is more efficient (e.g., passwords, safes, fences, locks).

Detective controls find errors or fraudulent incidents that escape the preventive controls. These controls are important when preventive controls are weak. For example, there are situations in which transactions are obtained from third parties, such as sales reports from franchisees or baggage claims reported by passengers at airports.

---

[21] Ibid.

[22] Ibid., p. 40.

[23] Crumbley, D.L., L.E. Heitger, and G.S. Smith. 2015. *Forensic and Investigative Accounting.* 7th ed. Chicago, IL: Commerce Clearing House.

Corrective controls are the actions taken to minimize further losses. They are there to correct errors, omissions, and frauds after detection. But internal controls can be broken, often by top management.

Types of Controls

Preventive Controls

- Segregating duties
- Requiring approvals
- Securing assets
- Password protection
- Using document control numbers
- Testing for drugs
- Positive pay system
- Rotating jobs and duties
- Backing up computers
- Lockbox system
- Preemployment testing
- Preemployment background checks

Detective Controls

- Written confirmations
- Reconciliations
- Management reviews
- Event notifications
- Surprise cash counts
- Counting inventory
- Hot lines

Corrective Controls

- Training personnel
- Redesigning processes
- Improving controls

- Budget variance reports
- Insurance
- Civil or criminal action

Government entities must balance risks and controls. Excessive controls create problems, but the absence of controls creates other problems.[24] Glenn E. Deck indicates that minimum controls in any operations exclude

- Competent personnel
- Adequate separation of duties
- Proper procedures for authorization
- Adequate documents and records
- Physical control over assets and records
- Independent checks on performance[25]

Management is to use a risk assessment or a similar process to help locate fraud. Regrettably, the report is esoteric and does not provide examples of fraud and red flags. The *Green Book* does provide 11 examples of common categories of control activities:[26]

1. Top-level reviews of actual performance
2. Reviews by management at the functional or activity level
3. Management of human capital
4. Controls over information processing
5. Physical control over vulnerable assets
6. Establishment and review of performance measures and indicators
7. Segregation of duties
8. Proper execution of transactions
9. Accurate and timely recording of transactions

---

[24] Ibid.
[25] Deck, G.E. 1988. "Controlling Fraud." In *Handbook of Government Accounting and Finance*, eds. N.G. Apostolou and D.L. Crumbley, ch. 26, 6. New York: John Wiley.
26 Op. cit., GAO-14-704G.

10. Access restrictions to, and accountability for, resources and records
11. Appropriate documentation of transactions and internal control.[27]

The Defense Department has a Contract Audit Manual that may be helpful for auditors, especially Chapter 8, "Cost Accounting Standards," and Chapter 9, "Audit of Cost Estimates and Price Proposals."[28] Cost accounting standards are designed to achieve uniformity and consistency in cost accounting practices that are applicable to all negotiated government contracts. In 1992, they were recodified into the Federal Acquisition Regulations, which deal with allocable and allowably of costs.

Searching for Fraud Symptoms

There are three major areas to look for fraud symptoms in the typical organization or individual:

- Source documents
- Journal entries
- Accounting ledgers

A red flag is like a ray of sunshine in a cloudy day. A forensic accountant or auditor may have to search numerous source documents to look for red flags. Appropriate source documents are:

- Cancelled checks and bank statements;
- Employee time cards (or time clock records);
- Employee contracts and buy or sale agreements;
- Sales invoices;
- Shipping documents;
- Expense invoices (e.g., travel and entertainment);
- Purchase documents;
- Credit card receipts;
- Cash register tapes;

---

[27] Ibid.
[28] Defense Contract Audit Agency, *DCAA Contract Audit Manual.*

- Revenue agent report;
- Loan documents;
- Tax returns of related parties;
- Payroll records, W2s, W9s, and 940s;
- Form 1099s;
- Form 990s and Form 941s;
- Insurance documents; and
- Brokerage account statements.

Red flags from source documents can include:

- Photocopies of missing documents
- False or changed documents
- Missing proper authorization
- Overstated voids or credits
- Second endorsements
- White outs and erasers
- Duplicate payments to same payee or for some product or service
- Large number of reconciling items
- Older items on bank reconciliations
- Overpayment of wages
- Ghost employees without necessary reductions
- Lost cash register tapes
- Lots of round numbers
- Too many beginning 9 digits (e.g., Benford's law)
- Bogus claims (e.g., health care, insurance)
- Many fictitious expense schemes (e.g., meals, mileage, sharing taxi, claiming business expenses never taken).

Forensic accountants should search for these journal entry fraud symptoms:

- Out-of-balance accounts
- Lack of supporting documentation (e.g., yellow post-it notes at WorldCom)

- Unexplained adjustments (e.g., suspense account)
- Unusual and numerous entries at end of period (e.g., top side entries)
- Written entries in computer environment
- Many round numbers
- Weekend entries
- Too many beginning 9 digits (e.g., Benford's law)
- Entries by unauthorized people.

To check for ledger fraud symptoms, a forensic auditor should look for:

- Underlying asset differences;
- Subsidiary ledger not agreeing with general ledger; and
- Inconsistencies between control accounts and supporting ledger. Fraud may cause differences.[29]

Four professors have outlined these fraud symptoms or red flags:[30]

- Items that should match, but do not (e.g., invoices for purchases, but no receiving report)
- Items that should not match, but do (e.g., vendor's address the same as an employee's address)
- Outliers and extremes (e.g., employee travels more than others or has significantly more overtime)
- Suspicious trends (e.g., invoice amounts do not match Benford's distribution)
- Accounting anomalies (e.g., employee asked by manager to process a transaction outside the system)

---

[29] Crumbley, D.L., L.E. Heitger, and G.S. Smith. 2015. *Forensic and Investigative Accounting.* 7th ed. Chicago: Commerce Clearing House.

[30] Albrecht,C.C., M.L. Sanders, and D.V. Holland. 2011. "The Debilitating Effects of Fraud in Organizations." In *Creative Accounting, Fraud and International Accounting Scandals,* ed. M. Jones, 179–80. West Sussex, England: John Wiley & Sons.

- Internal control weakness (e.g., no separation of duties)
- Analytical symptoms (e.g., unreasonable change in volume, mix, or price of products)
- Extravagant lifestyle (e.g., expensive clothes, jewelry, cars, vacations)
- Unusual behavior (e.g., no vacation, unusual work hours, or other eccentric behaviors)

The Institute of Internal Auditors (IIA) Practice Guide lists these employee red flags:[31]

- Employee acts unusably irritable
- Employee suddenly starts spending lavishly
- Overriding of controls by management or officers
- Consistently exceeds goals or objectives regardless of changing business conditions, competition, or both
- Preponderance of nonroutine transactions or journal entries
- Problems or delays in providing requested information
- Significant or unusual changes in customers or suppliers
- Transactions that lack documentation or normal approval
- Employees or management hand delivering checks
- Customers' complaints about delivery
- Poor IT access controls (e.g., passwords)
- Persons living beyond their means
- Person conveys dissatisfaction to fellow employees
- Unusually close association with suppliers
- Severe personal financial losses
- Addiction to drugs, alcohol, and gambling
- Change in personal circumstances (e.g., divorce)
- Develop outside business interest
- Employee consistently rationalizes poor performance
- Perceives beating the system
- Rarely takes vacations, and no one performs his or her work

---

[31] IIA Practice Guide. December 2009. *Internal Auditing and Fraud*, Altamonte Springs, Florida: IIA, 8–9.

Glenn E. Deck provides these audit principles to improve fraud deduction:

- Assume anyone can commit fraud
- Brainstorm the possibilities for fraud
- Inject an element of surprise in all audits
- Use both statistical and judgement sampling
- Examine unusual transactions
- Investigate exceptions rigorously
- Pay attention to the details of documents
- Verify documentation completely[32]

Proactive Is More Beneficial

There are two major types of fraud investigations: reactive and proactive. Reactive refers to an investigation after there is a reason to suspect fraud or occurs after a significant loss (e.g., predication). Being proactive is a preventive approach as a result of normal operations (e.g., reviewing of internal controls or identifying areas of fraud exposure). There is no reason to suspect fraud or there is an investigation to detect incidence of fraud.[33]

The threat of a future investigation reduces the occurrence of fraudulent behavior from 75 percent to only 43 percent. The larger the pay-off, the more likely it is that a person will commit fraudulent behavior.[28] In other words, give the fox a key to the hen house, and he or she is going to eat hens.

Some proactive and reactive approaches would be outlined as follows. *Proactive* approaches include:

- Effective internal controls
- Financial and operational audits
- Intelligence gathering

---

[32] G.E. Deck, op. cit., note 23, ch. 26, —9 and 10—.
[33] Davia, H.R. 2002. "Fraud Specific Auditing." *Journal of Forensic Accounting* 111, pp. 111–20.

- Logging of exceptions and
- Reviewing variances

Reactive detection techniques include:

- Investigating complaints and allegations
- Intuition and
- Suspicion[34]

G.E. Deck provides some common indicators of fraud found in government operations:

- Unusual checks, especially any check payable to cash or checks in even dollar amounts
- Any nonpayroll checks to companies endorsed by hand or payable to or endorsed by employees
- Any payment to a vendor that is cashed instead of being deposited
- Missing financial records, especially missing checks
- Altered entries or documents (especially checks and invoices)
- Using copies of invoices or receipts of originals as a basis of payment
- Unusually large quantities of a single-item purchase over a short period of time
- Unnumbered invoices or purchase orders
- Pattern of payments to individual or company that does not fit, especially if payments are similar or even amounts
- Document dates that do not make sense (e.g., same date on such related documents as purchase request, purchase order, invoice, receiving report, and canceled check)
- Document numbers and dates that do not make sense when set in date order (e.g., sequence of invoices is suspicious)

---

[34] Bologna, J., and R. Lindquist. 1995. *Fraud Auditing and Forensic Accounting*, 137. 2nd ed. New York: John Wiley.

- Vendor invoices with post office box numbers, typewriter numbering, rubber stamp heading, or no telephone number
- Item that does not make sense to have been bought from that vendor (e.g., consulting services from an electrical supply company)[35]

State and Local Government Susceptibility

Although bankruptcy of private sector entities is common, bankruptcy of government units has been rare. Government bankruptcy is an important issue for fraud prevention and detection because, like business corporations and organizations, governments facing severe financial difficulties can be fertile ground for fraud. Government bankruptcy also may trigger an investigative need to determine if fraud has contributed to such financial distress. Chapter 9 bankruptcies are used by municipalities, which includes cities, counties, townships, school districts, public improvement districts, bridge authorities, highway authorities, and gas authorities. Few municipalities file for bankruptcy.

For example, two former Massachusetts state employees pleaded guilty in October 2000 for eight different schemes of stealing $9.4 million in state funds. The employees were in charge of Massachusetts' Unpaid Check Fund, and they agreed to testify against five other employees (including two lawyers). These brazen and sophisticated embezzlement schemes exposed the lack of Internal Treasury controls to detect and prevent the largest theft of public money in the history of Massachusetts. The embezzlers agreed to help the attorney general follow the complex money trail in order to recover the stolen monies. The group devised a complex maze of real estate trusts and financial transactions to conceal the embezzlement before ultimately converting it to personal use.[36]

Governing, the States and Localities, is a website that tracks bankrupt cities and municipalities. Detroit, Michigan; San Bernardino and

---

[35] Op. Cit., note 22, pp. 26–9.

[36] News Release, Office of the Massachusetts Attorney General. October 18, 2000.

Stockton, California; and Central Falls, Rhode Island, are cities that have declared bankruptcy. Detroit is the largest city in the United States to file for bankruptcy.[37] Chapter 9 bankruptcy is really a means of reorganization.[38]

States face huge fiscal pressures, especially from a $1 trillion gap for unfunded pension obligations, but states cannot use the bankruptcy system to reorganize their debts. On January 14, 2014, the Final Report of the State Budget Crisis Tax Force outlined a troubling picture of the unsustainable and perilous position of many states and their local governments. The 28-page report's findings follow:

1. Obligations are rising faster than states' ability to meet them.
2. The rising costs of unfunded pensions continue to pressure state budgets and balance sheets.
3. The effect on the states of federal deficit reduction is unknown and often unconsidered.
4. Infrastructure funding remains inadequate.
5. The Affordable Care Act's impact is uncertain.[39]

Assessing the Financial Health of Governments

How does someone assess the financial health of a governmental entity? Following are early warning signs of possible trouble for municipal entities:

- Current year operating deficit
- Two consecutive years of Operating Fund deficit

---

[37] http://www.governing.com/blogs/by-the-numbers/municipal-bankruptcies-map.html

[38] Grossman, H.C., and T.E. Wilson. 1992 "Assessing Financial Health." In *Handbook of Governmental Accounting and Finance,* eds. N.G. Apostolou and D.L. Crumbley, 38.1–38.13. Somerset, NJ: John Wiley & Sons.

[39] Report of the State Budget Crisis Tax Force. January 2014. p. 10. http://www.statebudgetcrisis.org/wpcms/wp-content/images/SBCTF_FINALREPORT1.pdf

- Current year operating deficit that is larger than the previous year's deficit
- A General Fund deficit in the current year—balance sheet—current position
- A current General Fund deficit (two or more years in the last five)
- Short-term debt outstanding (other than Bond Anticipation Note) at the end of the fiscal year, greater than 5 percent of main Operating Fund Revenues
- A two-year trend of increasing short-term debt outstanding at fiscal year end
- Short-term interest and current year-end service greater than 20 percent of total revenues
- Property taxes greater than 90 percent of the tax limit
- Debt outstanding greater than 90 percent of the debt limit
- Total property tax collections less than 92 percent of total levy
- A trend of decreasing tax collections for two consecutive years or a three-year trend
- Declining market valuations for two consecutive years or a three-year trend
- Expanding annual unfunded pension obligations[40]

Ratio changes may be used to assess the financial health of governmental units, as shown by the office of the New York State Comptroller indicators in Table 10.1. A horizontal analysis over five to six years is recommended, and no single indicator is an indication of a positive or negative trend. Indicators 4, 5, and 6 may be the stronger measurements.[41]

State and local government retirement systems cover more than 19 million employees (one-sixth of the U.S. workforce). State public pension plans are underfunded by $4.7 trillion in 2014, up from $4.1 trillion in 2013. Thus, for each American, the unfunded liability is more than $15,000 per individual. The three largest underfunded states are California ($754 billion), Illinois ($331 billion), and New York

---

[40] Grossman, and Wilson, op. cit., pp. 38.13 and 38.14.

[41] Office of the New York State comptroller. www.osc.state.ny.us/localgov/muni

Table 10.1 Indicators of financial condition for municipal governments

Indicator	Indicator description	Ratios
1	Revenue and expenditures per capita	A. $\dfrac{\text{Gross Revenues}}{\text{Population}}$
		B. $\dfrac{\text{Gross Expenditures}}{\text{Population}}$
		C. $\dfrac{\text{Recurring Revenues (Gross Revenues} - \text{One - Time Revenues)}}{\text{Population}}$
	Recurring revenues capita	(Same as above)
Negative trend: Indicator b increasing faster than Indicator a and c		
2	Real property taxes receivable	$\dfrac{\text{ReaL Property Taxes Receivable}}{\text{Real Property Tax Revenue}}$
Negative trend: Percentage increases over time.		
3	Fixed casts—personal services and debt service	A. $\dfrac{\text{Salaries and Fringe Benefits}}{\text{Gross Expenditures}}$
		B. $\dfrac{\text{Debt Service Expenditures}}{\text{Gross Expenditures}}$
		C. $\dfrac{\text{Salaries and Fringe Benefits + Debt service}}{\text{Gross Expenditures}}$
Negative trend: Percentages increase over time.*		
4	Operating surplus or deficit	A. $\dfrac{\text{Gross Revenues} - \text{Gross Expenditures}}{\text{Gross Expenditures}}$
		B. $\dfrac{\text{Gross Revenues} - \text{Gross Expenditures} - \text{one- Time Revenues}}{\text{Gross Expenditures}}$

Notes:

* Some analysts use a variation of the 3b ratio based on debt service expenditures as a percentage of revenues. A ratio of 25 percent for debt service expenditures to "own source" revenues is considered a danger signal.[1]

---

[1] J.R. Razek, et al., op cit., p. 412.

Negative trend: Percentages decrease over time.		
5	Unreserved fund balance and appropriate fund balance	A. $\dfrac{\text{Unreserved Fund Balance}}{\text{Grass Expenditures}}$
		B. $\dfrac{\text{Appropriate Fund Balance}}{\text{Grass Expenditures}}$
Negative trend: Percentages decrease over time.[†]		
6	Liquidity—cash and investment as a percentage of current liabilities	$\dfrac{\text{Cash and Investments}}{\text{Current Liabilities}}$
	Liquidity—cash and investment as a percentage of gross monthly expenditures	$\dfrac{\text{Cash and Investments}}{\text{Grass Expenditures}/12}$
Negative trend: Percentages decrease over time.[‡]		
7	Long-term debt	$\dfrac{\text{Lone-Term Debt}}{\text{Population}}$

[†] Deficits in major funds in excess of 1.5 percent of fund expenditures or $50,000 (whichever is greater) are generally causes for concern. Some analysts use a variation of this ratio: the budgetary cushion. Here the fund balance is compared to revenues. The greater the fund balance as a percentage of revenues, the more likely a local government may weather hard times. A good rule of thumb is that a fund balance should be at least 5 percent of revenues.[2]

[‡] A government should generally have year-end cash equal to about 50 percent of current liabilities and 75 percent of average monthly expenditures. A governmental accounting textbook states that this quick ratio (or acid test) omits receivables and amounts due from other funds because of difficulties converting them into cash. They suggest that a large state government should consider a quick ratio of less than 50 percent as an indicator of financial stress.[3]

---

[2] Ibid, p. 411.

[3] Ibid, pp. 410–11.

Negative trend: Percentage increases over time.[§]		
8	Capital outlay	$\dfrac{\text{Capital Outlay}}{\text{Gross Expenditures}}$
Negative trend: Percentage decreases over time.[‖]		
9	Current liabilities	$\dfrac{\text{Current Liabilities}}{\text{Gross Revenues}}$
Negative trend: Percentage increases over time		
10	Intergovernmental revenues	$\dfrac{\text{Intergovernmental Revenues}}{\text{Gross Revenues}}$
Negative trend: Percentage increases over time.		
11	Economic assistance costs	$\dfrac{\text{Economic Assistance Cost}}{\text{Gross Expenditures}}$
Negative trend: Percentage increases over time.		
12	Public safety	$\dfrac{\text{Public Safety Cost}}{\text{Gross Expenditures}}$
Negative trend: Percentage increases over time.		
13	Tax limit exhausted	$\dfrac{\text{Tax Levy}}{\text{Tax Limit}}$
Negative trend: Percentage increases over time.[#]		
14	Debt limit Exhausted	$\dfrac{\text{Total Debt Subject to Limit}}{\text{Debt Limit}}$
Negative trend: Percentage increases over time.**		

[§] An increase in Indicator 7 would likely trigger a future increase in #3 formula as well as a decrease in Indicator 8.

[‖] This eighth indicator is an early warning sign of financial stress.

[#] The tax limit is the maximum amount of taxes that can be levied based on some statutory authority.

** Debt limit is the maximum amount of debt that can be issued under applicable statutory authority. Compare this ratio with indicators 3 and 7.

($307 billion). However, when the funding ratio is calculated, Illinois (at 22 percent), Connecticut (at 23 percent), and Kentucky (at 24 percent) are the worst. The most well-funded state, Wisconsin, is only 67 percent funded.[42]

The Final Report of the State Budget Crisis Task Force suggests that when the actuaries calculate a government's annual required contribution to the government's pension funds, the calculations should be made public contemporaneously. The report states that governments must make widespread changes in contributions, age of retirement, and benefits. Otherwise, these huge charges will crowd needed current services within existing revenue structures.[43]

Conclusion

Fraudulent financial statements may not be as great a problem in governmental entities as in the private and the not-for-profit sectors. However, governmental units face the same incentives, pressures, opportunities, rationalizations, and attitudes to achieve certain levels of performance to satisfy their stakeholders. Also, fraud, abuse, and corruption may be much greater in the public sector because it is not the employees' money. Accountants and auditors in government must be alert to the indicators and red flags of fraud, corruption, and abuse. Sound and effective internal controls must be in place and not overridden. More research is needed in the prevention and detection of fraud in governmental entities, because they are becoming such a huge part of our country.

---

[42] Luppino-Esposito, J. November 12, 2014. "Promises Made, Promises Broken 2014: Unfunded Liabilities Hit $4.7 Trillion." State Budget Solutions. http://www.statebudgetsolutions.org/publications/detail/promises-made-promises-broken-2014-unfunded-liabilities-hit-47-trillion

[43] Final Report of the State Budget Crisis Task Force. January 2014. http://www.statebudgetcrisis.org/wpcms/final-report-of-the-state-budget-crisis-task-force/sbctf_finalreport-2/

CHAPTER 11

Entrepreneurial Forensics: Assessing Customers and Decision Makers in the Biomedical Field

Michael S. Kinch and Emre Toker

Entrepreneurship has rightfully gained a reputation for being one of life's most challenging and rewarding paths. The magnitudes of the challenges and rewards are particularly steep when considering entrepreneurship in the biomedical sciences, which will be the focus of this chapter. We will emphasize particularly demanding aspects of entrepreneurship, namely those associated with an ever-changing definition of "the customer." We will define a customer in different ways, including the individual(s) or organization(s) that need to be persuaded to take a particular action or those who are impacted by the actions of the start-up product or company. This complex subject was selected based in part on the notoriously long and capital-intensive nature of biomedical product development, particularly in light of the need for governmental regulation. Yet, the lessons learned are not unique to biomedical products and comparably apply to other complex and technology-based products.

In our discussion of the changing array of customers throughout entrepreneurial development, we will define the different customers served by biomedical entrepreneurs, what each customer seeks, and the criteria they rely on for decision making. The complexity of the customer relationship expands as products and their sponsors mature. As we will see, this ever-expanding complexity is a common cause for new companies to fail to meet expectations and can convey catastrophic implications in terms of financial and literal life or death decisions.

The Investor as Customer

Arguably, the first customer for a budding entrepreneur comes in the form of the individuals or organizations that convey key enabling resources. Given the heavy capital requirement associated with biomedical product research and development, as measured either by dollars or time, early-stage investors seek a high multiple of return on their investment (Evens and Kaitin 2014; Kinch 2014). Compounding this, the risk associated with highly regulated products, based on their potential to convey efficacy or harm, has limited the pool of investors willing to invest in expensive and risky biomedical programs. That said, the reward for a successful product can be highly lucrative as the price and margins on new biomedical products routinely exceed those of many high-end luxury sports cars.

For a start-up company to be successful, it must carefully consider the importance of premoney valuations. A tendency of the founders to seek a high initial valuation can work contrary to the company's ability to survive. Despite taking the earliest risks, the ever-increasing need for more capital, first to prepare for and initiate clinical trials and later to expand the size and complexity of trials to compete for regulatory approval, often means that early investors do not have sufficiently deep pockets to continue investing. Repeated needs for financing invariably translates into multiple rounds of dilutions or potential *cram-downs*, which emphasize the need for favorable valuations at early stages of company formation (to make the initial investment attractive to early-stage investors).

Given the ever-increasing costs of development as a product progresses from preclinical exploration through clinical testing, among the chief goals demanded by the investor-customer is risk mitigation. Specifically, the savvy investor seeks to identify the fastest means to either double down on his or her investment to reach key value-driving inflection points or to kill the project at the earliest stage possible. The objective rationale for killing a project can include inferior (or nonsuperior) efficacy, safety or the ability to administer or manufacture a drug relative to the accepted standard as well as against competing products under development. Such a focus can put particular pressure on the management of biomedical start-ups as the overriding goal in the beginning is to eliminate as many underperformers as possible.

This risk aversion is coupled with a constant sense of urgency that is somewhat unique to biomedical products. In the well-understood case of new medicines, the period of marketing exclusivity is limited by key patent expirations, which undercuts at least 80 percent of future sales once generic drug competition commences. The time a pharmaceutical product spends in clinical trials alone routinely exceeds a decade and this translates into a relatively short period of time in which it can recoup its expenses, much less generate a profit. This urgency can often undermine the need for the extra time and cost needed to deploy the highest quality of product and may run contrary to the recent popularity of the "minimally viable product" that has been successfully applied to other technology fields such as software.

One way to minimize the risk of key patent expiries is to deploy a strategy of ever-widening patent fences. This well-established concept places the core drug or device at the center of an expanding number of patents that protect various aspects of drug efficacy, delivery, and manufacturing. For example, one can seek to protect the novel systems or dosing regimens used to deliver a drug, including the potential for protecting key combinations with existing medicines that may not be obvious to one practiced in the art. Likewise, intellectual property surrounding nonobvious means to manufacture or formulate the final product can provide a longer runway for a company to recover the extensive financial costs required to bring a biomedical product to market.

The Regulator as Customer

Biomedical products are understandably regulated based on their potential to either improve health (as intended) or to convey damage (from undesired side effects). As biomedical companies mature beyond an early concept or prototype into a potential new medical product, the regulator (in the United States, the Food and Drug Administration or FDA) transitions to become the primary customer. The regulator seeks to assure that only safe and efficacious products are introduced into the marketplace. Although this can place the FDA in the role of an adversary, successful companies instead regard the regulator as a collaborator, who helps identify the means to expedite and enhance product quality. An unfortunate issue that sometimes confounds this decision is rather unique to biomedical products and has been designated as the "Better than the Beatles" problem (Scannell et al. 2012). Specifically, a nod from the FDA requires that a new product be at least as good as the conventional standard of care. Were such a system to be applied to popular music, then only a song that was determined to be better than the Fab Five would be allowed to be played on the radio. This song, in turn, would then provide the hurdle that other songs would have to surpass.

To reach this lofty goal, more and more time and resources would need to be invested into the product to be sure that any new product has been evaluated as rigorously as possible. In enforcing the "Better than the Beatles" standard, the FDA must constantly balance itself against an ever-changing standard of protecting the public good. Protecting the public good means that all reasonable efforts must be deployed to prevent "bad drugs" from entering the marketplace. A prominent example is the refusal of FDA, and one of its junior reviewers, to bow to pressures from a major pharmaceutical to expedite the approval of a drug with the name Kevadon. The drug had been approved in Canada and many European countries, where it was on its way to the "blockbuster" status. Based on its widespread acceptance through much of the developed world, the application for approval for Kevadon, also known as thalidomide, was prevented by a single individual, Frances Oldham Kelsey, who was given the approval of Kevadon as an "easy" first assignment for a brand new employee (Bren 2001). Kelsey's actions minimized the impact of a drug that was later shown to be the definitive cause of a series of debilitating and gruesome birth defects experienced throughout the world (Mintz 1996). Other high-profile examples of "bad drugs" litter the airwaves (such as the painkiller Vioxx and the dietary supplement Phen-Fen) have at times led to criticisms that the FDA is overly lax in defending the public from dangerous chemicals (DiMasi, Milne, and Tabarrok 2014; Hamburg 2012; Hilts 2003; Woodcock 2009).

The FDA often must counter the perception as an overly loose regulator with the extreme opposite view as an overly critical one, often coming under criticism from both left and right at the same time. For example, the rapid emergence and notoriety of the HIV and AIDS epidemic in the early 1980s fomented emotional pleas to accelerate the approval of new medicines (Shilts 1987). Thus, the agency often finds itself in the unenviable position of not placing the bar too low to allow dangerous medicines to pass through while not so high that new medical breakthroughs are either impossible to achieve or can only do so after excessive delays. A general and defensible view is that high-profile failures (e.g., Vioxx) will trump low-profile improvements (e.g., has anyone heard of tenccteplase?).

For a biotechnology company aspiring to obtain an approval for a new medical breakthrough, the product sought by the customer changes over time. For example, the early stages of human clinical trials emphasize safety foremost over all other criteria (including efficacy against the disease). Particular emphasis must be placed on assessing potential toxicity concerns through the use of in silico investigation. Prior assessment of related drugs or the pathways they target can often identify signatures of potentially toxic drug candidates and allow the sponsor to identify the toxicity so that a strategy can be enacted to anticipate and minimize any risks. In silico studies provide a cost-effective means to assessing potential risk and should be buttressed through the use of more conventional laboratory-based investigation. Once this hurdle has been passed, the ability of the drug to mediate the desired effect is then evaluated and, based on the "Better than the Beatles approach," the product undergoes a grueling evaluation to support FDA licensure. Once an approval is obtained, the customer again changes and the FDA again transitions from emphasizing efficacy to safety as dictated by a requirement of the company sponsor to enact, and the FDA to oversee, a postapproval surveillance program to avoid tragedies such as those encountered with Vioxx or Phen-Fen.

Prescribers and Patients as Customers

As risks diminish and the product clears regulatory hurdles, biomedical companies often find themselves led by scientific or clinical leadership that has been rendered obsolete. For example, a persistent focus on diminishing risk by emphasizing the optimal efficacy-to-safety ratio and obtaining regulatory approval can further distract companies from focusing on commercial practicalities such as cost, distribution, sales, and marketing.

A prominent example includes the failure of Immunex, a Seattle-based biotechnology company, to develop a sufficient manufacturing and supply program for a popular arthritis drug, Enbrel, following its receipt of an FDA approval in 1998. A failure to meet consumer demand not only tarnished the product launch and adoption but ultimately undermined Immunex shareholder confidence to a point where they were eager to accept an offer from Amgen to acquire Immunex in 2001 (Bauman 2011; Money 2001).

Physicians walk a fine line between embracing new technologies that might benefit their patients and adhering to tried and true known treatments, where maintaining the status quo decreases potential liabilities and implications of malpractice. Thus, product adoption can be more sluggish than in other fields. A prominent example is reflected in the growth of so-called me-too or fast-follower drugs.

Mevacor was the first of a new class of cholesterol-lowering medications, known as statins (Stoy 1989). Launched in 1987, annual sales peaked at $1.3 billion in 1994 (the year before they succumbed to generic competition) (Cosgrove-Mather 2005). Since that time, at least four additional statin-based drugs have been approved. Over time, each newly introduced drug has eclipsed its predecessors in terms of sales. The sales of the fifth statin, Lipitor, launched in 1996, peaked at almost $14 billion in 2006 (the final year before generic competition eroded its dominance). The increasing sales of each new drug occurred despite the fact that the efficacy and safety profile of each of the five statins were essentially identical (Mills et al. 2010). However, as prescribers became more familiar with the new class of drugs, they were increasingly comfortable in recommending these to their patients. In some ways, the greater ease of adoption of innovative new products is an inverse of the conventional "innovator's dilemma" (Christensen 2013). As described in a recent popular book, the "innovator's dilemma" arises when the proponent of a new technology successfully launches a new product and becomes so focused and invested in it that they become susceptible to losing their markets to the next disruptive technology. With pharmaceuticals, it might be better considered that the innovator is susceptible to a lag in product adoption and that the lag allows either next-generation products or generic manufacturers to benefit from the changed mindset.

A variation in this theme is reflected in the failed launch of another innovative product in the early 2000s. FluMist was an intranasal influenza vaccine developed by Aviron of Mountain View, CA, a small biotechnology firm, purchased for $1.8 billion by Maryland-based MedImmune, Inc, in 2001 (MedImmune 2004). Clinical trials had demonstrated clear superiority of the vaccine, which was sprayed up the nose, and the lack of a needle was presumed to further increase the attractiveness of the product, particularly to children and a sizeable portion of the adult population, who regularly avoid flu shots due to needle aversion. Thus, MedImmune and its marketing partner, Wyeth Pharmaceuticals, assigned a premium price of $46 per dose, which was markedly higher than the $10 to $12 average price of its needle-based competitors (Pollack 2003).

To overcome some of the hesitations by physicians, FluMist was marketed using a novel return strategy. Most patients are unaware that physicians purchase influenza vaccine, mark it up, and then resell it to their customers. Thus, any unused material impacts the bottom line. To increase product appeal, MedImmune—Wyeth offered to buy back any unused FluMist. The unintended, but in retrospect obvious, consequence was that physicians instead ordered the same amount of injectable flu vaccine as normal, depleted all of their vaccine stocks, and only then prescribed FluMist. Consequently, FluMist sales barely reached 10 percent of the first year target (Rosenwald 2005). Even after eliminating the buy-back program, the high cost of FluMist drove away patients and prescribers and the product never lived up to the expectations created in the days before its approval.

Another example of a misunderstanding of the customer is conveyed by the experiences with Bexxar, an innovative new medicine for the treatment of leukemia and lymphoma (Horning et al. 2005). Bexxar targeted the same molecule on cancerous B cells as Rituxan, a proven billion-dollar antibody-based chemotherapy drug marketed by Genentech and Biogen-IDEC. However, the developer of Bexxar, Corixa

Pharmaceuticals of Seattle, WA, increased the efficacy of its antibody by adding radioactive iodine, which would literally burn up the tumor from within. The superiority of Bexxar was established in preclinical and clinical investigation. While focusing on the medical benefit of Bexxar, its developers overlooked the fact that most patients, who would receive the drug, were seen by medical oncologists. A different population of physicians, known as radiation oncologists, prescribed the radioactive Bexxar. As medical oncologists were less familiar with the drug and reluctant to lose their patients (and revenue streams) to other doctors, they largely ignored the launch of Bexxar and instead continued to prescribe drugs such as Rituxan. Consequently, Bexxar was ultimately withdrawn from the market and its innovative developer was acquired in a fire sale (GlaxoSmithKline 2013).

The individuals inhabiting the boardrooms of pharmaceutical companies have learned these lessons, which have lessened their interest in launching innovative new drugs and instead favored a safer and more lucrative approach of allowing others to innovate while reaping the benefits from a follow-on "me-too" product.

The Payer as Customer

It is not widely appreciated outside the medical profession that any physician, regardless of his or her specialty, can prescribe virtually any medicine (with a few exceptions for controlled and radioactive substances) for any patient. However, the payer, generally an insurance company, is not obligated to reimburse the patient for the costs of the medicine. Rather, insurers create a list of medicines, known as a "formulary," for which they will reimburse patients or health care providers. The drug components of the formulary are determined based on their safety, efficacy in treating disease, and cost-effectiveness.

As such, the payer is added to the list of customers served by a biomedical entrepreneur before and after FDA approval to begin marketing of a new medicine or device. Communication with and understanding of the payers will ideally allow the sponsor to identify an appropriate range of prices for their new medicine. Occasionally, one will read of examples in which a payer will refuse to reimburse the patient or provider. Well-known examples include the substitution of a generic equivalent in place of a branded drug but less well known is the fact that certain drugs are considered too expensive to warrant reimbursement.

The cost of new medicines has been growing and the U.S. market now captures more than a quarter trillion dollars in annual sales. In part, this trend reflects the fact that the development of a new drug has been rising at a logarithmic rate, presently standing at an average of $2.6 billion of investment to obtain approval for a single drug (CSDD Tufts 2014). Increasing prices also reflect the fact that certain unmet needs command a higher price, particularly medicines meant to treat catastrophic conditions (e.g., cancer) or rare diseases (known as orphan indications) (Kinch, Merkel, and Umlauf 2014). Increasingly, insurance companies and pharmaceutical manufacturers appear to be playing a game of "chicken" as increasing medicine costs appear to be approaching a practical ceiling. For example, payers in the United States now frequently refuse to pay for medicines that are marketed outside the country for considerably less money. Such differential pricing has created perceptions that American consumers and insurance companies are disproportionately underwriting the expenses of new medicines.

Providers are increasingly drawing the line by denying coverage for medicines that are deemed disproportionately expensive relative to their beneficial effects. A prominent example occurred in 2014 when a high-profile spat erupted between Gilead Sciences and Express Scripts, which oversees prescription benefits for more than 85 million Americans (Humer 2014). Gilead Sciences had gained FDA approval of Sovaldi for the treatment of hepatitis C virus (HCV) infection and priced the drug at an eye-watering $84,000 for a 12-week course of therapy, which translates into $1,000 per day (Sanger-Katz 2014). Appalled at the high price, Express Scripts removed Sovaldi from its formulary and instead negotiated an arrangement with Abbvie to purchase a competing medicine, which it deemed to demonstrate comparable efficacy for treatment of HCV but at a fraction of the price of Sovaldi.

The Express Scripts–Sovaldi incident could serve as a bellwether with powerful implications for the entire industry. Biomedical companies now face an additional challenge of adapting to a new type of financial regulation from payers, who are taking an increasingly firm stance against the escalating prices of new drugs. In particular, restrictions placed on pricing could have unforeseen implications in terms of recouping escalating sunk research and development costs. Such changes are occurring in a time when both investors and C-level staff of pharmaceutical and biotechnology companies evaluate the relative risks and rewards to determine how, and perhaps if, they will continue to develop new medicines.

Recommendation–Tools

From the beginning, companies focused on the biomedical sciences must consider all aspects of customers and how these change over time. The high capital costs and time require that the founders appreciate the risk–reward rationale practiced by investors for products that may require more than a decade and billions of dollars of investment before they could be launched. Likewise, the highly regulated nature of most biomedical products requires that the company acknowledge the viewpoint of the FDA (in the United States) in determining whether their product passes the "better than the Beatles" hurdle. Perhaps most importantly, the developer must at an early stage ask fundamental questions about who and how decisions will be made that affect the sales and marketing of the final product. To help provide an assessment of the questions confronting many biomedical start-ups, a review of the outcomes detailed earlier is provided (Table 11.1). This tool is intended to facilitate a self-assessment to identify and, ideally, to minimize the level and number of high-level risks that frequently confront biomedical entrepreneurs.

As a means of implementing the tool, a series of questions must be considered as the product or company is planned and initiated. These questions are broadly divided into the same categories and examples are shown in Table 11.2. The broad areas address each customer (investor, regulator, prescriber, patient, and payer) and consider key factors for each. For example, investors will require that the deployment of their limited resources into the company or product will convey a meaningful return and in a timely manner. A key for biomedical product development is the relatively high regulatory risk as most products are tightly controlled and the approval process can entail large amounts of capital investment and time. Therefore, the investor risk overlaps with the regulatory risk, where the questions largely focus on whether the

Table 11.1 Toolkit to assess customer risks and mitigation efforts

Customer	Risk	Mitigation
Investor	Excessive valuation	Assess long-term capital need and Return-on-Investment potential.
	Cumulative risks	Create roadmap of go or no decision points and value inflection points.
	Generic competition	Patent fence strategy
Regulator	Unanticipated toxicity	Exhaustive evaluation (in silico and laboratory based) of potential toxicities.
	"Better than the Beatles"	Distinguish the advantages of the new product versus existing and pending changes in the accepted standard of care.
Prescriber and patients	Failed product launch	Identify the key decision makers and educate these customers as to product superiority (and availability).
	Inverse innovator's dilemma	Assess the advantages and risks of positioning of the product as either an innovator or a "me-too" fast follower.
Payer	Assess payer role	Determine the degree of decision making practiced by the payer organization.
	Price sensitivities	Assess the past and future financial burdens the payer will invoke when pricing the final product.

Table 11.2 Evaluation questions for biomedical entrepreneurial forensics

Customer	Risk	Mitigation
Investor	Excessive valuation	How much investment will be needed and when? What is the net present value and how will this impact and be impacted by future needs for investment? What are the attractions for a potential investor in this opportunity (as compared to others in the same field or industry or in other industries)?
	Cumulative risks	What are the key value inflection points and what criteria will be used by investors to define success? What risk-mitigation strategies can be put in place to mitigate regulatory risks and market risks?
	Generic competition	What is the first or key intellectual property that precludes generic competition and when will this expire? What additional "patent fences" have been constructed to maximize market exclusivity (e.g., beyond the first or key patent)? Are there trade secrets (e.g., manufacturing techniques) that convey advantages and should these be protected?

(Continued)

*Table 11.2 Evaluation questions for biomedical entrepreneurial
forensics   (Continued)*

Customer	Risk	Mitigation
Regulator	Unanticipated toxicity	What potential toxicities are associated with the target being prosecuted? Is there clinical experience with similar drugs and what expected and unexpected toxicities were observed (and can this inform the design or deployment of your product? What models can be deployed to identify and preclude toxicities at a preclinical stage of development? At an early clinical stage of development?
	"Better than the Beatles"	What distinguishes the new product versus existing and pending changes in the accepted standard of care? Is the product transformational? A new mechanism? A follow-on?
Prescriber or patients	Failed product launch	Who makes the decision whether to utilize this new product? Physicians or patients? What criteria will they use to decide (e.g., efficacy, safety, cost, or all three)? Can you provide the product at the quantities needed at the time of launch? Is the supply chain sustainable?
	Inverse innovator's dilemma	What are the risks of "me-too" follow-ons capturing your market share. If the product is a follow-on, has the market been established, and what are the advantages of this particular follow-on relative to the innovator drug. Are there competitive advantages related to efficacy, safety, or cost?
Payer	Assess payer role	Who makes the decision whether to pay for this product (the patient or insurance company)? What considerations do they invoke in making the decision (efficacy, safety, cost, market size)? What are the risks that the payer will not select your product (even if the physician or patient prefers your product)?
	Price sensitivities	What are the costs associated with conventional interventions targeted by your product? Are your projected costs in-line with or superior to the current standard? If not, what is the justification for a higher price and will this be tolerated (by the patient, payer, or both)?

product is safe and efficacious in a manner that distinguishes it from the
conventional standard of care.

Two key considerations in the regulatory risk likewise overlap with
the market risks as defined by whether the patients or providers will adopt

the new product. Key questions to be considered include the time of exclusivity before generic competition fractures the market and the role of follow-on (me-too) drugs as potential existing or future competitors. A practical concern for a company transitioning from being a research and development organization to a sales and marketing organization is whether the manufacturing and distribution elements of the company are sufficiently organized to create and then meet the demand for the new product. The question of demand for biomedical products is somewhat unique in that the final user (i.e., the patient) is not necessarily the individual, who will decide whether to pay for the drug. Therefore, the new product must be considered in the context of the decision-making process and burdens associated with the payers including the potential that they will opt for a lower-cost alternative (even in the face of superior safety or efficacy of the new product).

Over the decades, entrepreneurship in the biomedical sciences has proven itself to be highly lucrative and rewarding in terms of introducing new products that fundamentally improve the health quality or quantity. The hurdles facing biomedical entrepreneurs are substantial, largely involving regulatory risk and the high costs of new product development. Nonetheless, the likelihood of achieving success can be increased substantially by identifying and assessing the needs and desires of the many customers and understanding how these change over the evolution of the new company.

References

Bauman, V. November 25, 2011. "Looking Back at Immunex and Ahead to Seattle's Biotech Future." http://www.bizjournals.com/seattle/blog/2011/11/looking-back-at-immunex-and-forward-to.html?page=all2011

Bren, L. 2001. "Frances Oldham Kelsey: FDA Medical Reviewer Leaves Her Mark on History." *FDA Consumer Magazine* 35.

Christensen, C. 2013. *The Innovator's Dilemma: When New Technologies Cause Great Firms to Fail.* Cambridge, MA: Harvard Business Review Press.

Cosgrove-Mather, B. January 13, 2005. "OTC Status Sought for Statin Drugs." http://www.cbsnews.com/news/otc-status-sought-for-statin-drugs/2005.

CSDD Tufts. 2014. "Cost to Develop and Win Marketing Approval for a New Drug Is $2.6 Billion." http://csdd.tufts.edu/news/complete_story/pr_tufts_csdd_2014_cost_study2014

DiMasi, J.A., C.P. Milne, and A. Tabarrok. April 7, 2014. "An FDA Report Card: Wide Variance in Performance Found Among Agency's Drug Review Divisions." http://www.manhattan-institute.org/html/fda_07.htm

Evens, R.P., and K.I. Kaitin. 2014. "The Biotechnology Innovation Machine: A Source of Intelligent Biopharmaceuticals for the Pharma Industry—Mapping Biotechnology's Success." *Clinical Pharmacology & Therapeutics* 95, no. 5, pp. 528–32.

GlaxoSmithKline. 2013. "GSK to Discontinue Manufacture and Sale of the BEXXARÂ® Therapeutic Regimen (Tositumomab and Iodine I 131 Tositumomab)." Washington, DC.

Hamburg, M. February 7, 2012. "50 Years after Thalidomide." http://blogs. fda.gov/fdavoice/index.php/2012/02/50-years-after-thalidomide-why-regulation-matters/2012

Hilts, P.J. 2003. *Protecting America's Health: The FDA, Business, and One Hundred Years of Regulation.* New York: Alfred A. Knopf.

Horning, S.J., A. Younes, V. Jain, S. Kroll, J. Lucas, D. Podoloff, and M. Goris. 2005. "Efficacy and Safety of Tositumomab and Iodine-131 Tositumomab (Bexxar) in B-cell Lymphoma, Progressive after Rituximab." *Journal of Clinical Oncology* 23, no. 4, pp. 712–19.

Humer, C. 2014. "Express Scripts Drops Gilead Hep C Drugs for Cheaper AbbVie Rival." http://www.reuters.com/article/2014/12/22/us-express-scripts-abbvie-hepatitisc-idUSKBN0K007620141222

Kinch, M.S. 2014. "The Rise (and Decline?) of Biotechnology." *Drug Discovery Today* 19, no. 11, pp. 1686–90.

Kinch, M.S., J. Merkel, and S. Umlauf. 2014. "Trends in Pharmaceutical Targeting of Clinical Indications: 1930-2013." *Drug Discovery Today* 19, no. 11, pp. 1682–85.

MedImmune. 2004. "FDA Approves Needle-free FluMist—First Flu Vaccine Delivery Innovation in Over 50 Years." http://www.drugs.com/news/fda-approves-needle-free-flumist-first-flu-vaccine-delivery-innovation-over-50-years-3314.html

Mills, E., P. Wu, G. Chong, I. Ghement, S. Singh, E. Akl, O. Eyawo, G. Guyatt, O. Berwanger, and M. Briel. 2010. "Efficacy and Safety of Statin Treatment for Cardiovascular Disease: A Network Meta-analysis of 170 255 Patients from 76 Randomized Trials." *QJM: An International Journal of Medicine* 104, no. 2, pp. 109–24

Mintz, M. 1996. "Remembering Thalidomide." *Washington Post National Weekly Edition.*

Money, C. December 17, 2001. "Amgen to Buy Immunex." http://money.cnn.com/2001/12/17/deals/amgen_immunex

Pollack, A. 2003. "THE MEDIA BUSINESS: ADVERTISING; Anatomy of a Failed Product Introduction: How a Nasal Spray Flu Vaccine Flopped in the Marketplace." *The New York Times*. http://www.nytimes.com/2003/11/19/business/media-business-advertising-anatomy-failed-product-introduction-nasal-spray-flu.html

Rosenwald, M.S. 2005. "FluMist Sales Falling Short, Survey Finds." *The Washington Post*. http://www.washingtonpost.com/wp-dyn/articles/A51955-2005Jan5.html

Sanger-Katz, M. 2014. "Why the Price of Sovaldi Is a Shock to the System." http://www.nytimes.com/2014/08/07/upshot/why-the-price-of-sovaldi-is-a-shock-to-the-system.html?_r=0&abt=0002&abg=02014

Scannell, J.W., A. Blanckley, H. Boldon, and B. Warrington. 2012. "Diagnosing the Decline in Pharmaceutical R&D Efficiency." *Nature Reviews Drug Discovery* 11, no. 3, pp. 191–200.

Shilts, R. 1987. *And the Band Played on: People, Politics, and the AIDS Epidemic*. New York: St Martins.

Stoy, D.B. 1989. "Controlling Cholesterol with Drugs." *The American Journal of Nursing* 89, no. 12, pp. 1628–31.

Woodcock, J. 2009. "A Difficult Balance—Pain Management, Drug Safety, and the FDA." *The New England Journal of Medicine* 361, no. 22, pp. 2105–07.

CHAPTER 12

International Forensics

J. Mark Munoz

Introduction: A Foray into the Unknown

Company A is a Philippine-based manufacturer and trader of sporting goods and defense equipment. Starting from a single retail store in the 1940s, the company has grown into a national icon with over 50 stores nationwide.

While enjoying a significant market share in the industry, the company faced mounting competition. The company managed to thwart competitive attacks by constant reinvention. In the 1950s to 1970s, they focused on branding and aggressive retail expansion. In the 1980s, they started manufacturing targeted product lines to gain cost advantages and improve profitability. In the 1990s, they became one of the early movers in online trading.

When the Asian economic crisis happened in the late 1990s, the company struggled financially. As sales slumped, they found it harder to financially support the extensive manufacturing and retail infrastructure that they built. Moreover, with a workforce of over 1,000 people, salary-related expenses became a serious financial burden.

In order to cope with their financial difficulties, the management team decided to venture into a market space that offered high margins and limited competition. They embarked into a new world of small-arms manufacturing. The management team rationalized that this move would allow them to sell a new product to their captive sports and outdoors-oriented clientele. They had been selling a limited line of small arms in their stores. They envisioned a scenario where the new small arms brand would lower costs by consolidating marketing and logistics operations.

It would also lead to higher profitability since the product would move from the factory to the retail store—eliminating the middlemen.

The manufacturing of small arms proved to be a daunting challenge. It was not a core competency of the company. The learning curve was steep and they had to hire even more people with specialized expertise to keep the operations going.

While the process required a heavy investment and took several years to come to fruition, the management team prevailed. They successfully launched a small-arms product line that they marketed in their stores.

Within a few years, the company's financial strength grew. However, with growingly efficient manufacturing operations, they found themselves having thousands of units in excess inventory each year.

The management team decided that the best way to address the issue was to sell the products internationally. They promptly hired a new international sales manager to sell the excess small arms overseas.

The foray into foreign shores led to unprecedented corporate activities such as international trade show participation, identifying international distributors, and entering into international contracts.

Through the efforts of the company's international sales manager, most of the excess inventory were sold in different parts of the world.

The largest opportunity was seen in the United States, and the firm decided to focus all their attention in this market. They appointed a large distributor to handle the sales of all their products in the United States.

Within the first year of operations, several problems emerged. While the distributor's purchase volume was large, the competitive terrain in the United States required that the products had to be sold cheap. Additional investment was required since products had to be modified to conform with United States small-arms standards and regulations. Furthermore, some products had to be occasionally shipped back to the Philippines from the United States for repairs, which added to business expenses. The management team thought about setting up a U.S. office just to manage repairs but opted not to pursue it due to the high investment cost. The distributor's product purchases fluctuated heavily based on market inclinations and changing laws. Since the volume of sales to the U.S. market was hard to predict, the profit angle became questionable.

Meanwhile, the company's finances were in a tailspin. The management team had incurred a huge debt to upgrade the factory and was paying high monthly amortizations. Furthermore, they had hired several new workers and had the highest payroll in their history.

The company relied heavily on the U.S. distributor and the U.S. market; the only country the company exported to. The company had only one exclusive distributor for the United States.

With a fragile business situation, a series of events led to a near corporate demise. With high-profile cases of gun violence occurring in the United States, policies and regulations in the country changed leading to new requirements for gun ownership. There had been some changes in market psychology. These changes led to limited purchases from their sole U.S. distributor. For three months, the distributor opted not to purchase products due to a high volume of unsold inventory. With high factory operating costs and no export sales, the company could no longer pay its bills. The management team frantically tried to sell its inventory in new foreign locations. However, since international export sales need time to develop and their brand was unrecognized, they were unsuccessful. The company faced a real prospect of bankruptcy.

Scope

In the aforementioned case, the company's challenge was a result of a lack of regard of business considerations essential in international business development. These factors include the following.

Market Environment

An internationalizing company needs to carefully examine, in advance, the environment in which it intends to operate. Aside from understanding business basics such as common business practices, infrastructure, and competition, it would need to assess the sociopolitical and legal terrain carefully. In the discussed case, Company A's lack of anticipation of U.S. standards and changing laws compounded their problems.

Relationships and Networks

A company's international success is oftentimes reliant on the quality of existing on-ground relationships. In many cases, firms with trustworthy and influential local partners make considerable headways in their business development efforts. Limitations on the quality and quantity of local relationships can lead to problems. In the case of Company A, working with only one distributor for the entire U.S. market was a risky proposition. Should the relationship fail, they would have to start from scratch. Furthermore, relying exclusively on one distributor limits opportunities for tapping the full market potential of a target location. For example, If the U.S. distributor geographically covers the East Coast, its coverage of the West Coast may be limited. In this case, partnering with just one distributor would mean that its business coverage of the country is limited.

Country Compatibility

Companies expanding in international locations where the culture and practices are somewhat compatible have higher likelihood of success. A company expanding in a new location that presents language barriers and extensive cultural differences will likely encounter cross-cultural misunderstandings and business conflict. In the case of Company A, while the Philippine and U.S. culture tend to be largely compatible, attitudes toward business differ. For instance, Asian cultures provide a high emphasis on face. People of this culture would go to great lengths to avoid saying no or embarrassing a colleague. As such, clear communication can be clouded. In the Philippines, the term *hiya* (feeling of shamefulness) leads to a shy and socially sensitive behavior. While not determined as a certainty in this case, it is possible that *hiya* and the Filipino propensity for *utang na loob* (feeling of indebtedness) prevented Company A from exploring other distributors in the United States. It is possible that the indirect Filipino business approach may have led to misunderstandings with the direct business approach of the U.S. distributor.

Organizational Preparedness

Many international organizations fail because they are not prepared for foreign expansion. There have been several cases where companies enter foreign markets to solve domestic business problems. Company

A wanted to expand overseas because of inventory excess in the local market. While exporting a product overseas makes good business sense, companies need to carefully consider their level of preparation. Overseas business development means investment of financial and other corporate resources for activities such as recruitment and training of additional personnel; travel and communication expenses; sales and marketing–related expenses (i.e., brochures and catalogs, mailings, trade show participation); distribution and logistics expenses; and product modification among others. Lack of preparation and planning, along with resource limitations, can lead to problems. Company A did not anticipate ahead of time the financial requirement needed to expand overseas. They were not aware that the product had to be modified and a large investment in the factory operations would be necessary. The poor planning of the management team along with resource limitations set the stage for business failure.

Analysis

From the aforementioned case, it is evident that international business development should never be a spur-of-the-moment initiative for companies. It requires a well-planned and researched course of action. Venturing overseas is similar to embarking on an expedition. One does not just find a ship and sail. In business, just like a voyage, extensive preparation is needed such as examining maps and weather patterns, preparing for food supplies, finding a competent crew, gathering as much information on the destination as possible, rallying investors and stakeholders, and forging important friendships and alliances.

Managerial forensics is anchored on the premise that proper evidence needs to be gathered to best understand the corporate status in regard to its health and ailments. It is a method for identifying areas of potential weaknesses so that corrections can be made, solutions found, and turnaround strategies identified prior to the pursuit of new business opportunities.

International forensics may be viewed as a subset of managerial forensics. As managers carry out diverse management activities, including business internationalization, there is a need to gather evidence and use it as the foundation for strategic action.

Figure 12.1 Model for international forensics

International forensics requires the examination of at least four key factors when developing businesses overseas. These four factors include: market environment, relationships and networks, country compatibility, and organizational preparedness. Figure 12.1 illustrates a Model for International Forensics.

The Model for International Forensics outlined in Figure 12.1 suggests the following:

- *Orientation*—Two of the four attributes relate to external orientation (i.e., market environment and country compatibility) and the other two pertain to an internal orientation (i.e., relationships and networks and organizational preparedness). External orientation means a regard for the outside environment in which the organization operates. Internal environment refers to organizational architecture in which the company operates. In the practice of international forensics, corporate attributes relating to its external and internal orientation need to be examined.
- *Timing*—The even dispersion of the attributes suggests that the mentioned factors are of equal relevance and importance. Depending on the corporate competencies and market conditions, one or two of the factors may require more immediate attention. For example, if Company A's major competitor makes a pioneering entry in a country where both firms are not present, an immediate market investigation of the country would be necessary. In such a case, Country A should prioritize its effort in understanding the market environment. While all factors are important, in certain

cases, one or two would require more immediate action and prioritization.

- *Depth and scope*—An important consideration relates to how extensively should efforts be placed on the four mentioned factors. The appropriate answer depends on diverse factors such as organizational goals, competencies, competitive threats, and emerging opportunities. It makes sense for a company to invest considerable time and effort on the activity that would offer the highest potential for return. When Company A explores an overseas location with very high market potential, but one where they have no existing relationships, a significant effort should be placed on relationship building. In such a case, it is beneficial for them to go wide and deep when attempting to find the best relationship in this location.

In essence, managerial forensics and international forensics are a function of research, analysis, and execution. In seeking to find the best corporate solutions, these three activities need to be well executed and tightly linked. A misstep in one can lead to corporate blunders and failures.

Expert Opinion

Academic literature supports the relevance of market environment, relationships and networks, country compatibility, and organizational preparedness when expanding overseas.

Need for Market Environment Examination

Foreign business environments need to be carefully assessed, since they are driven by external change propagators (Bilkey 1978).

Government policies along with institutional factors have an impact on a company's internationalization efforts (Child, Ng, and Wong 2002; Wrigley and Currah 2003). In some locations, the legal environment can cause challenges since, in some cases, laws are not as clear cut (Khanna and Palepu 1997). Bureaucracy hinders success in foreign shores (Datta, Rajagopalan, and Rasheed 1991).

There are considerable risks associated with a company's internationalization (Johanson and Vahlne 1977). Instability in the operating economic environment can pose as a threat (Svetlicic, Jacklic, and Burger 2007). Political, financial, and currency exchange issues pose as serious barriers to international expansion (Bae and Jain 2002; Bhattacharya and Wheatley 2006).

The business environment requires careful scrutiny. Internationalization can result in misunderstanding of consumer preferences, government laws and policies, competitive activity, and country infrastructure, which can lead to problems (Mitchell, Shaver, and Yeung 1992). Furthermore, the competitive landscape is often intense (Wrigley and Currah 2003).

Importance of Relationships and Networks

Relationships and networks facilitate international market entry (Kotha, Rindova, and Rothaermel 2001; Lechner and Dowling 2003; Woolcock and Narayan 2000). There is merit in linking human and relational capital (Hitt et al. 2006). Interorganizational linkages tend to expand knowledge of international locations and lead to internationalization (Hakanson 1982; Steensma et al. 2000). A firm's previous network formation experiences facilitate the internationalization process (Chetty and Campbell-Hunt 2004).

Firm relationships enhance international business operations. When firms have inadequate knowledge or weak capabilities, cooperation with other firms with relevant capacities can help the firm manage its weaknesses (Madhok 1997). Cooperative engagements allow effective resource mobilization and facilitate prompt strategic action of firms (Chang 1995). A firm's social capital can provide a unique competitive advantage (Hitt, Lee, and Yucel 2002).

Relationships and networks open the door to finding the appropriate ally in the foreign location. Social linkages play a role in international partner selection and the enhance the formation of international alliances (Kogut and Singh 1988; Wong and Ellis 2002).

Relevance of Country Compatibility

Cultural differences are a reality in international business. Distance as a result of cultural differences exists (Kogut and Singh 1988).

Extensive country differences can deter investment and business expansion decisions. Cultural disparity poses as a barrier to international entry with viewed risk negatively impacting internationalization intent (Arbaugh, Camp, and Cox 2008). Expansion decisions are sometimes driven by psychic distance, the extent to which a target country is different from the home country in terms of culture, politics, education, and other factors (Johanson and Vahlne 1977).

There are diverse business implications on the lack of compatibility between the home and host countries. The more different the market is from the home country, the more difficult it will be for internationalizing firms to learn from the international expansion (Eriksson et al. 2000). Product offering has to be excellent for companies to manage the challenges of foreignness (Caves 1982). When locations are culturally incongruent, possibilities of long-term survival are lower (Delios and Henisz 2003).

Need for Organizational Preparedness

Business expansion into international locations can be demanding for any organization. Firms that lack know-how, adequate resources, and internationalization competencies are bound to face numerous challenges. There are challenges in acquiring relevant information and integrating them into a viable strategy (Bijmolt and Zwart 1994). A typical risk associated with internationalization pertains to the identification of suitable agents in the host country (Reeb, Kwok, and Baek 1998).

Several firm attributes aid in the internationalization process. Corporate history is essential. Because of their exposure to unstable political environments, new multinational enterprises tend to be more prepared to succeed in diverse business terrains (Cuervo-Cazurra and Genc 2008). Factors such as skills and past experiences impact internationalization (Child, Ng, and Wong 2002). A dynamic organizational architecture is also essential. Matthews (2006) observed that innovative organizational forms have been used in internationalization. Corporate diversity is helpful. Diverse top management team compositions lead to heightened ability for internationalization (Carpenter and Frederickson 2001).

When a firm internationalizes, new business demands emerge. Firms will have to have setup costs, and monitor and modify products to thrive in foreign markets (Hitt, Hoskisson, and Ireland 1994). Firm internationalization suggests that firms possess nontangible assets such as know-how, proprietary information and materials, and corporate image (Caves 1996). In some cases, international diversification led to performance decline (Geringer, Beamish, and duCosta 1989). A firm that is new and also simultaneously entering a foreign market needs to create dual sets of solutions, image, and build credibility in both domestic and foreign locations (Pakes and Ericson 1998). Firms need to stay in control of the internationalization process (Vermeulen 2001).

The size of the organization can pose as a challenge for internationalizing firms (Mugler and Miesenbock 1986; Wilson 2000). Key challenges in the internationalization of small organizations are (1) limited market power, (2) small market segment served, and (3) having the leadership infrastructure to respond to challenges (Aldrich and Auster 1986). Small firms tend to be financially challenged as they expand internationally and may not be able to pursue suitable opportunities (Smallbone and Wyer 1995).

Management competencies are critical in the internationalization process. Managers may have to deal with heightened complexity (Grant 1987). When an organization does not have the adequate number of managers or skilled and experienced leaders, the effort to internationalize is hindered (d'Amboise and Muldowney 1988; Reuber and Fischer 1997). Foreign investments are typically limited by management factors such as types of skills available, organizational capacity, politics, and growth transitions (Buckley 1989).

The ability of the management team to plan effectively and manage conflict is essential. Companies that make sequential market entry tend to do better than those that enter markets in an unplanned or unsystematic manner (Barkema, Bell, and Pennings 1996). Unplanned ventures into foreign locations will lead to failure (Monti and Yip 2000).

As firms internationalize, diverse viewpoints on markets to enter, how to enter, and resource commitments lead to conflict, misunderstandings, and heightened communication costs (O'Reilly, Snyder, and Boothe 1993).

Internationalizing firms need to be well prepared for new financial demands. As firms generate new operational arrangements in foreign shores, significant resource commitment is necessary (Zott 2003). There are high related costs in internationalization (Burgman 1996; Servaes 1996). Internationalization had profit reduction due to the learning curve (Lu and Beamish 2004). Internationalization is affected by financial mismanagement by managers (Stulz 1990).

Recommendations

Based on the mentioned case and research literature, there is merit in considering the four factors of international forensics when companies contemplate international expansion. Firms that are already overseas and are experiencing challenges should find the Model for International Forensics useful in identifying where problem areas exist and what solutions may be viable. Consulting firms looking to assist companies that are struggling in their internationalization efforts should be able to derive fresh perspectives and turnaround approaches from the model. International organizations, government entities, NGOs, investment firms, and policy makers could use the model in crafting policies and strategies that support firm internationalization and development. The academe could use the model as a medium of understanding the art and practice of managerial forensics and open new avenues for further research in the future.

As a tool in managerial forensics, the Model for International Forensics is useful in self-assessment and strategy formation. Specific questions that can be asked are featured in Table 12.1.

A company's planned international expansion can pose new threats to the firm's organizational health. A poorly planned international market entry can cause errors in management judgments and drain financial resources. International expansion cannot be done in haste nor should it be viewed as a solution to existing domestic problems.

Table 12.1 Evaluation questions for international forensics

Market environment	With regard to the target country, what is the political, business, sociocultural, and economic environment like? How do these impact the enterprise? How efficient is the legal system? Are property rights and contract laws well enforced? Is there corruption? What policies and regulatory matters could affect the firm? How stable is the financial system? Have foreign investments been pouring in the country or is there capital flight? How well developed is the infrastructure, logistics, and distribution systems? What is the level of sophistication of consumers? What media channels are available? How extensive is the competition?
Relationships and networks	What existing networks and relationships are in place? Are these adequate? What relationships need to be developed in the target location to best further the business? What are the best ways to tap into these relationships and cultivate them? What short-, medium-, and long-term strategies can be developed to optimize networking and relationship building?
Country compatibility	What are the cultural similarities and differences between the home country and host country? How can these impact the planned enterprise? What preparation needs to be done to address cross-cultural differences? What types of cultural training and developmental programs would be best for executives and employees?
Organizational preparedness	Does the organization have the executive, manpower, and financial resources to succeed in the international expansion? If not, what needs to be done to enhance the level of preparedness? Does the company have an International Business Development Plan? What organizational adjustments need to be implemented to effectively pursue the internationalization efforts? What changes in the organizational architecture is necessary?

Much like the basis of success of historic expeditions, business internationalization requires extensive research, preparation and planning, cultivation of relationships, and cross-cultural understanding.

References

Aldrich, H.E., and E. Auster. 1986. "Even Dwarfs Started Small: Liabilities of Size and Age and Their Strategic Implications." *Research in Organizational Behavior* 8, no. 1, pp. 165–98.

Arbaugh, J.B., S.M. Camp, and L.W. Cox. 2008. "Why Don't Entrepreneurial Firms Internationalize More?" *Journal of Management Issues* 20, no. 3, pp. 366–82.

Bae, C.S., and V. Jain. 2002. "Multinationality: Evidence from US Industrial Firms." Paper presented at the 2002 Academy of International Business Meeting, 28 June–1 July, Puerto Rico.

Barkema, H.G., J.H. Bell, and J.M. Pennings. 1996. "Foreign Entry, Cultural Barriers, and Learning." *Strategic Management Journal* 17, no. 2, pp. 151–66.

Bhattacharya, M., and K.K. Wheatley. 2006. "Organizational Risk and Capital Investments: A Longitudinal Examination of Performance Effects and Moderating Contexts." *Journal of Management Issues* 18, no. 1, pp. 62–83.

Bijmolt, T.H.A., and P.S. Zwart. 1994. "The Impact of Internal Factors on the Export Success of Dutch Small and Medium-Sized Firms." *Journal of Small Business Management* 32, no. 2, pp. 69–83.

Bilkey, W.J. 1978. "An Attempted Integration of the Literature on the Export Behavior of Firms." *Journal of International Business Studies* 9, no. 1, pp. 33–46.

Buckley, P.J. 1989. "Foreign Direct Investment by Small and Medium Sized Enterprises: The Theoretical Background." *Small Business Economics* 1, no. 2, pp. 89–100.

Burgman, T. 1996. "An Empirical Examination of Multinational Capital Structure." *Journal of International Business Studies* 27, no. 3, 553–70.

Carpenter, M.A., and J.W. Frederickson. 2001. "Top Management Teams, Global Strategic Posture, and Moderating Role of Uncertainty." *Academy of Management Journal* 44, no. 3, 533–45.

Caves, R.E. 1996. *Multinational Enterprise and Economic Analysis*, 2nd ed. Cambridge, UK: University Press.

Caves, R.R. 1982. *Multinational Enterprise and Economic Analysis*. Cambridge, UK: Cambridge University Press.

Chang, S.J. 1995. "International Expansion Strategy of Japanese Firms: Capability Building through Sequential Entry." *Academy of Management Journal* 38, no. 2, 383–407.

Chetty, S., and C. Campbell-Hunt. 2004. "A Strategic Approach to Internationalization: A Traditional Versus a 'Born-Global' Approach." *Journal of International Marketing* 12, no. 1, pp. 57–81.

Child, J., S.K. Ng, and C. Wong. 2002. "Psychic Distance and Internationalization: Evidence from Hong Kong Firms." *International Studies of Management & Organization* 32, no. 1, pp. 36–56.

Cuervo-Cazurra, A., and M. Genc. 2008. "Transforming Disadvantages into Advantages: Developing Country MNE's in the Least Developed Countries." *Journal of International Business Studies* 39, pp. 957–79.

d'Amboise, G., and M. Muldowney. 1988. "Management Theory for Small Businesses: Attempts and Requirements." *Academy of Management Review* 13, no. 2, pp. 226–40.

Datta, D.K., N. Rajagopalan, and M.A. Rasheed. 1991. "Diversification and Performance: Critical Review and Future Directions." *Journal of Management Studies* 28, no. 5, pp. 529–48.

Delios, A., and W. Henisz. 2003. "Political Hazards, Experience, and Sequential Entry Strategies: The International Expansion of Japanese Firms, 1980–1998." *Strategic Management Journal* 24, no. 11, pp. 1153–64.

Eriksson, K., J. Johanson, A. Majkgard, and D. Sharma. 2000. "Effect of Variation on Knowledge Accumulation in the Internationalization Process." *International Studies of Management & Organization* 30, no. 1, pp. 26–44.

Geringer, J.M., P.W. Beamish, and R.C. duCosta. 1989. "Diversification Strategy and Internationalization: Implications for MNE Performance." *Strategic Management Journal* 10, no. 2, pp. 109–19.

Grant, R.M. 1987. "Multinationality and Performance among British Manufacturing Companies." *Journal of International Business Studies* 18, no. 3, pp. 79–89.

Hakanson, H. 1982. *International Marketing & Purchasing of Industrial Goods: An Interaction Approach.* Chichester, UK: John Wiley & Sons.

Hitt, M., L. Bierman, K. Uhlenbruck, and K. Shimizu. 2006. "The Importance of Resources in the Internationalization of Professional Service Firms: The Good, the Bad, and the Ugly." *Academy of Management Journal* 49, no. 6, pp. 1137–57.

Hitt, M., R.E. Hoskisson, and R.D. Ireland. 1994. "A Mid-range Theory of the Interactive Effects of International and Product Diversification on Innovation and Performance." *Journal of Management* 20, no. 2, pp. 297–326.

Hitt, M.A., H. Lee, and E. Yucel. 2002. "The Importance of Social Capital to the Management of Multinational Enterprises: Relational Networks among Asian and Western Firms." *Asia Pacific Journal of Management* 19, no. 2–3, pp. 353–72.

Johanson, J., and J.E. Vahlne. 1977. "The Internationalization Process of the Firm—A Model of Knowledge Development and Increasing Foreign Market Commitments." *Journal of International Business Studies* 8, no. 1, pp. 23–32.

Khanna, T., and K. Palepu. July–August 1997. "Why Focused Strategies May Be Wrong for Emerging Markets." *Harvard Business Review* 75, no. 4, pp. 41–51.

Kogut, B., and H. Singh. 1988. "The Effect of National Culture on the Choice of Entry Mode." *Journal of International Business Studies* 19, no. 3, pp. 411–32.

Kotha, S., Rindova, V., and F. Rothaermel. 2001. "Assets and Actions: Firm Specific Factors in the Internationalization of US Internet Firms." *Journal of International Business Studies* 32, no. 4, pp. 769–91.

Lechner, C., and M. Dowling. 2003. "Firm Networks: External Relationships as Sources for the Growth and Competition of Entrepreneurial Firms." *Entrepreneurship & Regional Development* 15, no. 1, pp. 1–26.

Lu, J.W., and P.W. Beamish. 2004. "International Diversification and Firm Performance: The S-curve Hypothesis." *Academy of Management Journal* 47, no. 4, pp. 598–609.

Madhok, A. 1997. "Cost, Value, and Foreign Market Entry Mode: The Transaction and the Firm." *Strategic Management Journal* 18, no. 1, pp. 39–61.

Matthews, J.A. 2006. "Dragon Multinationals: New Players in 21st Century Globalization." *Asia Pacific Journal of Management* 23, no. 1, pp. 5–27.

Mitchell, A., M. Shaver, and B. Yeung. 1992. "Getting There in a Global Industry: Impacts on Performance of Changing International Presence." *Strategic Management Journal* 13, no. 6, pp. 410–32.

Monti, J.A., and G.S. Yip. July–August 2000. "Taking the High Road When Going International." *Business Horizons* 43, no. 4, pp. 65–72.

Mugler, J., and J. Miesenbock. 1986. "Determinants of Increasing Export Involvement of Small Firms." In *International Council of Small Business* 9, pp. 189–205. Proceedings World Conference.

O'Reilly, C., R. Snyder, and J. Boothe. 1993. "Effects of Executive Team Demography on Organizational Change." In *Organizational Change and Redesign: Ideas and Insights for Improving Performance*, eds. G. Humber and W. Glick. Oxford, UK: Oxford University Press, 147–75.

Pakes, A., and R. Ericson. 1998. "Empirical Implications of Alternative Models of Firm Dynamics." *Journal of Economic Theory* 79, no. 1, pp. 1–45.

Reeb, D., C.Y. Kwok, and Y. Baek. 1998. "Systematic Risk of the Multinational Corporation." *Journal of International Business Studies* 29, no. 2, pp. 263–79.

Reuber, A.R., and E. Fischer. 1997. "The Influence of Management Team's International Experience on the Internationalization Behaviors of SME's." *Journal of International Business Studies* 28, no. 4, pp. 807–25.

Servaes, H. 1996. "The Value of Diversification During the Conglomerate Merger Wave." *Journal of Finance* 51, no. 4, pp. 1201–25.

Smallbone, D., and P. Wyer. 1995. "Export Activity in SMEs." Working Paper Series, No 9, Center for Enterprise and Economic Development Research (CEEDR).

Steensma, H.K., L. Marino, M. Weaver, and P.H. Dickson. 2000. "The Influence of National Culture on the Formation of Technology Alliances by Entrepreneurial Firms." *Academy of Management Journal* 43, no. 5, pp. 951–73.

Stulz, R. 1990. "Managerial Discretion and Optimal Financing Policies." *Journal of Financial Economics* 26, no. 1, pp. 3–27.

Svetlicic, M., A. Jacklic, and A. Burger. 2007. "Internationalization of Small and Medium Enterprises from Selected Central European Economies." *Eastern European Economics* 45, no. 4, pp. 36–65.

Vermeulen, F. 2001. "Controlling International Expansion." *Business Strategy Review* 12, no. 3, pp. 29–36.

Wilson, H. 2000. "Internationalization of Small and Medium-Sized Enterprises (SMEs)." In *International Business Theories, Policies and Practices*, eds. M. Tayeb and M. Harlow, 198–220. UK: Pearson Education.

Wong, P.L., and P. Ellis. 2002. "Social Ties and Partner Identification in Sino-Hong Kong International Joint Ventures." *Journal of International Business Studies* 33, no. 2, pp. 267–89.

Woolcock, M., and D. Narayan. 2000. "Social Capital: Implications for Theory, Research, and Policy." *The World Bank Observer* 15, no. 2, pp. 225–49.

Wrigley, N., and A. Currah. 2003. "The Stresses of Retail Internationalization: Lessons from Royal Ahold's Experience in Latin America." *International Review of Retail, Distribution and Consumer Research* 13, no. 3, pp. 221–43.

Zott, C. 2003. "Dynamic Capabilities and the Emergence of Intraindustry Differential Firm Performance: Insights from a Simulation Study." *Strategic Management Journal* 24, no. 2, pp. 97–125.

PART III

Strategies for Corporate Revival

Executive Interview on Managerial Forensics: Keith Cooper, Senior Managing Director, FTI Finance and FTI Consulting

J. Mark Munoz

Management concepts and perspectives are best validated by real-world practitioners. In this chapter, Dr. Mark Munoz, professor of management and international business at Millikin University, poses a few questions on managerial forensics to Keith Cooper, senior managing director of FTI Finance and FTI Consulting. FTI Consulting is a global business advisory firm that provides solutions to diverse business challenges. Anchored on the philosophy of providing "critical thinking at a critical time," the firm leverages its expertise across 27 countries on six continents.

(Start of interview)

1. **What is your understanding of managerial forensics? How is it practiced in the field?**

   Managerial forensics is not used a lot in industry. At FTI, we do restructuring, performance improvement, and operations revitalization. We do our analysis using a data-driven approach. Managerial forensics is about analyzing the root cause of underperformance, inability to achieve business plans or objectives, or why the business lags behind peer organizations. It's largely about using data for business analytics and diagnostics.

In one case, a poultry processor that I know was on the brink of bankruptcy. After a thorough data analysis, strategies for a turnaround were identified and the company was able to achieve substantial returns. This case is indicative of how managerial forensics is practiced. We need to look beyond financial statements and take a hard look at past performance. There has to be a close examination and detailed analysis of the customers, products, and operational costs.

In the case of the poultry business, to determine the best strategy we looked at its entire business processes in close detail. We examined the contract growers, processing, and the types of products created.

Oftentimes, it takes a lot of work to meticulously examine the details. However, it leads to the identification of opportunities to improve profitability.

In business, hard decisions have to be made. Key questions need to be asked: Should operations be curtailed? How? Should volume be reduced to reduce cost? Should cash be conserved? Should the plant be kept open? How can production be optimized?

Managerial forensics is the tool one can use to identify the most profitable product, reduce losses, and implement the best managerial action.

Another key consideration is benchmarking. Companies want to benchmark their performance against established standards. At FTI, while we appreciate benchmarks, we prefer to have open conversations with people. We have conversations with plant managers to discuss the challenges and issues that confront them in the workplace.

We do look at benchmarks, but we also look at data of their operations, Six Sigma practices, and specific performance by line by machine. We identify the dispersion of performance to determine efficiency improvement opportunities. We compare volume produced against labor and manpower. Patterns often emerge and we can do cost analysis. It is important to look at high performers versus laggards. Where are the pinch points?

Data is important because it guides discussions, and helps focus on issues in the value chain of the business. It identifies areas in the process where there are bottlenecks. It helps provide answers to questions such as: What's happening? Where is waste created? Where are the delays? Data points consultants and managers toward the right direction.

2. Do you believe the practice of management can be enhanced with evidence-based organizational analysis?

Absolutely. In one case, a client in the oil services industry thanked our team for embarrassing him. That is, providing him with new information and analysis of the business that he never thought of before. Many businesses often think they have everything when they have the best technologies and systems in place. In reality, when data is cut and presented in a certain way an entirely new picture may emerge. Managers should ask themselves key questions such as: How does this data help us focus on key management issues? How can the company acquire the right data in a timely manner in order to make the right management decisions?

When companies do the same things over and over again, it becomes helpful to step back, evaluate, and take on a fresh approach.

Evidence-based data can help drive better decisions. It sets the stage for new questions and a new way of looking at the business.

Unfortunately, many businesses operate by the seat of their pants. They don't gather the right data or don't value data at all. This approach can lead to many challenges. Firms can benefit from an evidence-based organizational analysis.

3. In your view, what are the best models in organizational analysis?

The best models utilize key performance indicators that drive profitability and cash flow. It is important to understand cash flow and what investment is needed to maximize cash flow over the long term.

Additionally, service to customers is important. How can customer bases be managed in order to enhance profitability and preserve business longevity?

Companies need to have a dashboard of key performance indicators. It would have elements pertaining to the overall goals and objectives of the business. It grounds them to their essence. What are the mission, vision, and goals of the organization? What will I try to be?

What niche should I compete in? All of these issues need to be considered. It should drive decision making and help the firm focus on its essence.

Measurability is important as well. It should be based on facts.

The best models are ones where work is aligned with proper systems and have clear performance indicators that are in line with business objectives.

4. In your experience, what are common causes for corporate failures?

One, lack of foresight and action. Management didn't recognize changes in the market place. As a result, they were slow to respond to market changes, such as commodity swings and pricing issues.

Two, misdirected focus. Some businesses are founded by entrepreneurs who built the enterprise through sales generation. When new competitors came in, they did not transform the business. They failed to have a holistic perspective of the market place and missed out on the new realities.

Third, technological upheaval. This happens to mature companies where the advent of new technology caused their business to go away. They didn't anticipate the impact the new technology had on consumer buying behavior as an example.

5. If you were to introduce a managerial forensics "surgical kit" what tools should be in there? What managerial tools are necessary to accurately diagnose and heal ailing corporations?

There are many organizational analysis tools that companies can use. It largely depends on the industry. ERP systems can help extract data and facilitate analysis.

Critical to this effort though is the actual analysis. The right questions need to be asked. One has to examine implications to profitability, the customer, and the product.

Logical groupings need to be considered. What are the best combinations of services and customer segments in order to optimize profit? Where should the focus be?

The managerial forensics toolkit should include measurement instruments for profitability and profit margins. They should provide the means for companies to understand where profit dollars are much higher and what aspects of the business would have the greatest profit impact.

CHAPTER 14

Team Health: Measuring It, Understanding It, and Improving It

Colin Price and Sharon Toye

The ability of an organization to operate with energy is critically dependent on whether or not teams, at every level, are healthy. In their desire and efforts to improve cultural health, organizations tend to focus on either organization-wide interventions—such as cost-cutting programs, organization redesign initiatives, or performance management system revamps—or programs that build individual leadership capability through classroom and on-the-job learning. However, in most organizations, team health gets forgotten.

After forensically studying health and performance data from more than 2,000 teams, there is some bad news but also some good. First, the bad news: Most teams are unhealthy, and senior teams are especially so. But on the upside, the energy that can be released by increasing the health of teams is enormous. Taking bonus as a proxy for performance, we find that leaders of healthy teams have a whopping 23.6 percent increase in bonus compared to unhealthy teams. If every team in your organization were healthy, just imagine the possibilities for the performance of your organization. And the prescription for the improvement of team health is clear. Teams are amenable to forensic analysis and pragmatic action. This chapter describes what questions to ask when diagnosing the health of teams; reveals the results of our research with more than 2,000 teams; and offers six recommendations on how to improve team health.

Literature Review and Cases

What Makes Teams Effective—and What Makes Them Ineffective

Senge (1990, 10) believed that:

> The discipline of team learning starts with "dialogue", the capacity of members of a team to suspend assumptions and enter into a genuine "thinking together" … [It] also involves learning how to recognize the patterns of interaction … that undermine learning.

Ancona and Caldwell (1992) felt that, too often, the focus around making a team healthy was on *internal* dynamics (such as having a clear structure and goals), and they found that the amount and type of *external* communication teams engage in determine performance. Almost a decade after writing *The Wisdom of Teams* and a significant gain in experience and insight around teams, Katzenbach and Smith (2001, vii) established that "the most important characteristic of teams is discipline; not bonding, togetherness, or empowerment." Hackman (2002) viewed teams as performing well when they had clear goals, team members with the right skills and experience to perform well-defined tasks, adequate resources, and access to coaching.

So what makes teams ineffective? Lencioni (2002) outlined five dysfunctions of teams: absence of trust; fear of conflict; lack of commitment; avoidance of accountability; and inattention to results. Lencioni (2005) wrote on how to diagnose the five dysfunctions, how you can address these dysfunctions, and within this process, how you can build a healthy team. The key points included the importance of knowing whether a team really is a team; that teams should focus on results and monitor performance; that teams do not have to agree on every decision but they must understand exactly what has been decided and commit to supporting the decision; and that building functional teams requires dedication and hard work.

Unhealthy Teams Can Destroy the Value of an Organization

Consider this example of how unhealthy teams can affect the organization as a whole.

Supermarket giant Morrisons suffered five years of slumping profits, which led, in 2015, to the chief executive officer (CEO) and half of the senior management team being sacked (Butler and Rankin 2015).

As the new CEO, David Potts, who carried out the sacking, stated, "I will now be constructing a leaner management board, with the aim of simplifying and speeding up the business." However, observers have said,

> If Potts had merely wanted to slim down the management board, he could have done so without asking anybody to leave …. If it was just streamlining, he could have stripped the jobs from the management board. There was clearly an element of performance. (Goodley 2015)

Healthy Teams Create Great Value

Now consider the consistent finding over decades of how great teams can add value:

- Mavrinac and Siesfeld (1998, 4) collected data from more than 250 institutional portfolio managers to show a direct link between top team effectiveness and company valuation. They found that approximately 35 percent of the decisions regarding whether or not to invest in a company are driven by nonfinancial data pertaining to a top team.
- West, Patterson, and Dawson (1999) tracked 160 companies over a 10-year period and found that the quality and effectiveness of the top team accounted for a 43 percent variation between profitability in different businesses.
- Hackman and Wageman (2005) studied executive teams at major international organizations and found that when the CEO of a major U.K. telecommunications company reorganized and revitalized his executive group, within three years, the company increased its market capitalization from £10 billion to £30 billion and increased its customer base several-fold.

To date, the understanding of what creates health or ill health in teams has been anecdotal. Very little real data have been collected on any significant sample of teams. The closest we have to that is the work provided by Wageman et al. (2008), who analyzed data from more than 120 top teams around the world; their work included observations, in-depth interviews of leaders and members, and quantitative and qualitative data about teams. Wageman et al. developed five prescriptions:

1. Decide if you want—and need—a team. They found that poorly performing senior leadership teams were not real teams. Fewer than 7 percent of the teams studied agreed on who was on the team.
2. Create a compelling purpose for your leadership team that is challenging as well as clear. They found that leaders frequently overchallenged individual members but underchallenged the team. They also found that challenge without clarity harms performance.
3. Get the right people on your team—and the wrong ones off. They identified certain team member characteristics that can enhance—or undermine—top team effectiveness that include: the right skills and experience, executive leader self-image, conceptual thinking, empathy, and integrity.
4. Give your leadership team the structure it needs to work. They found that effective senior leadership teams have a sound structure (i.e., the right size, meaningful tasks, and clear norms) and get more support.
5. Coach your team: effective senior leadership teams have more helpful leaders, get more coaching from their leaders, and coach each other more.

This was a definite advance in the science of teams and the understanding of what creates or dissipates energy in teams. However, these findings are essentially drawn from *qualitative data* and *generalized* for all teams. We wanted to know more.

Our Research Methodology: Analyzing Data from 2,000 Teams

Our work over the last three years has been focused on closing this gap in our collective knowledge about teams. We set out with the aim of researching multiple teams, gathering robust data on them, and then tussling with those data to understand what studying teams in this way, at this depth, and with this robustness, tells us about what enhances their health and therefore their performance and the performance of their organizations.

To do this, we have analyzed data from a significantly larger sample of teams than those completed by researchers to date—2,000 teams across a wide number of organizations, functions, and geographies, in industries as diverse as banking, private equity, insurance, engineering, telecommunications, health care, and charitable institutions. We have measured team health through the application of a tried and tested questionnaire—the Team Accelerator Questionnaire (TAQ), which takes only 15 to 20 minutes to complete online—http://minitaq.cocompany. com/—with robust reliability and validity. We have used this instrument to take the temperature of teams from within the team, as well as through the perceptions of key stakeholders outside the team. Crucially, we have also correlated team health with financial performance data from organizations.

What Our Research Revealed about Healthy Teams

The detailed, robust, and rigorous analysis of the teams we have studied means we now have a clear picture of how to build healthy teams and how much value healthy teams create for their organizations. There is a lot of value on the table, and this does not get realized by accident. As with any athletic team, prowess and performance take work, targeted muscle building, and skill development. Yes, you can shine a light—any light— on teams, and teams will improve, but as soon as the light diminishes, so do the gains in the health and performance of the team. Our research shows what the health plan needs to focus on.

When we meet healthy teams in organizations (whatever their sector, industry, functional specialism, or place in the hierarchy), we meet teams that build on each other's energies and talents, generating synergy to deliver a shared purpose. We can recognize them as healthy because they:

- Align, execute, and renew better than their competitors;
- Create a shared agenda that produces competitive advantage;

- Execute with a metabolic rate that drives outstanding levels of achievement;
- Renew continuously, setting stretching objectives; and building improvement capabilities that outpace others;
- Have high levels of trust and productive conflict;
- Operate in a high-challenge, high-support mode; and
- Focus on both performance and health.

Five Arenas of Team Performance

Our research, our experience, and our work with both healthy and unhealthy teams have revealed that teams that truly are at the peak of their performance develop five arenas, shown in Figure 14.1 and described in the following sections. For each of the five arenas, we use three specific tests to diagnose the health of teams we work with, which are listed after the description of each arena:

1. *Mandate.* Teams need to be very clear on why they exist—what their shared purpose is. For unhealthy teams, that purpose is not clear and agreed on by members of the team. This is a crucial step to enable focused attention and effort. To determine if a particular team has a clear mandate, it should meet these three criteria:
    - *Unique commission*: The team has a deep and shared understanding of the expectations of its stakeholders.
    - *Shared purpose*: Team members are mutually accountable for, and collectively committed to, a shared purpose. Focusing on work only the team can do, the team members leverage their unique position as integrators.
    - *Coherent direction*: Both the vision and the strategy are agreed, tightly integrated, and clearly articulated.
2. *Governance.* Once a mandate is clear, the job of the team is to ensure its governance processes are in good shape to enable the team to deliver that mandate. How do team members share information, communicate with each other, allocate and integrate tasks, and align their incentives to ensure smooth operating? It does not matter

whether the team is leading an airline or running an Olympic relay race. Healthy high-performing teams know how to make effective decisions at pace, how to manage hand offs, and how to decide on the best team members for the job. To determine whether a team has strong governance, it should meet these three criteria:

- *Tight composition*: The team contains the right "fact holders" with the right skills and mix of perspectives, while avoiding the burden of excessive size.
- *Aligned incentives*: The team is incentivized to deliver its strategy, achieve targeted outcomes, and role-model behaviors, balancing collective and individual accountability.
- *Agile processes*: The team interacts flexibly with effective cadence and with clear individual and collective decision rights.

3. *Behaviors*. Team members need to behave in ways that support the achievement of their commission and purpose. Here, the work of the team is to identify behaviors that help and hinder and to develop ways of working that share leadership responsibilities across the team and that enable productive use of talent, robust dialogue, and productive conflict. To determine whether a team has healthy behaviors, it should meet these three criteria:

- *Distributed leadership*: The team leader operates as a *first among equals*, leveraging the full capabilities of the team.
- *Productive conflict*: Empathy trumps ego and the team is able to rupture and repair, support, and challenge.
- *Explicit standards*: Team members support each other when it counts, and the foundations of respect, disclosure, and directness are in place. They role model this for the organization.

4. *Connections*. It is a rare team that can deliver what it needs to without involving, engaging, and managing stakeholders and employees. To determine whether a team has healthy connections, it should meet these three criteria:

- *Compelling story*: The team translates its strategy into a compelling story and uses it to powerfully engage key audiences.
- *Focused grip*: The team follows through and drives for impact, commissioning work that results in competitive advantage.

- *Stakeholder influence*: The team actively considers and then consciously shapes the wider context in which it operates through managing key relationships.

5. *Renewal.* Like healthy people, teams never stand still. Just as healthy living is a continuous process that requires paying attention to the latest research, the newest techniques, and a whole-system approach, the same applies to teams. Team members come and go, organizational contexts and industry drivers shift, and stakeholders change. To assess the renewal capability of a healthy team, it should meet these three criteria:

- *Foresight*: The team has sufficient focus on the future and avoids short termism.

- *Learning*: The team takes time to reflect and learn, drawing on external and varied perspectives, translating them into productive improvement.

- *Energy*: The team works in a way that creates rather than saps energy. It channels and leverages the energy of the organization in pursuit of performance and health.

These five arenas come together to deliver team health, as shown in Figure 14.1.

The Data Analyzed: A Spectrum of Team Health

Team health was calculated based on the number of respondent groups rating above 3.8 on a 5-point scale across the 15 tests of brilliant teams (three tests for each of the five arenas). The respondent groups included all four of the interested and affected parties of teams—the members, the leader, the line manager of the leader, and the team's stakeholders.

Teams with varying levels of health have been categorized as *Healthy* when all four respondent groups score above 3.8; *Within reach* when three respondent groups scored above 3.8; *Middling* when two respondent groups scored above 3.8; *Troubled* when only one respondent group scored above 3.8; and *Unhealthy* if none of the respondent groups scored above 3.8.

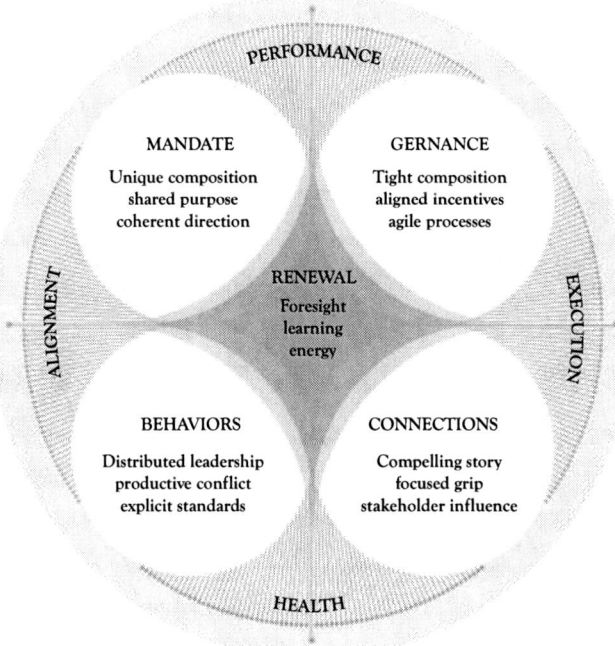

Figure 14.1 The five arenas of team performance

Figure 14.2 shows the distribution of teams within these five levels. As you can see, using 3.8 as the standard produces a beautifully clear normal distribution curve. We also found that, true to our prediction that team health is not a naturally occurring phenomenon for most teams, only 14 percent of the teams we studied could be defined as *Healthy* whereas 27 percent were *Troubled* or downright *Unhealthy*.

Healthy Teams Outperform *Unhealthy* Teams

The critical question here is: Do healthy teams deliver superior performance? Organizations use various measures to assess performance of their people, teams, functions, and the enterprise as a whole. In our research, we looked at bonuses as a proxy for economic performance. We applied a multiple regression analysis and found that being a healthy team compared to an unhealthy team explains 13 percent of the variance in performance.

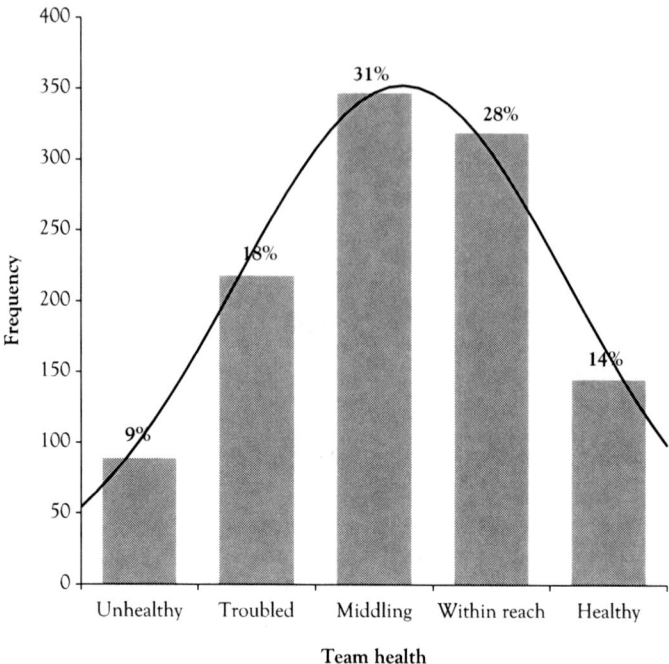

Figure 14.2 Distribution of teams within each level of team health
(n = 1,262)

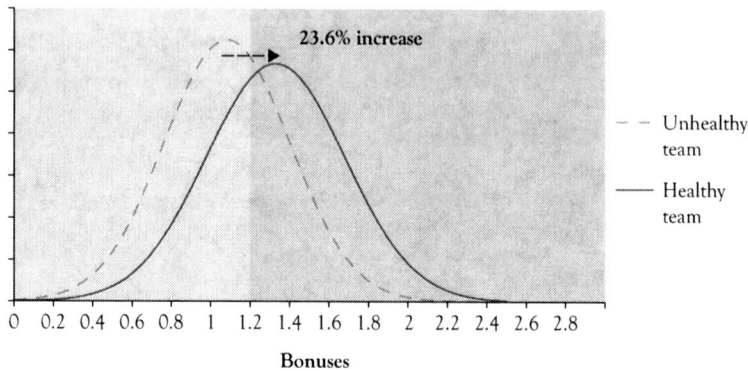

Figure 14.3 The distribution of bonuses for unhealthy and healthy
teams (n = 149)

Figure 14.3 shows that, on average, across all the teams we studied, healthy teams had an economic impact 23.6 percent higher than unhealthy teams. They reduced costs quicker, got to market more effectively, and launched products more smoothly.

So we have established that there is room for improvement in most teams in most organizations—particularly in senior teams. We also now have a definitive link between healthy teams and the performance of those teams and their leaders. But does that make a difference to the performance of the organization?

Healthy Teams Make for Healthy Organizations

Different organizations measure health in different ways—for example, engagement scores, enablement scores, and culture assessments. One of the most robust measures of organizational health has come from the work of Price and Keller. In *Beyond Performance,* they found these results for companies in the top quartile of organizational health:

- They are 2.2 times more likely than lower-quartile companies to have above-median EBITDA (earnings before interest, taxes, depreciation, and amortization) margin,
- They are 2.0 times more likely to have above-median growth in enterprise value to book value, and
- They are 1.5 times more likely to have above-median growth in net income to sales.

Across the board, correlation coefficients indicate that roughly 50 percent of performance variation between companies is accounted for by differences in organizational health (Keller and Price 2011, 6). The compelling research conducted by Price and Keller shows that organizational health drives performance. We now know that teams drive organizational health. How do we know this? We took various ways of measuring organizational health, and they all showed the same thing. *Teams matter.*

On average, the health score of an organization is 16 percent higher for *Healthy* teams compared to *Unhealthy* teams for the part of the organization that team leads. Think of Russian dolls: A healthy senior team will see the whole organization as 16 percent healthier, and the team below this healthy team will also rate its organization as healthier, and so on down through the organization of teams nested under the healthy senior team.

We found that being a *Healthy* team compared to an *Unhealthy* team explains 30 percent variance in organizational health scores. To give a sense of how stark the findings are, 80 percent of healthy teams are also rated in the top quartile of organizational health as compared to 0 percent of unhealthy teams. Being a *Healthy* team has an impact on five of the organizational health dimensions: *Healthy* teams are 22 percent higher in *leadership*, 23 percent higher in *innovation and learning*, 15 percent higher in *accountability*, 13 percent in *capabilities*, and 12 percent higher in *external orientation* compared to *Unhealthy* teams.

This indicates that *Healthy* teams are experiencing better leadership, which is the most important dimension for being in the top quartile for organizational health; in addition, *Healthy* teams are clearer about what is required of them (accountability), which we have found to be the most important dimension for *not* being in the *bottom* quartile for organizational health.

Leaders of healthy teams truly contribute to organizational health by:

- Producing an increased capability for learning and innovation, which are so essential for surviving and thriving in today's uncertain and fast-changing world
- Increasing the likelihood that they will have higher engagement scores by 3.25x. They are also likely to have superior senior management perceptions and well-being ratings.

As a result, healthy teams win out in the war for talent.

Six Ways to Improve Team Health

So irrespective of how big they are, or what the gender and age of the team leader are, what in particular do healthy teams do to build their health? Here, we revisit our five arenas and 15 tests of a healthy team. We found that all the tests are foundational; however, it pays for teams with different levels of health to focus on different arenas.

1. *Focus at the Top*. We found that senior teams are the most *Unhealthy* teams in organizations. As you can see from Figure 14.4, we found

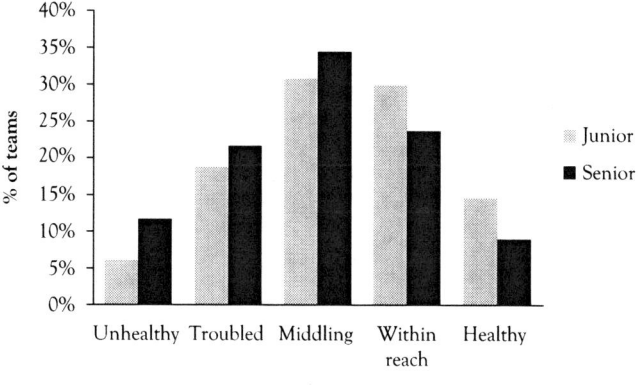

Figure 14.4 *Distribution of teams within each level of team health for senior and junior teams (n = 1,262)*

1.9 times more *Healthy* junior teams than senior teams. In addition, we found that senior teams rate their team lower on 13 of the 15 tests of brilliant teams. The bottom line? You are more likely to be a good team if you are more junior, and just as the responsibility and impact of teams at the senior levels become most critical—just when organizations need their teams at the top levels to be finely tuned and delivering their best in a competitive field—they are not! The upside of this finding? The sheer scale of opportunity for organizations to train and coach their teams to become healthier.

2. *Connect with customers.* Teams that are not customer facing have yet another hurdle to jump. Our research shows that the further a team is away from facing the customer, the harder that team needs to work to become a healthy team. In organizations, teams that have their purpose for existence more *in their faces* (i.e., customer-facing teams) have 1.4 times more chance of being a *Healthy* team and 1.3 times more chance of being *Within reach* compared to noncustomer-facing teams. In addition, customer-facing teams score significantly higher on *14 of the 15* tests of brilliant teams than noncustomer-facing teams. The bottom line: Connecting with customers is important for team health. Figure 14.5 elaborates.

Healthy customer teams have cracked the code on how to put in place agile processes, influence their stakeholders, and communicate compellingly to interested constituencies. For noncustomer-facing

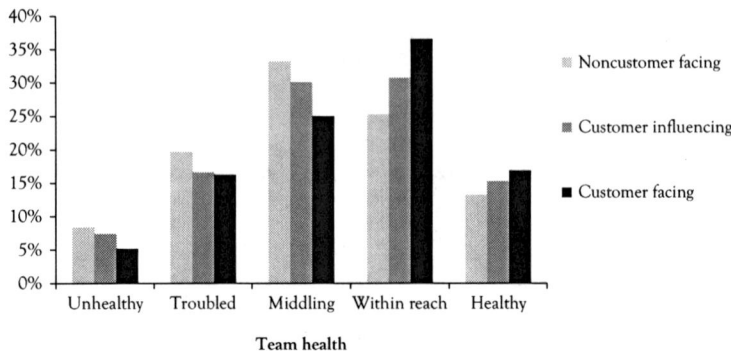

Figure 14.5 Customer-facing teams are healthier

teams, the story becomes familiar: shared purpose, foresight, and unique commission are what make the difference. Added to this mix is a focused grip on the work they set their organization to do. Focusing on these areas will help to shift the health indicators for noncustomer teams.

3. *Start with mandate.* The top four constraints felt by all teams relate to mandate and so spending time on clarifying a team's mandate is well spent. Although all teams have constraints, *Healthy* teams report having significantly fewer constraints compared to *Unhealthy* teams on 40 of the 45 constraints. As shown in Figure 14.6, the top four constraints are felt to a lesser extent the healthier the team is, and so it would be wise for an *Unhealthy* team to focus on tackling:

- Allowing too many priorities to pull it in competing directions
- Becoming caught up in *troubleshooter* mode to focus only on today's problems
- Finding it difficult to integrate the different portfolios of each team member into a coherent purpose
- A tension between the team's priorities and the expectations of its stakeholders

The bottom line is that *Unhealthy* teams have more constraints and they particularly allow too many priorities to get in the way of performance.

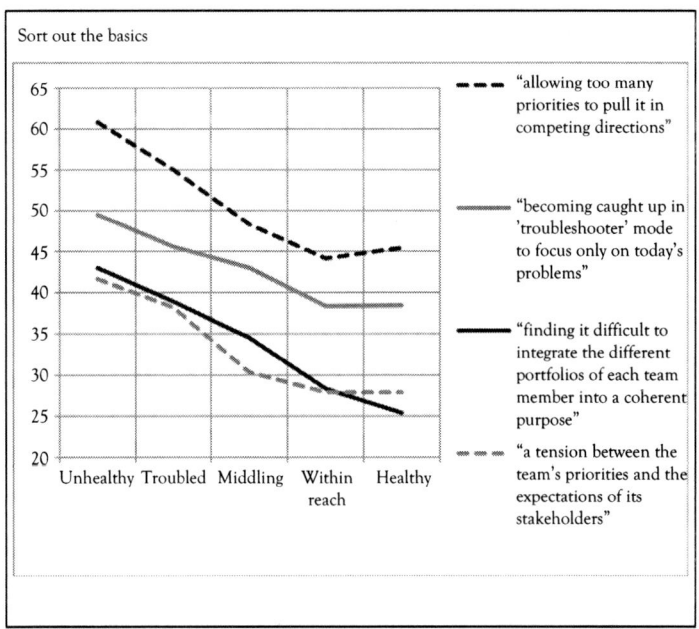

Figure 14.6 *Top four constraints across the five levels of team health* (*n* = 1,262)

4. *Tailor your priorities.* All tests are foundational for *Healthy* teams; however, it pays for teams with different levels of health to focus on different levels. We looked at the average scores of the 15 tests of brilliant teams across all respondent groups and found that a team that wants to remain *Healthy* should focus on aligning the team around a shared purpose and building stakeholder influence (i.e., one clear message) from all the team members to all the different constituencies the team interacts with. A team that collectively increases all aspects of shared purpose and stakeholder Influence is more likely to be a *Healthy* team.

In contrast, *Unhealthy* and *Troubled* teams are better off spending their energy on the basics of why they exist (Unique Commission), what's the future plan to deliver (Foresight); and communicating key messages powerfully across the organization (Compelling Story), as increasing these factors makes them less likely to be an *Unhealthy* team.

5. *Hold up the mirror.* Gathering an outside-in view of the team is critical to ensuring that teams meet the needs of their stakeholders, who view teams differently from the way the team sees itself. Team members have a *rosier* view of their team health, as their average scores are significantly higher than the three other constituency groups; team members, on average, score the highest on 10 of the 15 tests of brilliant teams, as shown in Figure 14.7. Also, team leaders and commissioners are more aligned, as there is no difference in their mean ratings of 13 of the 15 tests.

6. *Question optimism.* Do not believe the predictions of *Unhealthy* teams, which have a larger gap between how they view their current performance and predictions for performance. It is important to question this optimism, as without intervention, these teams are unlikely to achieve their performance ambitions. *Healthy* teams rate their current performance as higher than that of *Unhealthy* teams; a 37 percent variance in the team's overall current performance is explained by the health of a team. As shown in Figure 14.3, we have seen that *Healthy* teams do, in fact, perform higher. Furthermore,

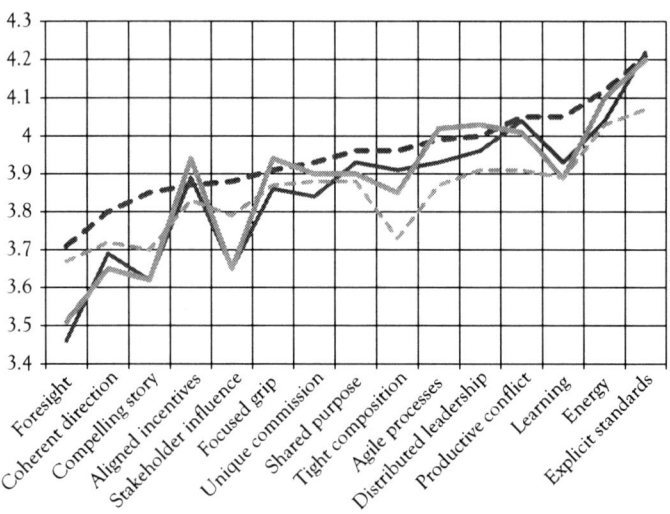

Figure 14.7 Mean scores for the four respondent groups for each of the 15 tests of brilliant teams (n = 812)

although both *Healthy* and *Unhealthy* teams tend to rate their current performance in line with the performance data, *Unhealthy* teams tend to be optimistic in their future performance predictions. When you ask teams to predict what their overall performance will be in a year's time, the ratings across all levels of team health are more positive: team health explains only a 5.2 percent variance in future performance ratings. This is known as the *Optimism Bias* outlined by Kahneman (2011), which describes how most of us have an unrealistically optimistic view about predicting the future.

Conclusions and Implications to Managerial Forensics

Our research with more than 2,000 teams has revealed what makes healthy teams. We believe that managers can derive great benefits from using a tool similar to our TAQ to assess the current health of their own teams. This questionnaire can benchmark teams by assessing responses from four different groups: the members of the team, the team leader, the line manager of the team leader, and stakeholders invested in the team's work. We evaluate each team according to five criteria: the team's *mandate*, how a team is *governed*, the team's *behaviors*, the team's *connections*, and team *renewal*.

We have found that healthy teams outperform unhealthy teams—and although that may sound obvious, our research has shown that healthy teams had an economic impact of 23.6 percent higher than unhealthy teams. Healthy teams reduce costs faster, get to market more effectively, and launch products more smoothly. Finally, we have found six ways that organizations can improve the health of their teams. First, organizations should *focus at the top*. Second, organizations need to ensure that teams *connect with customers*. Third, organizations should ensure that teams *start with purpose*. Fourth, organizations should *choose their recipe carefully*. Fifth, organizations should *hold up a mirror*. Sixth and finally, organizations should *guard against optimism bias*.

The soft stuff is the hard stuff. The intangibles of team dynamics are hard to manage but contain enormous value and are amenable to forensic examination and disciplined development.

References

Ancona, D., and D. Caldwell. 1992. "Bridging the Boundary: External Activity and Performance in Organizational Teams." *Administrative Science Quarterly* 37, no. 4, pp. 634–65.

Butler, S., and J. Rankin. 2015. "Morrisons Sacks Supermarket Boss Dalton Philips." http://www.theguardian.com/business/2015/jan/13/morrisons-chief-dalton-philips-quits-supermarket

Goodley, S. 2015. "Morrisons Sacks Half Its Senior Management Team." http://www.theguardian.com/business/2015/mar/24/morrisons-sacks-half-its-senior-management-team

Hackman, J.R. 2002. "Leading Teams: Setting the Stage for Great Performances—The Five Keys to Successful Teams." *Harvard Business School Archive.* http://hbswk.hbs.edu/ archive/2996.html

Hackman, J.R., and R. Wageman. 2005. "Top Teams: Why Some Work and Some Do Not." Working Paper Hay Group.

Kahneman, D. 2011. *Thinking, Fast and Slow.* New York: Farrar, Straus and Giroux.

Katzenbach, J., and D. Smith. 2001. *The Discipline of Teams. A Mindbook-Workbook for Delivering Small Group Performance.* New York: John Wiley & Sons.

Keller, S., and C. Price. 2011. *Beyond Performance: How Great Organizations Build Ultimate Competitive Advantage.* New York: John Wiley & Sons.

Lencioni, P. 2002. *The Five Dysfunctions of a Team: A Leadership Fable.* San Francisco, CA: Jossey-Bass.

Lencioni, P. 2005. *Overcoming the Five Dysfunctions of a Team: A Field Guide for Leaders, Managers, and Facilitators.* San Francisco, CA: Jossey-bass.

Mavrinac, S., and T. Siesfeld. 1998. Measures That Matter: An Exploratory Investigation of Investors' Information Needs and Value Priorities. *Ernst & Young Centre for Business Innovation and the Organisation for Economic Cooperation and Development.*

Senge, P. 1990. *The Fifth Discipline: The Art and Practice of the Learning Organization.* London: Random House.

Wageman, R., D.A. Nunes, J.A. Burruss, and J.R. Hackman. 2008. *Senior Leadership Teams. What It Takes to Make Them Great.* Boston, MA: Harvard Business School Press.

West, M., M. G. Patterson, and J. Dawson 1999. "A Path to Profit? Teamwork at the Top." *Centre Piece: The Magazine of Economic Performance,* pp. 7-11.

CHAPTER 15

Protecting Value via Information Management

Al Naqvi

Overview

By the time boards and management teams realize that their companies are in insurmountable trouble, it is often already too late. In today's business environment, stability is not a guarantee. Short product life cycles, intense global competition, technological knowledge exchange, and capital flow have all redefined the rules of the game. In this new environment, capital is attracted to those who perform, and is quick to retract from those who destroy value. Rapid reaction of capital upon bad news, when combined with other operational and environmental factors, creates conditions where speed of decline and demise is often intensified. In a world marked by accelerated speed of demise, companies need diagnostic systems, methods, and processes that can provide early warning signals of trouble. Two interrelated issues impact why management teams and boards often fail to monitor and diagnose problems: (1) Inability of boards to receive irrefutable and precise information such that they can overcome their own biases and intervene at the earliest signs of trouble; (2) Having models, metrics, and systems that can provide monitoring and diagnostics information. This chapter presents an epistemologically based model that connects board responsiveness and value loss of companies with diagnostic information that boards receive. In addition, a metric known as Instant Valuation is identified as the diagnostic and a new type of monitoring information technology system is introduced in the chapter.

The Need for Preemptive Information

Before the *point of no return* is reached in their decline and demise, companies issue all kinds of signals. These signals often go unnoticed by managers, analysts, and boards. Identifying and tracking these signals are usually not practical, or even possible. The problem comes from the fact that our information systems were designed to process transactions, provide historical information, and give us some view into the future via predictive analytics. They are not designed to act as risk monitors or diagnostic systems to constantly take the pulse of an organization and provide ongoing feedback to assess the relative health of the firm. In other words, our systems either focus on the past or the future, but fail to provide a view of the current state of a business. This paper presents a model for designing an information system that can provide preemptive and actionable information to a firm about its *current state*. Such information can give managers and shareholders the ability to make better decisions and provide preemptive information to avoid the decline and demise of companies.

The Bliss to Bust Cycle

Good companies can lose value fast. Consider the following examples:

- In the mid-June 2014 time frame, SanDisk stock traded at above $100. A year later, it has lost nearly 50 percent of its value.
- Ralph Lauren, one of the world's most prominent brands in the lifestyle products industry, was trading at over $180 in the beginning of 2015. By the middle of the year, the stock had lost nearly 28 percent of its value.
- Between mid-2014 and mid-2015, Chesapeake Energy lost nearly 60 percent of its market value.
- Mattel, a toy manufacturing and distribution company, lost nearly 30 percent of its value between July 2014 and July 2015.

- Fossil Group, a consumer fashion goods company, lost nearly 30 percent of stock value in one year preceding July 2015.
- Hewlett-Packard Co. lost over 20 percent value in 2015.

These firms play in different industries, are large companies, and have long and established operating histories. Yet, when their value loss began, these firms spiraled into decline rapidly.

In his book *How the Mighty Fall* Jim Collins[1] presents five stages of decline in a firm:

Stage 1: Hubris born of success
Stage 2: Undisciplined pursuit of more
Stage 3: Denial of risk and peril
Stage 4: Grasping for salvation
Stage 5: Capitulation to irrelevance or death

Each of these stages of decline is in effect a representation of the managerial mindset, decisions, and actions. Implied in the definition of the stages is that at each stage, managers make decisions, and boards approve such decisions, with certain type and amount of information. Recent research suggests that boards are falling short to provide appropriate guidance and oversight for companies.[2] *Harvard Business Review* points out that

> Boards aren't working. It's been more than a decade since the first wave of post-Enron regulatory reforms, and despite a host of guidelines from independent watchdogs such as the International Corporate Governance Network, most boards aren't delivering on their core mission: providing strong oversight and strategic support for management's efforts to create long-term value.

---

[1] Collins, J. 2009. *How the Mighty Fall: And Why Some Companies Never Give in.* New York: Jim Collins.
[2] Barton, D., and M. Wiseman. 2015. "Where Boards Fall Short." *Harvard Business Review.* https://hbr.org/2015/01/where-boards-fall-short (accessed July 15, 2015).

Boards are often slow to move and to take the steps necessary to avoid the demise and decline of companies. Outside of malicious intentions (board members intentionally destroying value) or cultural constraints (e.g., the inability to overcome personal bias due to friendship with the chief executive officer [CEO]) the irresponsiveness or lack of early intervention from boards can be a function of not having clear and irrefutable information about the factors leading to the demise of a firm. This becomes even harder during the early stages of the problems since the signs of trouble are so insignificant that it is not possible to connect the dots of demise. As such, the structure, format, and comprehensiveness of information presented to the boards are critical to avoid decline and demise.

Thus, a framework is needed that classifies the stages of decline and demise in terms of information availability and its comprehension—and that links such factors with the speed and intensity of responsiveness of boards to act. This epistemologically oriented model can provide the foundation for a new type of information model. Unlike the Collins model, which presents a managerial view of the decline, an alternative way to think about the decline stages is board of directors' awareness level about decline and demise of a firm. We can call it B-to-B Cycle and it too has five stages. These stages can be described as follows:

Stage 1: No obvious trouble signs
Stage 2: Some indications of trouble
Stage 3: General knowledge of issues
Stage 4: Clear and undeniable
Stage 5: Full acceptance

Stage 1: No Obvious Trouble Signs: At this stage, everything seems fine and there are no visible signs of trouble. A firm is producing cash and achieving milestone. Boards are happy and investors are satisfied. Despite the outward tranquility, problems have begun to take shape. This is usually the time when earliest signs of trouble arise; however, they are usually ignored. Present success conceals the emergence of problems. Even though at this stage there are no obvious signs of trouble, some farsighted expert professionals can see signs of problems into the future. Usually, such awareness results from connecting the dots, observing patterns, or having a truly visionary insight.

Stage 2: Some Indications of Trouble: At this stage, analysts notice some signs of trouble. Boards generally bring up the issues in board meetings, but do it as cautiously and gently delivered question or two. When asked, management teams typically reject these indications in a cavalier fashion. Some analysts begin covering the trouble signs—but they are typically a minority.

Stage 3: General Knowledge of Problems: At this stage, the company's problems become widely known. Management teams and boards resolve to fight the problems and make claims to turn the tide in their favor. Analysts sharpen their tone and stock value takes a hit. Even though the problems are known, a sense of optimism prevails. This is typically the stage in which the state of delusion can be most powerful. This is also the time when activist investors start flocking around like hawks.

Stage 4: Clear and Undeniable Proof: At this stage, trouble signs become extremely obvious. There exists no doubt that the company is heading in the wrong direction. Stock plummets and management teams begin feeling the heat from every angle. At this stage, some type of informal restructuring plans are developed. Mostly, such plans are internally developed, typically with the help of strategy consulting firms. Boards want leaders to be held accountable but delusional hopefulness persists. In many cases, even at this stage, some board members choose to wait and give management more chance to perform. The cash position of companies becomes challenging and management teams are left with fewer options.

Stage 5: Full Acceptance: At this stage, finally, all parties come to a conclusion that the situation is hopeless and that formal restructuring is needed to turnaround a firm. At this stage, specialty restructuring firms are hired, management teams are ousted, and turnaround specialists are brought in.

A Behavioral System

It should be clarified that the B-to-B cycle is derived from human cognitive and behavioral angle rather than a financial or operational metrics angle. It is based on how boards and managers process information about the firm, how this information flows through the mental filters and biases, and how each stage of denial is overcome, often reluctantly, to the reality that confronts a firm. The key question is that is it possible to speed up the acceptance of reality without actually suffering the demise and having all the negative consequences?

The Topology of Bliss to Bust Cycle

As shown in Figure 15.1, the *y*-axis represents the value of a firm and the *x*-axis represents the time duration. Although it is a simple curve, the topology of the B-to-B cycle provides incredible insights into the potential demise and decline of companies.

> *Trouble Inflection Point*: Trouble Inflection point or TIP is the point where first signs of trouble begin in a stable state.
>
> *Full Acceptance and Comprehensive Turnaround*: Full Acceptance and Comprehensive Turnaround (FACT) is the point where there is absolutely no uncertainty about the troubles and a restructuring or turnaround team is put in place to lead the turnaround.
>
> *B-to-B Cycle Time*: Bliss to Bust Cycle Time is the time it will take for a company from TIP inflection point till the FACT point.

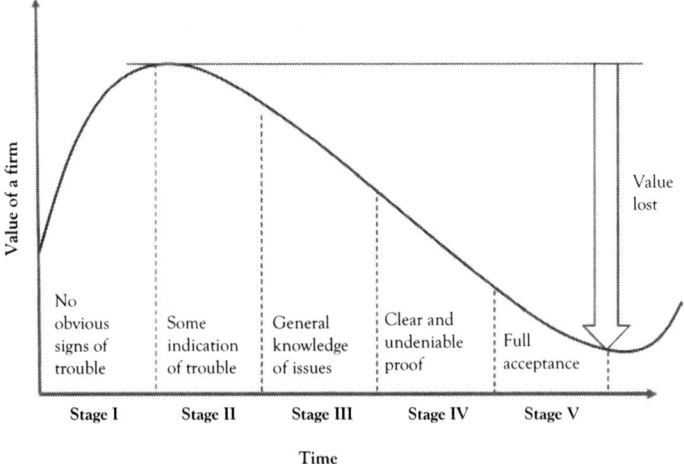

Figure 15.1 B-to-B Cycle

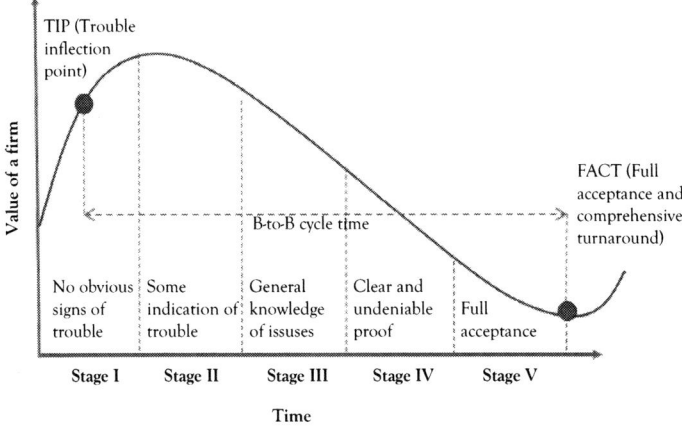

Figure 15.2 B-to-B Cycle

Obviously, this can be a factor of the industry, company's position in the market, its resources, decision style of the management teams, board's ability to overcome its own biases, and many other factors.

In order to manage the cycle a company needs to: (1) Identify the TIP event. Not all TIPs will result in rapid or guaranteed value loss. Some TIPs can be handled easily and value destruction can be reversed. Knowing when the TIP event has transpired improves the ability to intervene; (2) for each TIP event, estimate the B-to-B cycle time, that is, expected duration till a meaningful value destruction reversal plan can be put in place; (3) for each TIP event, determine what value loss may happen; and (4) build a plan to reduce the value loss impact, and accelerate the recovery time.

The goal is to lessen the value loss and shorten the duration of the negative impact of problems. In this case, the curve (value loss) becomes shallower and narrower versus deeper and longer (Figure 15.3). It should be pointed out that for managers and boards responsible for protecting and increasing a company's value, there are two paths available if the Tipping Inflection Point has been reached. The company needs to minimize the value loss, shorten the time duration of the B-to-B cycle in order to accelerate recovery, or perform both.

Although the duration of each stage in Figure 15.3 is assumed to be equal, each stage can have different durations. The last stage may arrive

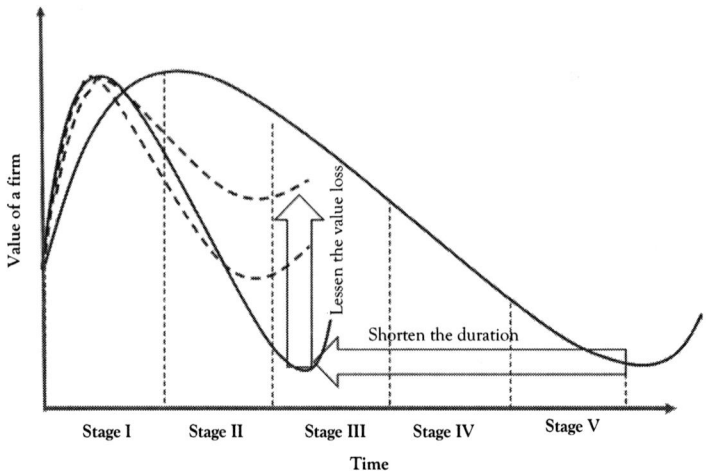

Figure 15.3 Reduce the impact

faster, or slower, depending on the quality of information available and the ability of the boards to understand the situation and make decisions accordingly. It is important to understand that faster recognition of problems can enable faster response to fix them.

Other factors that can also be studied in the B-to-B cycle are how much capital will be lost before an official turnaround begins and how many alternative options would be lost as you move from stage to stage.

Building a Stage 1 Diagnostic

How can we identify a TIP? When you visit an ER department, the first level of triage operation or clinical intervention is to attach various kinds of diagnostic equipment to your body. These patient-monitoring devices allow ER (Emergency Room) physicians and clinical staff to evaluate the present or current state of a patient. At that point it is usually less relevant what the patient's vital signs were six months ago or what they would be in a year's time. The most relevant and important information is the patient's current state. If we apply the same concept to companies, we would want to develop metrics that can provide us a solid understanding of a company's current state. In other words, we would need to develop tools and metrics that can help us identify a TIP event. In order to develop those metrics, we would need to understand the two primary sources of problems.

To be successful companies need two types of excellences. First, they need to be excellent investors of capital provided to them by the shareholders. Investment excellence means that companies operate with solid strategies, scan the markets for right opportunities, and invest responsibly in projects that return greater than the expected cost of capital. Second, they need to be excellent operators so that they can maximize the potential of the investments the company has made. Operating excellence includes all operational capabilities such as marketing, distribution, sales, supply chain management, sourcing, manufacturing, service management, and so on.

In fact, the aforementioned strategy can be viewed as a twofold strategy whereby companies invest in areas in which they can apply their unique operational strengths as they develop unique operational strengths in the areas in which they have invested. For example, it will be prudent for Toyota to invest in areas in which it can apply its unique operating methodology (e.g., Lean Production) to exploit greater efficiencies. Similarly, once investment is made it should make sure that it continues to develop and expand its operating models to accommodate the unique requirements of the new investments.

Hence, we can study the genesis of problems in a firm from two angles: (1) Investment problems—where a company fails to make successful investments; and (2) operating problems—where a company fails to execute and deliver upon the investment promise due to operating issues.

If we canvass the two problems, we would get a two-by-two as shown in Figure 15.4.

Talking about a failed offshoring project, Robert McDonough, COO of United Technologies, said, "I think we failed on both the planning and the execution side."[3] Such a paralyzing situation is reflected in the top-right quadrant of Figure 15.4 where a firm lacks both investment and operational excellences. The other two problems can arise when a company with solid operational capabilities is unable to make good

---

[3] Mann, T. 2014. "Otis Finds Reshoring Manufacturing Is Not Easy." *Wall Street Journal.* http://www.wsj.com/articles/sb10001424052702304518704579519432946574424 (accessed June 10, 2015).

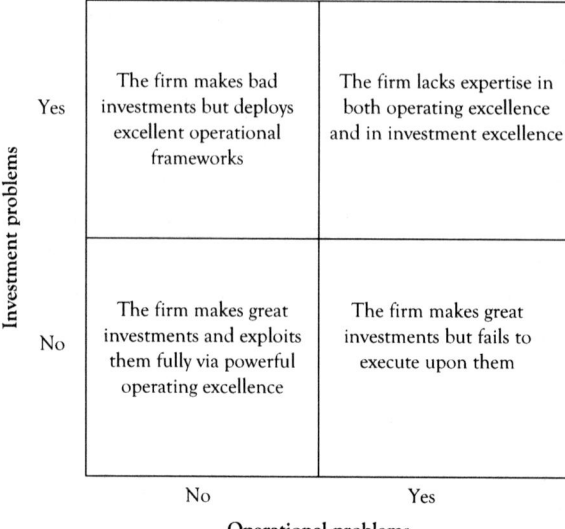

Figure 15.4 Strategic problems matrix

investment decisions or when a company that makes good investments fails due to underdeveloped operating models.

As such we need a measure, or a group of metrics, that help assess both investment excellence and operating excellence. This metric must provide ongoing information about a firm and its operating and investment decisions. It should also give an assessment of the health of a firm. Most importantly, it should also provide a good understanding if a TIP event has happened.

Instant Valuation—The Key Metric

In order to determine the current state of affairs of a company, it is critical that shareholders and managers receive the latest information on the Investment Excellence capabilities and the Operating Excellence capabilities of a company. The former will determine if the company will be able to find the right opportunities to invest in and the latter determines if the firm will be able to monetize or capitalize on those opportunities.

While accountants and financial experts debate what the most relevant measures of performance of a firm are, the value of a company is a reflection of the combination of all of the critical measures in a single number. After all, there is nothing in the operating or financial dynamics of a company that is not included in its valuation number. Hence, the only measure that contains information about the operating and investment performance of a company is its valuation.

The single most important measure of a company's current state is its valuation. This number sums up the value of its cash flows from the future and brings it to the present moment. It links a company's present with its future. It is established on the basis of a company's cash flows as well as the relative riskiness of the cash flows. Since this one number incorporates a company's investment strategy and its operating strategy, it is the best indicator for a company's health.

Some of the main attributes of valuation include the following:

1. Contains both operational and investment information
2. Primarily focuses on expected profits, cost of capital, and how much capital is plowed back into the company
3. This number sums up literally all activities and risks of a firm in a single number.

While public companies are constantly being valued by public markets, such valuations happen on the basis of the information available to the markets. What is needed is for companies to develop internal valuation numbers and to monitor them constantly.

Designing the Value Monitoring Information System

To overcome the risk of rapid decline and demise, and to manage value loss preemptively, a company should have a system that constantly monitors its intrinsic value. Intrinsic value is based on the fundamentals of a firm.

This should be done in a manner such that the leadership team and the boards should be able to know a company's valuation on an ongoing basis—even as a constantly updated dashboard item. This dynamically available valuation metric can be termed as Instant Valuation. It is important to understand that the information setup to calculate Instant Valuation will have the following key features:

1. All of company's projects, their status, and their impact on the value of the company will directly contribute to the Instant Valuation number.
2. Each project will have its risk attributes understood and its cost of capital identified.
3. Each product line will also roll into the valuation number.
4. Each business unit will be reflected in the Instant Valuation number.

In such a system, future cash flows and options (including real options) will flow into the valuation number. Thus, the Instant Valuation number would encapsulate all the operational and investment information of a firm. Such a valuation number will change constantly, as new information comes in from a company's forecasting and accounting systems. It will also be able to show the impact on the value of a firm by cash flows on an annualized basis.

The fundamental concept behind this methodology is that as soon as new information becomes available on any operational issues (e.g., failed marketing, operational problems, R&D failure, supply chain issues, etc.) or investment issues (e.g., inability to reinvest capital in new opportunities, investments made in value destroying opportunities, overpaying for an acquisition, etc.) it will immediately impact the valuation number. Since the valuation calculation is driven by future cash flows and risks, any strategic or tactical changes will be reflected in the number immediately. This will also signify a cultural shift where line managers will have to estimate and provide empirical estimates of the impact of any operational or investment issues. For boards and executive teams, any significant unfavorable changes in the Instant Valuation number can indicate looming problems.

Such a system can provide truly interesting analysis for public companies. They will be able to dynamically monitor their market-determined stock value with the Instant Valuation, study the variances, and understand the differences between shareholder expectations and management expectations. Such a system can also be used to identify opportunities to buy back stock or raise capital through stock offerings.

How Does It Work?

A system of Instant Valuation that links the operational performance of a firm with its value is presented in Figure 15.5. As shown in the figure, the formulae for determining the value of a firm are based on the return on invested capital (ROIC), weighted average cost of capital (WACC), and growth. In simple terms, the formula is saying that the expected cash flows produced by the firm, the capital invested to date, the growth factor, and the relative riskiness of the investment determine the value of the firm. Cash flows of a firm are derived by revenues and expenses; cost of capital is a function of capital structure and risk; and growth depends on the reinvestment of capital into the company (which is based on factors such as growth rate, dividends, and stock buyback policies). In effect, the valuation number really encapsulates the entire spectrum of the business.

As Figure 15.5 shows, stemming out from the base valuation is the primary determinant—that is, ROIC, WACC, and growth. As we branch out further, we can break ROIC into revenues, expenses, and the invested capital (Level 1). Note that, in this case, we have nicely blended the information from both balance sheet and income statement. In Level 2 analysis, we can use the traditional metrics from both income statement and balance sheet (e.g., revenues, expenses, capital invested, etc.) to understand what is impacting the value of a firm at a high level. In Level 3 analysis, we break down revenues, expenses, and other factors into lower-level drivers of value (e.g., price per unit, volume, etc.). In Level 4 analysis, we can take each driver of value and link it with market, business environment, and lower-level operational factors that determine the value of drivers (e.g., market share and competitive factors). Lastly, Level 5 analysis links specific corporate initiatives and projects with their impact on the value of a firm. In other words, specific actions of management team (e.g., implementing strategic sourcing or launching a new social media campaign) can now be directly connected with the value of the firm. To learn more about valuation methods and value trees refer to the book *Valuation, Measuring and Managing the Value of Companies.*[4]

Please note that the example shown in Figure 15.5 is not comprehensive and is only used for illustrative purposes. A comprehensive tree will have hundreds, possibly thousands, of branches. The Instant Valuation

---

[4] McKinsey & Company Inc., T. Koller, M. Goedhart, and D. Wessels. 2010. *Valuation: Measuring and Managing Value of Companies.* Hoboken, NJ: John Wiley & Sons.

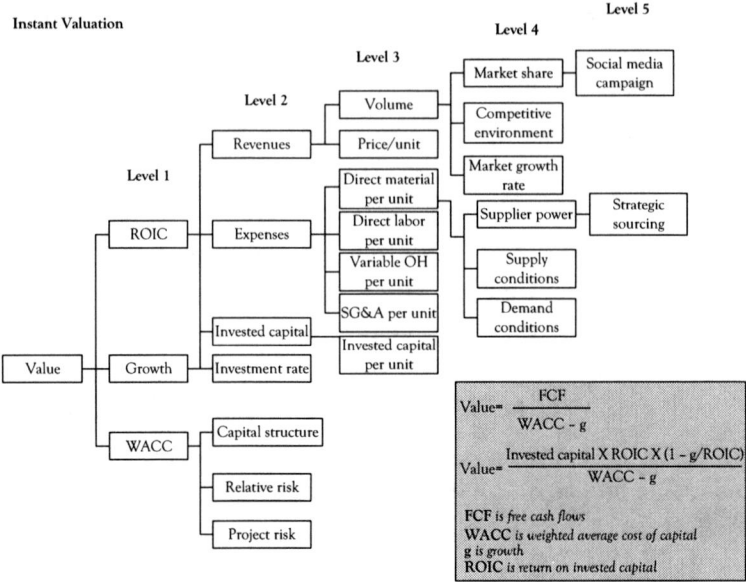

Figure 15.5 Instant valuation

Model not only builds the model from core financial metrics, but also links up the corporate initiatives and projects with the value of a firm. Level 5 analysis is the most distinguished part of the model.

Notice how this system can provide preemptive and early warning signals to the management team. For instance, as soon as management discovers that a specific corporate initiative or project will not be able to achieve its goal, its variance will show up in the operational and market factors, which in turn will show up in value drivers, and value drivers will impact the ROIC, growth, or WACC—and all of these factors when combined will show the impact on the valuation of a firm. These trees can be built for business units, product lines, and projects. With a tool as such, boards will have the visibility at the lowest level of details and will be able to see the factors that impact the value of a firm.

Who Manages the Instant Valuation?

While the management team has significantly greater information than external analysts or external shareholders, their own biases may interfere with their assumptions for growth, risk, or the retention rate of capital (some of the determinants of valuation). These assumptions may create significant disparity between the company's internal numbers and the valuation as determined by the public markets. If a company fails to truly estimate the impact of an operational or investment issue on the future earning potential of a firm or fails to empirically determine the riskiness of a project, an illusion of stability can be created. Hence, it is advisable that such a system be implemented and managed by independent auditors or external entities.

What Would It Take to Implement Such a System?

Advances in technology such as big data can now enable companies to collect and process massive data streams. With real-time data streaming and ability to process large data sets, companies now have the option to implement such valuation-based systems.

Key Points

Rapid value loss in companies is too big a problem to ignore. In many cases, significant wealth is destroyed in the process. The key question is that how is it possible that large companies, with established business models, can fall so quickly. This chapter solves the problem by introducing two tools: (1) B-to-B Cycle Analysis and (2) Instant Valuation. Bliss to Bust Analysis is an epistemological tool that estimates the speed and intensity of responsiveness of the boards of a company when trouble lands. This tool can be used to assess how efficiently and effectively a board will respond to a company's problems. Instant Valuation is a tool that gives boards the ability to drill down and preemptively determine the potential upcoming loss of value. The core assumption being made here is that if boards have such information, they will be able to intervene and impact change quickly—and hence save companies from rapid wealth destruction. Notice that one tool provides information to the boards so they can act quickly and the other tool is a behavioral tool that studies the efficiency and effectiveness of a board's response. With these two tools in place boards should have no excuse as to why they are not able to respond to looming crises rapidly. These two tools also establish higher accountability. Companies that either fail to provide meaningful and actionable information to the boards or boards that fail to decisively and rapidly act on the information are both critical failure points in management of value of firms—and the aforementioned tools help solve that problem. Further research should focus on empirically measuring the responsiveness of boards and relate that to value loss of firms.

CHAPTER 16

Changing Direction

Diana Heeb Bivona

Introduction

For nearly half a century, researchers and practitioners have pondered the question as to how and why businesses fail. As a result, numerous predictive failure models have been created. Why is there such an obsession with creating these models? The belief is that better predictive failure models could be the key to creating early-warning systems for businesses. If businesses are better able to understand the origins of failure, managers could take specific corrective actions to prevent failure and turn a company around.

Researchers have attempted to put forth a predictive model to explain why businesses fail. However, no definitive predictive model has been created that is both highly accurate and applicable to companies across all industries, regardless of size, location, or age (Balcaen and Ooghe 2006). The quest to find answers as to why businesses fail has taken researchers down a winding path, yielding a fragmented field of literature.

An Evolving Approach

The earliest framework proposed in the literature was the quantitative approach. Beaver (1967) and Altman (1968) were among the first researchers to build predictive business failure models based on financial ratio analysis using statistical methodology. The underlying premise of this quantitative framework assumed that the failure process was a static event, and predictive models could be built based on specific financial data. Other researchers followed Beaver and Altman's quantitative approach favoring financial ratios and statistical models to predict business failure

(Dimitras, Zanakis, and Zopounidis 1996; Laitinen 1993; Lussier 1995; Ooghe, Spaenjer, and Vandermoere 2009; Wang and Campbell 2010). However, the similarity ends there. Each researcher adopted his or her variation of Beaver or Altman's quantitative framework using different financial variables, sample characteristics, and validation techniques. As a consequence, several alternative methodologies were developed.

While the earlier quantitative framework concentrated on creating a diagnostic tool for *predicting* business failure, some researchers suggested that a more useful objective was to focus on *preventing* failure. Thus, attention shifted to identifying the causes (why) and processes (how) of business failure, and the framework changed. Argenti, an early pioneer of the qualitative approach, suggested that the reasons for corporate collapse could be found within a company, in the management, and the management process undertaken. For Argenti (1976), the blame for the failure of a business was placed squarely on the doorstep of the leadership.

Several other researchers adopted a qualitative framework (Balcaen and Ooghe 2006; Dimitras, Zanakis, and Zopounidis 1996; Sharma and Mahajan 1980; Van Caillie 1999) but focused on a variety of organizational management aspects in failing companies. Three primary categories drew the attention of researchers: corporate governance, strategic management, and operational management (Van Caillie 1999).

Crutzen and Van Caillie (2007, 15) proposed an integrative model on how and why companies fail. This model adopted the resource-based view of a firm that sees businesses as "heterogeneous bundles of idiosyncratic, hard-to-imitate resources and capabilities." Firm performance is directly tied to the ability to utilize its resources, capabilities, and deployment effectively in such a way as to create a sustainable competitive advantage. Superior performance occurs when those resources and capabilities are successfully aligned with strategic industry factors—features of the competitive environment that determine a company's profitability (Amit and Schoemaker 1993). Conversely, failure ensues when a company's resources and capabilities are no longer (or never were) aligned with its competitive environment (Thornhill and Amit 2003).

Phases of Failure

Crutzen and Van Caillie (2007), to date, have created the only integrative model of its kind. It is unique in its attempt to integrate commonly identified quantitative and qualitative factors found from nearly 50 years of research on this topic. The model incorporates, in an interpretative perspective, the findings of previous studies, combining failure factors and events recognized as the *most often evoked* in the literature (Crutzen and Van Caillie 2007).

Crutzen and Van Caillie (2007, 18) viewed the business failure process as a misalignment between a company's resources and distribution and its external environment. A company enters the failure process when its "set of resources and its deployment are inadequate and not adapted to the requirements of its environment." If a firm's resources and deployment are inadequate, it cannot react to pressures, both internal and external, and the firm can neither create nor sustain a valuable strategic position (Thornhill and Amit 2003). This failure to take corrective action to bring in line the resources of the company and its distribution with the requirements of the environment starts the firm on a spiral into failure (Crutzen and Van Caillie 2007).

Failure symptoms become visible if corrective actions are not taken to align a firm's resources and deployment with the environment. Often, these symptoms developed during the first phase but were hidden and not immediately recognized by management. Those symptoms that are often financial emerge in the second phase. Symptoms include, but are not limited to, insufficient sales or revenue, a decrease in profitability, a decline in competitiveness, reduced market share, a lack of cash flow and liquidity, an increase in external debts, and a greater need for more external financing (Crutzen and Van Caillie 2007; Laitinen 1992; Thornhill and Amit 2003).

Left unaddressed, firms enter into the third phase of business failure. Rapid deterioration marks this stage. Critical warning signals emerge. The firm's liquidity and solvency are critically low leading to mistrust among management and stakeholders. Creditors become more diligent and less swayed. Often, when a company enters this phase, it is too late to reverse the failure trend, and bankruptcy is imminent (Argenti 1976; Crutzen and Van Caillie 2007; Laitinen 1991, 1992).

Possibilities and Limitations

The integrative model provided by Crutzen and Van Caillie (2007) identifies a potential pathway to business failure and signals that can alert a firm to problems. To date, it is the most comprehensive model attempting to integrate and build upon the existing literature in the field. Crutzen and Van Caillie's (2007) research is relevant because it indicates that if a set of failure causes and a series of events can be identified along a business failure pathway, then corrective actions can be determined and specific corrective actions undertaken.

The integrative model proposed by Crutzen and Van Caillie (2007) is promising but does have limitations. First and foremost, it has not been empirically validated. The model does not address the time dimension, that is, when does a firm enter and move through each phase of failure. Finally, more details as to how organizational and financial factors combine and contribute to each phase of failure are needed. However, despite these concerns, the research does suggest that even when entering the failure process, companies still have a window of opportunity under which they can commence a successful turnaround strategy.

Undertaking Corrective Actions

As previously discussed, effectively identifying and diagnosing the problems of a business is just the first (and most important) step in preventing business failure. Assuming that an accurate differential diagnostic of the underlying issues has accurately pinpointed the case, the next step is to undertake a successful turnaround. Turnarounds are a multistage process that often requires overcoming both external circumstances and internal organizational constraints. Numerous theories exist in the business literature about how to revive a poorly performing business, but proven strategies for an effective turnaround are scant.

The track record for companies attempting a turnaround is sobering. Only about one-third of companies intending to turn around their dire situation can do so. Of these, only 40 to 50 percent can position themselves for long-term growth (Yandava 2012). The turnaround process can be painful and does not guarantee success. It is, however, the best alternative to failure.

Turnaround approaches are hampered by several factors including limited resources, lack of stakeholder support, and time constraints. The method of turnaround selected often depends on an organization's resources, the causes of the failure, and the nature and extent of the failure itself (Spremo and Prodanović 2013). Historically, turnarounds have focused on taking either a strategic or an operational approach, but as with the differential diagnosis approach to managerial forensics, a more systematic approach to a turnaround is warranted.

Conditions in the external environment and internal organizational constraints are much more complex than what companies have experienced in previous decades. As discussed earlier, the causes of business failure typically are just as complex and integrated. This requires a turnaround approach that necessitates a systematic analysis and realignment of capabilities throughout the company. Therefore, a successful turnaround needs to focus on the strategy, leadership, organization, operations, marketing, and financials of a company.

Several notable companies found themselves at one point or another at a crossroads needing to undertake a turnaround—a turnaround that required a similar systematic approach. For example, Kraft Foods witnessed a decline in its revenues of 9 to 11 percent when consumers began to demand healthier food options. The *quick fix* might suggest simply offering healthier product offerings. Instead, Kraft returned to its core competencies, undertook an operational restructuring, centralized its core processes, and reevaluated the performance and accountability of its senior management team (Yandava 2012) to place it squarely on the road to recovery and subsequent profitability once again.

Retailer Talbots Inc. ignored a change in its consumer demographic, which led to an annual loss of $189 million in 2007—a wake-up call after enjoying several years of stable revenues. Talbot's restructuring, like Kraft, took a holistic approach. Talbots reconnected with its loyal core customer and focused on ensuring that its product offerings supported what its core customer base wanted. The company worked on ways to strengthen its centralized inventory control mechanisms and distribution system to increase sale management efficiencies and created greater synergy between its online store catalog and bricks and mortar locations. Talbots also undertook an aggressive financial restructuring. With the procurement of BPW Acquisitions Corporation, the company was able to reduce its outstanding debt and stock repurchasing program. After two years of loss, the company was able to achieve an operating income of around $60 million in 2010. The increased liquidity in stock also allowed the company to focus further on additional long-term strategies (Yandava 2012).

Table 16.1 offers a synopsis of the business areas to assess, possible focus questions, and potential turnaround solutions. This is by no means an exhaustive list, merely illustrative of the methodical approach to a turnaround that should be undertaken.

Table 16.1 Assessment by business area

Area of focus	Focus question	Potential turnaround solutions
Strategy	How can we utilize our key strengths or core competencies to continue differentiating ourselves from the competition?	• Reevaluate core competencies • Adopt a growth strategy • Invest in R&D • Implement a flexible value chain to respond to customer demand
Leadership	How can we ensure the leadership team remains committed to our strategy?	• Reaffirm commitment of senior management team to turnaround strategy • Ensure that departments are focused on the integrated solution • Commit to leading change • Advocate accountability in all actions • Incorporate sustainability and corporate social responsibility
Organization	How can we create a positive and dynamic environment in which our employees can flourish?	• Ensure strategy moving forward is integrated with organizational performance • Create a positive dynamic workplace that attracts and retains top talent • Foster an environment of innovation, collaboration, and continuous improvement

Operations	How can we improve the efficiency of our business processes?	• Improve key efficiencies in the value chain • Strategically invest in key technologies • Focus on product or service quality
Marketing	How can we enhance our marketing and sales activities?	• Reevaluate competitive environment • Create innovative market offers • Build or improve customer relationships
Financial	How can we change our capital structure to achieve financial efficiencies?	• Dispose of nonperforming assets • Restructure debt and equity • Exchange equity for debt • Evaluate opportunities to improve cash flow both in the short and long term

After an assessment of the business areas has been completed, a turnaround plan defining the specifics: how, cost, where, and when can be undertaken. This is then followed by the implementation and management of the plan of action and the final step, a thorough evaluation of the turnaround plan.

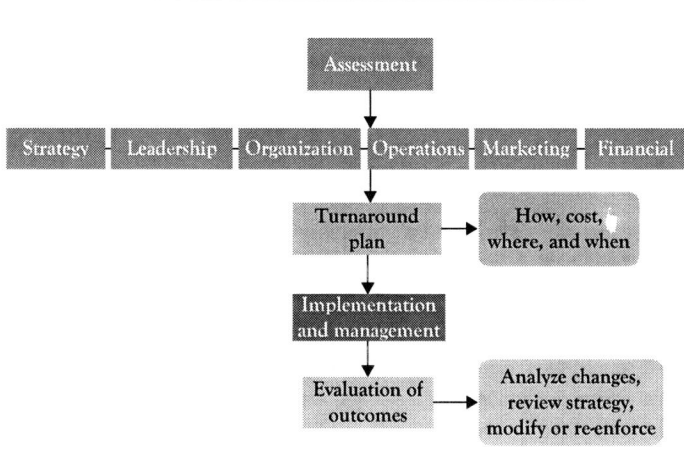

Conceptual framework of a successful turnaround

Conclusion

Predicting business failure and effecting a successful turnaround remain two of the most investigated topics in management literature. The research over the last few decades has grown exponentially, but significant gaps remain. A uniform analytical framework for predicting failure and a successful turnaround strategy remain elusive.

It can be easy to miss early warning signs of business failure as they are often weak, or the evidence needed to *prove* there is a problem is hard to pin down (Yandava 2012). However, the longer we ignore the signals that there is a problem, the more limited is our ability and resources to undertake an effective turnaround strategy. While a turnaround typically arises from urgent necessity, it may prove beneficial to view it as an opportunity to redefine our business strategically to ensure it remains relevant and sustainable over the long term. In simplistic proverbial terms, if we are given lemons we need to work to make lemonade.

References

Altman, E.I. 1968. "Financial Ratios, Discriminant Analysis and the Prediction of Corporate Bankruptcy." *Journal of Finance* 23, no. 4, pp. 589–609.

Amit, R., and P.J. Schoemaker. 1993. "Strategic Assets and Organizational Rent." *Strategic Management Journal* 14, no. 1, pp. 33–46.

Argenti, J. 1976. *Corporate Collapse: The Causes and Symptoms.* London, UK: McGraw-Hill.

Balcaen, S., and H. Ooghe. 2006. "35 Years of Studies on Business Failure: An Overview of the Classic Statistical Methodologies and Their Related Problems." *The British Accounting Review* 38, no. 1, pp. 63–93.

Beaver, W.H. 1967. "Alternative Accounting Measures as Predictors of Failure." *Accounting Review* 43, no. 1, pp. 113–22.

Crutzen, N., and D. Van Caillie. 2007. "The Business Failure Process: Towards an Integrative Model of the Literature." *EIASM Workshop on Default Risk and Financial Distress.* http://www.hec.ulg.ac.be/sites/default/files/workingpapers/WP_HECULG_20070502_Crutzen_VanCaillie.pdf

Dimitras, A., S. Zanakis, and C. Zopounidis. 1996. "A Survey of Business Failures with an Emphasis on Prediction Methods and Industrial Applications." *European Journal of Operational Research* 90, no. 3, pp. 487–513.

Laitinen, E. 1991. "Financial Ratios and Different Failure Processes." *Journal of Business Finance & Accounting* 18, no. 5, pp. 649–73.

Laitinen, E.K. 1992. "Prediction of Failure of a Newly Founded Firm." *Journal of Business Venturing* 7, no. 4, pp. 323–40.

Laitinen, E.K. 1993. "Financial Predictors for Different Phases of the Failure Process." *Omega* 21, no. 2, pp. 215–28.

Lussier, R.N. 1995. "A Nonfinancial Business Success versus Failure Prediction Model for Young Firms." *Journal of Small Business Management* 33, no. 1, pp. 8–20.

Ooghe, H., C. Spaenjers, and P. Vandermoere. 2009. "Business Failure Prediction: Simple-Intuitive Models versus Statistical Models." *IUP Journal of Business Strategy* 6, no. 3/4, pp. 7–44.

Sharma, S., and V. Mahajan. 1980. "Early Warning Indicators of Business Failure." *Journal of Marketing* 44, no. 4, pp. 80–89.

Spremo, T., and S. Prodanović. 2013. "The Turnaround Strategy and Courses of Action of Companies in the Crisis." *Zbornik Radova Ekonomskog Fakulteta U Istocnom Sarajevu* no. 7, pp. 101–11. doi: 10.7251/ZREFIS1307101S.

Thornhill, S., and R. Amit. 2003. "Learning about Failure: Bankruptcy, Firm Age, and the Resource-Based View." *Organization Science* 14, no. 5, pp. 497–509.

Van Caillie, D. May 1–14, 1999. "Business Failure Prediction Models: What Is the Theory Looking for?" Paper presented at the Second International Conference on Risk and Crisis Management, Liege, Belgium.

Wang, Y., and M. Campbell. 2010. "Business Failure Prediction for Publicly Listed Companies in China." *Journal of Business & Management* 16, no. 1, pp. 75–88.

Yandava, B. 2012. "A Capability-Driven Turnaround Strategy for the Current Economic Environment." *Journal of Business Strategies* 29, no. 2, pp. 157–85.

PART IV

Conclusion

CHAPTER 17

Conclusion

J. Mark Munoz and Diana Heeb Bivona

The business world faces diverse challenges and opportunities at an ever-increasing pace. Some management teams readily find the appropriate business solutions, while others do not. In many cases, fresh perspectives both from internal and external sources are necessary.

Business failures have become a common reality in contemporary business. In 2014, about 963,739 business and nonbusiness filings were submitted to U.S. Bankruptcy Courts (U.S. Courts 2014). Gallup reported that the death rate of U.S. businesses surpassed the birth rate of new businesses for the first time in 2008 since these metrics were tracked, and that trend has continued (Clifton 2015).

Managers have the ability to determine the course and the fate of their companies. For instance, rates of business failures vary across industries and countries and require the implementation of strategic approaches such as continuous monitoring, price increases, and reduced country exposure to manage risk in high insolvency-risk locations (Dun and Bradstreet 2012).

More than ever, practitioners of management need to develop skills in organizational introspection, research, and analysis in order to identify the best strategies. Ultimately, the companies that best capture and manage data and information will gain advantages in innovation, competition, and productivity (McKinsey 2011).

The challenges and opportunities brought about by the contemporary business environment will make managerial forensics an important practice in the coming years. It underscores the notion that the fate of an enterprise is shaped by the concerted action of its leaders. It supports the thinking that well-planned, methodical, and efficiently executed courses of action can lead to favorable business results. It highlights the need to continuously strive for optimal organizational health.

The featured chapters in this book highlight the key attributes of the managerial forensics approach.

Holistic Approach

Managerial forensics requires a comprehensive examination of the entire organization. Similar to the way a physician examines an entire human body, an effective forensics examination requires a holistic approach.

Historically, many managers have successfully identified problems associated with their particular business units (marketing, operations, finance, etc.). Unfortunately, these business units have often followed the silo approach limiting the communication across all functions of the company, more importantly, limiting the ability to identify multidimensional organizational problems. By using this holistically focused approach, the interconnection of various business units are noted, and nonfunctioning or dysfunctional areas within those functions are identified. The best solutions then can be found when a firm is methodically evaluated in its totality.

When attempting to overcome a problem that is causing the decline of our business, we become solely focused on *fixing* the problem in isolation. We fail to look at how our decisions related to addressing that issue will potentially impact the health and well-being of the company as a whole. Adopting a holistic approach reminds managers to maintain focus. To remember as Aristotle suggested "the whole is greater than the sum of its parts."

Detail Oriented

Diagnosing decline is never easy. If it were, we would never see global companies led by very savvy executives go through long periods of poor performance. However, the reality is that understanding the causes of corporate malaise can evade even the smartest of managers. Managerial forensics supports the notion of *never leaving a single stone unturned*. This means that, in finding solutions to corporate malaise or demise, the forensic analyst needs to objectively dig deep to uncover root causes of problems.

Making sense of reams of complex information in a large organization is riddled with hazards that can detour or throw the problem-discovery process into chaos. Wading through reams of data is challenging because our abilities as human beings to handle and process information is limited. As a result, we frequently take mental shortcuts when looking for the reasons why our company is in decline. These shortcuts may mislead us into believing that we are handling the overwhelming amount of information, but often these shortcuts lead us to misperceive or misinterpret the reasons for decline.

Similar to a clinical examination, this means going beyond the health history and intake of a patient, and using other diagnostic tools to gain an understanding of the problem. With the differential diagnostic process, x-rays and MRIs as well as blood and tissue samples are extracted to better understand the patient's condition. In managerial forensics, the use of diverse diagnostic and technological tools can lead to similar results. For example, big data analytics can help extrapolate information unknown to the executive team. This acquired knowledge can help shape management decisions.

Scientifically Grounded

Managerial forensics underscores the notion that scientific methodologies can be applied to the practice of management. The more reliable the information and the more thorough the research, the higher the chances of making an effective business decision.

As seasoned managers, the tendency is to rely on our past experiences and intuition to fill in what we do not know when dealing with limited information. This can lead us to accept misinformation because cognitively it is easier to accept a readily available piece of information than to evaluate it for its accuracy. Misinformation is dangerous because it is sticky and resistant to correction. Managerial forensics does not remove intuition from the equation, but forces us to question the facts and evaluate the accuracy of the information we use to make decisions.

Appropriate Timing

Gathering information and diagnosing the problem in a timely manner are critical to the practice of managerial forensics. Similar to a patient in critical care, a struggling organization needs immediate and appropriate intervention to get through a crisis.

Set Turnaround Plan

An organization cannot be nursed back to health in a day. The pathway to recovery requires a series of carefully planned and monitored steps. The effective practice of managerial forensics includes the identification of a well-defined strategic plan toward organizational health.

A model for the practice of managerial forensics is shown in Figure 17.1.

Figure 17.1 shows that managerial forensics is not about the pursuit of a singular action but rather a combined set of simultaneous actions geared toward the achievement of optimum results. In the case of managerial forensics, success is measured not solely by the efforts placed but the results delivered. It is about identifying an appropriate set of management tools that assist firms in deciphering salient information to ensure that problems and solutions are uncovered.

The managerial forensics approach used in tandem with the tools discussed in this book—governance evaluation, leadership assessment, ethical review, marketing autopsy, forensic accounting in the private sector and government context, entrepreneurial forensics, and international

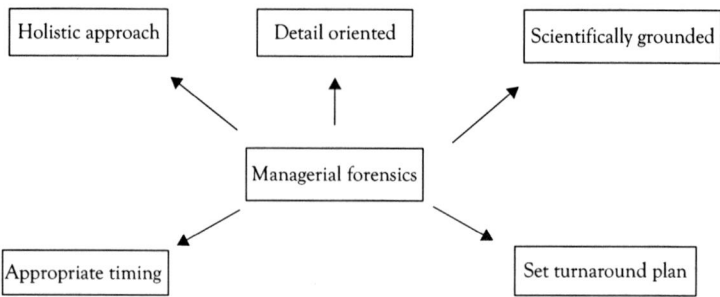

Figure 17.1 The managerial forensics approach

forensics—provides a strong diagnostic framework to uncover what is ailing the organization and how corporate health may be restored. Managers and consultants can select what they believe to be the viable tools to make their assessment. These tools, much like surgical tools, can be used singularly or collectively when striving to uncover and investigate operational malaise. In the case of Blockbuster, a marketing forensic investigation during the height of its troubles could have helped address pressing strategic issues they faced. In the case of Enron, leadership assessment combined with the effective use of forensic accounting early on could have made a difference.

In applying these tools, asking the right questions can make a big difference. Examples of a few key questions to start the diagnostic process are illustrated in Table 17.1.

Table 17.1 Tools and questions for managerial forensics

Assessment area	Questions
Governance	• Are governance structures in place? • What needs to be changed and by when? • Does the company structure support decision making and resource allocation? • Are the various units managed in such a way as to promote the organization's mission? • Are governance risks managed through transparency, monitoring, and clear accountability?
Leadership	• Is there a clearly defined and communicated vision in place? • Is the strategy flawed? Is it sustainable? • How effective is the current leadership in implementing the vision, mission, and strategy? • Are the goals measurable? • Are plans actionable?
Ethics	• Is the organization ensuring that employees and management are in compliance with existing laws, regulations, and the organization's own policies? • Are compliance efforts coordinated effectively company-wide? • Is the organization aware of its organizational culture, ethical culture, and ethical climate? • Is the organization operating ethically? • Has organizational preparedness been evaluated through internal and external assessments?

(Continued)

Table 17.1 Tools and questions for managerial forensics   (Continued)

Assessment area	Questions
Operations or production	• Does the operations strategy support company objectives? • Have competitive priorities been successfully translated into operational capabilities? • Is excellence in product, service, and system execution consistently being achieved?
Marketing	• Is the firm competing effectively within its market space? • Have the target market been correctly identified? • Has the company clearly defined and communicated its value proposition to its clients? • Is the firm's strategy for retaining customers working? • Have assessments in customer value added, marketing strategy, and marketing program been made?
Financial or accounting	• Does the capital structure support the organization's need for financial efficiencies? • Does the accounting system support existing operations? • Is the financial information received reliable and free of *creative accounting*? • Are preventative measures in place? • Is cash flow regularly monitored? • Has a thorough financial review including internal control audits been made?
Entrepreneurship	• Has the start-up clearly differentiated itself in the market? • Has the company created or added value? • Has the start-up correctly achieved product–market fit such that profitability occurs? • Does the organization understand its customers and stakeholders well, including their expectations and ability to generate revenue? • Has customer risk assessment been made for key stakeholders?
International	• Could an international angle help address organizational challenges? (i.e., cost reduction, market expansion, profitability improvement) • Is there a specific plan and strategy in place for expanding operations internationally? • Has the organization successfully adapted to foreign markets as needed? • Are there other international variables left unaddressed? • Have considerations been made on market environment, country compatibility, relationships and networks, and organizational preparedness?

In a situation where a company is experiencing a major decline in profitability, asking and answering some of these questions can help identify the root causes of the problem. Utilizing the tools mentioned in this book helps managers and consultants identify solutions.

Ironically, the best scenario for managerial forensics is not to ever need to use it. This means that the organization is in perfect shape and no weaknesses have to be addressed. However, these cases are the exceptions rather than the rule. With diverse challenges such as economic recessions, rapidly evolving markets, heightened consumer sophistication, disruptive technological changes, and intense competition, companies are likely to experience some difficulty in at least one operational area. In such cases, the ability of managers and consultants to execute managerial forensics well and at the right time would have a huge impact on organizational health. The chapters on team health, information management, and directional change offered remedial options for organizational challenges. These chapters prescribed the need for a proactive rather than a reactive approach to attaining corporate health.

In the end, managerial forensics is anchored in methodical corporate examination, diagnosis, and revival. With proper and reliable investigations undertaken, managers can craft viable strategies that pave the way for a successful recovery and the achievement of optimal organizational health.

There are diverse viewpoints on the most effective corporate turnaround strategies. Strategies need to be focused on people, by helping employees engage in strong relationships with customers (Reiss 2013). Clarity of strategy and speed of execution are important considerations (DuBois 2011). A combination of attributes such as financial acumen, effective control and monitoring, change management, innovation, and a strong digital presence contributes to success (Todrin 2012).

Whatever the strategy one uses in managing corporate problems and attempting a turnaround, it is apparent that the acquisition of reliable information in a timely manner helps in the corporate assessment and strategy formulation. In essence, the practice of effective managerial forensics is the key.

Managerial forensics, while supportive of evidence-based management principles, goes a step further in advocating the use of science, and specifically the differential diagnostic tool (SOAP) to assist in identifying the root causes in stagnating or declining businesses. As previously mentioned, the practice of managerial forensics is not about finding standard solutions, but rather the application of a well-defined process that increases the probability of finding the right solution to the right problem.

The authors hope that the approach and tools introduced by the contributors, who are experienced management experts and consultants in their respective fields, will be helpful to managers, entrepreneurs, and consultants as they seek remedies to corporate malaise or near demise. Hopefully, students and educators will take interest in this exploratory and pioneering work and encourage additional research. Lastly, it is hoped that government officials and policy makers use the insights offered in this book to support the practice of management and entrepreneurship.

The world of business is certainly dynamic and evolving. As business models change in response to economic and market changes, competition, and shifting consumer preferences, managerial forensics will likely evolve along with it. This book is not the end of this pioneering exploration on managerial forensics, but rather the beginning of endless possibilities that lie ahead.

References

Clifton, J. January 13, 2015. "American Entrepreneurship: Dead or Alive?" Gallup. http://www.gallup.com/businessjournal/180431/american-entrepreneurship-dead-alive.aspx

DuBois, S. 2011. "Why Turnaround CEO's Must Race the Clock." http://fortune.com/2011/09/16/why-turnaround-ceos-must-race-the-clock/ (accessed January 7, 2015).

Dun and Bradstreet. 2012. "Global Business Failures Report." http://www.dnbcountryrisk.com/FreeSamples/ICI/ICI_06.12.pdf (accessed January 6, 2015).

McKinsey. 2011. "Big Data: The Next Frontier for Innovation, Competition and Productivity." http://www.mckinsey.com/insights/business_technology/big_data_the_next_frontier_for_innovation (accessed January 6, 2015).

Reiss, R. 2013. "Top CEO's Say the Secret of the Great Turnaround is Focus on People." http://www.forbes.com/sites/robertreiss/2013/10/28/top-ceos-cite-the-secret-of-the-great-turnaround-is-focus-on-people/ (accessed January 7, 2015).

Todrin, D. 2012. "Five Steps to a Successful Business Turnaround." http://www.entrepreneur.com/article/223955 (accessed January 7, 2015).

U.S. Courts. 2014. "Filings by Chapter and Nature of Debt." http://www.uscourts.gov/uscourts/Statistics/BankruptcyStatistics/BankruptcyFilings/2014/0914_f2.pdf (accessed January 6, 2015).

Index

OTHER TITLES IN THE CORPORATE GOVERNANCE COLLECTION

John A. Pearce II, Villanova University and Kenneth Merchant,
University of Southern California, Editors

- *A Director's Guide to Corporate Financial Reporting* by Kristen Fiolleau, Kris Hoang and Karim Jamal
- *Blind Spots, Biases, and Other Pathologies in the Boardroom* by Kenneth Merchant and Katharina Pick
- *A Primer on Corporate Governance, Second Edition* by Cornelis A. de Kluyver
- *A Primer on Corporate Governance: Spain* by Felix Lopez-Iturriaga and Fern Tejerina-Gaite
- *A Primer On Corporate Governance: China* by Jean Jinghan Chen

Announcing the Business Expert Press Digital Library

Concise e-books business students need for classroom and research

This book can also be purchased in an e-book collection by your library as

- a one-time purchase,
- that is owned forever,
- allows for simultaneous readers,
- has no restrictions on printing, and
- can be downloaded as PDFs from within the library community.

Our digital library collections are a great solution to beat the rising cost of textbooks. E-books can be loaded into their course management systems or onto student's e-book readers.
The **Business Expert Press** digital libraries are very affordable, with no obligation to buy in future years. For more information, please visit **www.businessexpertpress.com/librarians**. To set up a trial in the United States, please email **sales@businessexpertpress.com**.

CPSIA information can be obtained
at www.ICGtesting.com
Printed in the USA
FSOW02n1628150117
29497FS